T0214250

Lecture Notes in Computer Science 12119

More information about this series at http://www.springer.com/series/7412

Abderrahim El Moataz ·
Driss Mammass · Alamin Mansouri ·
Fathallah Nouboud (Eds.)

Image and Signal Processing

9th International Conference, ICISP 2020
Marrakesh, Morocco, June 4–6, 2020
Proceedings

 Springer

Editors
Abderrahim El Moataz
GREYC, University of Caen Normandie
Caen, France

Alamin Mansouri (iD)
ImViA
University of Burgundy
Dijon, France

Driss Mammass
IRF-SIC, Faculty of Sciences
Ibn Zohr University
Agadir, Morocco

Fathallah Nouboud
Math - Info
University of Quebec
Trois-Rivières, QC, Canada

ISSN 0302-9743   ISSN 1611-3349   (electronic)
Lecture Notes in Computer Science
ISBN 978-3-030-51934-6   ISBN 978-3-030-51935-3   (eBook)
https://doi.org/10.1007/978-3-030-51935-3

LNCS Sublibrary: SL6 – Image Processing, Computer Vision, Pattern Recognition, and Graphics

This Springer imprint is published by the registered company Springer Nature Switzerland AG
The registered company address is: Gewerbestrasse 11, 6330 Cham, Switzerland

Preface

The 9th International Conference on Image and Signal Processing (ICISP 2020) was scheduled to take place in Marrakesh, Morocco, during June 4–6, 2020. Unfortunately, the coronavirus pandemic and the ensuing confinement, as well as the uncertainties around when a return to normal life would ensue, led us to cancel the physical holding of the ICISP conference. Fortunately, LNCS agreed to publish the proceedings of the 40 best papers and as a result we are grateful to present the ISCISP 2020 proceedings.

Historically, ICISP is a conference resulting from the actions of researchers from Canada, France, and Morocco. Previous editions of ICISP were held in Cherbourg-Octeville, France (2008, 2014, and 2018), in Trois-Rivières, Quebec, Canada (2010 and 2016), and in Agadir, Morocco (2001, 2003, and 2012).

ICISP 2020 is sponsored by EURASIP (European Association for Image and Signal Processing), CNRST (National Center for Scientific and Technical Research), STIC pole (National STIC competence pole), and ARPRI (association of research on patter recognition and imaging).

For this 9th edition, in addition to the ICISP session, we scheduled three special sessions: Data and Image Processing for Precision Agriculture (DIPPA 2020), Digital Cultural Heritage (DCH 2020), and Machine Learning Application and Innovation (MALAI 2020).

From 84 full papers submitted, 40 were finally accepted. The review process was carried out by the Program Committee members; all are experts in various image and signal processing areas. Each paper was reviewed by three reviewers and also checked by the conference co-chairs. The quality of the papers in these proceedings is attributed first to the authors and second to the quality of the reviews provided by the experts. We would like to thank the authors for responding to our call, the reviewers for their excellent work, and the organizers of the three special sessions. We would also like to thank the members of the Local Organizing Committee for their advice and help. We are also grateful to Springer's editorial staff for supporting this publication in the LNCS series.

We hope this publication provides a good view into the research presented at ICISP 2020, and we look forward to meeting you at the next ICISP conference.

This edition is in memory of Pr. Driss Aboutajdine, who was part of the team responsible for thinking and organizing ICISP since 2001. He contributed to the success of all ICISP editions. Prof. Aboutajdine was the ex-Director of the National Center for Scientific and Technical Research of Morocco (CNRST), a member of the Academy Hassan 2 of Science, and national coordinator of the pole of competence STIC.

June 2020

Abderrahim El Moataz
Driss Mammass
Alamin Mansouri
Fathallah Nouboud

Organization

General Chairs

Abderrahim El Moataz	University of Caen Basse-Normandy, France
Fathallah Nouboud	University of Quebec à Trois-Rivières, Canada

Program Committee Chairs

Driss Mammass	Ibn Zohr University, Morocco
Alamin Mansouri	University of Bourgogne, France

Special Sessions Chairs

Dippa 2020

Mohamed El Hajji	CRMEF-SM Agadir, Morocco
Youssef Es-Saady	Ibn Zohr University, Morocco
Adel Hafiane	INSA Centre Val de Loire, France
Raphaël Canals	University of Orléans, France

DCH 2020

Alamin Mansouri	University of Bourgogne, France

MALAI 2020

Abdelaziz El Fazziki	Cadi Ayyad University, Morocco
Jihad Zahir	Cadi Ayyad University, Morocco

Local Organizing Committee

Hassan Douzi	Ibn Zohr University, Morocco
Mouad Mammass	Ibn Zohr University, Morocco
Mohamed Salim El Bazzi	Ibn Zohr University, Morocco
Soufiane Idbraim	Ibn Zohr University, Morocco
Mustapha Amrouch	Ibn Zohr University, Morocco
Hasna Abioui	Ibn Zohr University, Morocco
Taher Zaki	Ibn Zohr University, Morocco
Ali Idarrou	Ibn Zohr University, Morocco

Web Chair

Mouad Mammass	Ibn Zohr University, Morocco

Proceedings Chair

Mohamed El Hajji   CRMEF-SM Agadir, Morocco
Youssef Es-Saady   Ibn Zohr University, Morocco

International Associations Sponsors

European Association for Signal Processing (EURASIP)
National Center for Scientific and Technical Research (CNRST)
National STIC Competence pole (STIC pole)
Association of Reasearch on Pattern Recognition and Imaging (ARPRI)

Sponsoring Institutions

Ibn Zohr University, Morocco
University of Quebec à Trois-Rivières, Canada
University of Caen Basse-Normandie, France
University of Bourgogne, France
Faculty of Sciences, Agadir, Morocco

Program Committee

Tarik Agouti   Cadi Ayyad University, Morocco
Abdellah Ait Ouahman   Cadi Ayyad University, Morocco
Mustapha Amrouch   EST, Ibn Zohr University, Morocco
Jilali Antari   FPT, Ibn Zohr University, Morocco
Aissam Bekkari   ENSA, Cadi Ayyad University, Morocco
Mostafa Bellafkih   INPT Rabat, Morocco
Yannick Benezeth   University of Bourgogne, France
Djamal Benslimane   Lyon 1 University, France
Giuseppe Boccignone   University of Milan, Italy
Frank Boochs   Mainz University of Applied Sciences, Germany
Stéphanie Bricq   University of Bourgogne, France
Pierre Buyssens   IRISA, France
Raphaël Canals   University of Orleans, France
Pierre Chainais   École Centrale de Lille, France
Pamela Cosman   UC San Diego, USA
Jose Crespo   Universidad Politécnica de Madrid, Spain
Meurie Cyril   University Gustave Eiffel, France
Christian Degrigny   Haute École Arc, France
Hassan Douzi   FSA, Ibn Zohr University, Morocco
Mohamed El Hajji   CRMEF-SM Agadir, Morocco
Azeddine El Hassouny   ENSIAS, Mohamed V University, Morocco
Jessica El Khoury   University of Bourgogne, France
Abderrahim El Moataz   University of Caen Basse-Normandie, France

Abdelmounîm El Yacoubi	Institut Mines-Télécom SudParis, France
Hasna Elalaouhi-Elabdallaoui	Cadi Ayyad University, Morocco
Abdelaziz Elfazziki	Cadi Ayyad University, Morocco
Abdel Ennaji	University of Rouen, France
Youssef Es-saady	FPT, Ibn Zohr University, Morocco
Sony George	Norwegian Colour and Visual Computing Laboratory, Norway
Abel Gomes	University of Beira Interior, Portugal
Adel Hafiane	INSA Centre Val de Loire, France
Rachid Harba	University of Orleans, France
Jon Yngve Hardeberg	Norwegian University of Science and Technology, Norway
Markku Hauta-Kasari	University of Eastern Finland, Finland
Aymeric Histace	ETIS, ENSEA, France
Khalid Housni	Ibn Tofail University, Morocco
El Hassane Ibn Elhaj	INPT Rabat, Morocco
Ali Idarrou	ESTG, Ibn Zohr University, Morocco
Soufiane Idbraim	FSA, Ibn Zohr University, Morocco
Jérôme Idier	University of Nantes, France
M'bark Iggane	Ibn Zohr University, Morocco
Mustapha Kardouchi	University of Moncton, Canada
Ali Khenchaf	ENSTA Bretagne, France
Mohammed Lamine Kherfi	University of Quebec à Trois-Rivières, Canada
Alexandre Krebs	University of Bourgogne, France
Zakaria Lakhdari	University of Caen Basse-Normandie, France
Hajar Lazar	Cadi Ayyad University, Morocco
Francois Lecellier	University of Poitiers, France
Ludovic Macaire	University of Lille, France
Richard Macwan	University of Bourgogne, France
Amal Mahboubi	University of Caen Basse-Normandie, France
Driss Mammass	FSA, Ibn Zohr University, Morocco
Mouad Mammass	ENCG, Ibn Zohr University, Morocco
Alamin Mansouri	University of Bourgogne, France
Franck S. Marzani	University of Bourgogne, France
Jean Meunier	University of Montreal, Canada
Cyrille Migniot	University of Bourgogne, France
Pascal Monasse	LIGM, University of Paris-Est, France
El Mustapha Mouaddib	University of Picardie, France
Hajar Mousannif	University Cadi Ayyad, MA
Neeta Nain	Malaviya National Institute of Technology Jaipur, India
Fathallah Nouboud	University of Quebec à Trois-Rivières, Canada
Jean-Marc Ogier	University of La Rochelle, France
Marius Pedersen	Norwegian University of Science and Technology, Norway
Alessandro Rizzi	University of Milan, Italy

Contents

Machine Learning Application and Innovation

Biomedical Imaging

Deep Learning and Applications

Pattern Recognition

Segmentation and Retrieval

Mathematical Imaging and Signal Processing

Digital Cultural Heritage and Color and Spectral Imaging

Approach to Analysis the Surface Geometry Change in Cultural Heritage Objects

Sunita Saha[1]([⊠]) ⓘ, Piotr Foryś[1] ⓘ, Jacek Martusewicz[2],
and Robert Sitnik[1] ⓘ

[1] Faculty of Mechatronics, Warsaw University of Technology, Warsaw, Poland
Sunita.Saha@pw.edu.pl
[2] Faculty of Conservation and Restoration of Works of Art,
Academy of Fine Arts in Warsaw, Warsaw, Poland

Abstract. The three-dimensional digitization of the cultural heritage objects during different stages of the conservation process is an important tool for objective documentation. Further data analysis is also important to monitor, estimate and understand any possible change as accurately as possible. In this work, the cultural heritage (CH) objects were selected for 3D scanning, analysis and visualisation of the change or degradation on their surface over time. The main goal of this work is to develop analysis, and visualization methods for CH object to assess local change in their surface geometry to support conservation processes documentation. The analysis was based on geometrical analysis of change in global distance between before and after chemical cleaning for a chosen object. The new local neighborhood distance histogram has been proposed as a local measure of surface change based on optimized k-neighborhood search algorithm to assess the local geometry change of a focus point.

Keywords: 3D scanning · Surface data analysis · Monitoring of conservation process · Cultural heritage digitisation · 3D visualization · Surface geometry

1 Introduction

3D imaging is a technology that can be accomplished through various methods [1]. It can be used for disparate applications, ranging from quality and process control in factory automation to scientific imaging and research and development of algorithms in 3D model acquisition [2, 3]. The image related application is the natural fit for an algorithm course. They can be demonstrated graphically and have the property that they usually contain large enough number of surface points. There are various virtual reality applications [4] for which a number of researchers are working for its further development in terms of 3D imaging technology and acquisition technology like remote sensing, GIS, cultural heritage, etc. The use of geometric tools in analysis of detection change information i.e. spatial information works efficiently in solving the 3D problem [5].

The understanding of 3D structures is essential to many scientific endeavors. Recent theoretical and technological breakthroughs in mathematical modeling of 3D data and data-capturing techniques present the opportunity to advance 3D knowledge

© Springer Nature Switzerland AG 2020
A. El Moataz et al. (Eds.): ICISP 2020, LNCS 12119, pp. 3–13, 2020.
https://doi.org/10.1007/978-3-030-51935-3_1

into new realms of cross-disciplinary research [6]. 3D knowledge plays an important role in archaeology [7]. In conservation science, the restoration phases of cultural heritage objects can be digitized with various techniques [8]. Cultural heritage analysts and scientists are continuously trying to develop the more objective and automated techniques that could document of different phases of the conservation pipeline and could give the scientific visualization for the changes detected [1]. The work shows the application of structured light techniques to the analysis of a CH object before and after conservation. In order to detect and visualize changes the authors applied a local geometry analysis and presented the outcome through a global and local distance histogram. The goal of the work is in line with the application of optical metrology techniques to CH questions.

2 Related Study

The study on 3D imaging reveals applications in the area of surface change analysis. Robert Sitnik and *et al.* [9] researched on monitoring the marker-based surface degradation process using the 3D structured light method of scanning and visualize the 3D surface geometry. They analyzed the mutual overlapping of neighboring point clouds. The simulation of the work was validated using CloudCompare (v2.9.1) software based on the distance parameter. The main drawback of this work is the necessity of attaching physical markers to the object surface.

Pintus Ruggero and *et al.* [10] presented the recent techniques for performing geometric analysis in cultural heritage applications focusing on the factors of shape perception enhancement, restoration, and preservation support, monitoring overtime, object interpretation, and collection analysis. The survey was based on the geometric scale at which the analysis was performed and the cardinality of the relationships among object parts exploited during the analysis.

Manfredi Marcello and *et al.* [11] developed a reliable RTI method for monitoring changes in cultural heritage objects to get detailed information on the object surface. In this work, they captured the RTI and compared the normal vectors to the limits: the method was able to detect the damage automatically. The RTI method of analysis works well on collecting the detail information of the object's surface but for geometrical analysis of change, it is not convenient. Also in [12] E. Marengo and *et al.* presented another development technique based on multi-spectral imaging for monitoring the conservation of cultural heritage objects which is based on the change in color information rather than geometric analysis of change.

3 Measuring System

Due to the requirements concerning resolution, accuracy and illumination conditions, we developed a custom measurement set-up. We decided to use the structured light technique (SL) [13, 14] for geometry measurement. The phase-shifting method combined with Gray codes for improved phase unwrapping has been chosen as the most accurate SL implementation. We select six phase-shifts after careful assessment of the developed measurement head intensity transfer nonlinearities. We used an SL calibration method based on modeling of phase distribution concerning detector coordinates [15].

The 3D scanner we used was designed and developed in the ZTRW laboratory and it is a completely functional prototype that is ready to perform valid measurements of objects surface (Fig. 1). The main units in the 3D scanner are the projector and detector. They are mounted on a special base made from carbon fiber composites. Using such composites prevents the deformation of the scanner construction caused by thermal expansion and ensures that the projector-camera system can be treated as a rigid body. As a projector, we have used a LightCrafter 4500 from Texas Instruments [16]. The matrix detector – a color camera from IDS [17] captures images of projected patterns (each single measurement consist of 14 frames). The fringes projection and image grabbing is synchronized by external microcontroller build on Atmega (we've used Arduino UNO) [18]. Custom made software for calibration and measurement [14] is saved on microcomputer IntelNUC. The Maximum Permissible Error for the scanner has been estimated as EMPE = 0.25 mm [19].

Fig. 1. Custom-designed 3D scanner.

4 Experimental Object

To analyze the change that appeared on an object surface, an object from the cultural heritage field was chosen from Studio of Conservation and Restoration of Ceramic, Faculty of Conservation and Restoration of Work of Art, Academy of Fine Arts in Warsaw Poland. The object is a ceramic tile and the area of interest was decided as the front of the object shown in Fig. 2(a) at the initial state and in Fig. 2(b) there is presented its photo after the cleaning process.

The chosen object which is a ceramic tile is a part of an element of the top of the tile stove which is from the end of the 19th century. The origin of the tile is demolished stove from one of Warsaw's tenements houses. The tile was made in one of the ceramic factories in Velten near Berlin. It was formed in plaster mould from a ceramic mass. It

Fig. 2. The chosen object and area of interest (Front) before (a) and after (b) cleaning.

was burnt to biscuit at temperatures up to 1000 °C. This type of tiles in the 19th century was not originally covered with enamel or painted. The basic tiles in those tile stoves were covered with a white glaze, but the decorative elements were in the natural color of fired clay.

The floral ornament decorating of the tile is a precise carving decoration. During about a hundred years of use of the tiled stove, the apartment owners covered the upper part of the tiled stove with several layers of paint (oil and emulsion painting) as a way of unprofessional renovation treatment - instead of cleaning the tile surface. In this way, the visibility of sculptural decoration details was lost. During the conservation works, it was decided to remove the over painting. This made it possible to regain the original ethical values of antique decoration.

5   3D Scanning and Analysis

5.1   3D Scanning

The 3D scanning of the chosen object, with our measuring system, was done in Studio of Conservation and Restoration of Ceramic, Faculty of Conservation and Restoration of Work of Art, Academy of Fine Arts in Warsaw. The scanning process was done after setting the measuring system in day light condition. In Fig. 3 the scanning process is shown. To scan the area of interest for the chosen object we needed a total of 13 scans to get the complete area chosen.

The second scan after cleaning the object was done choosing the same environment and in the same lighting condition as before. The object was scanned at the same studio and the complete scanning of the object it needed a total of 20 scans for the entire area of interest.

5.2   Reconstruction of the Object

The obtained point clouds from each scan were stitched to reconstruct the 3D model for the chosen object. For each scan, some noisy points appeared where the surface of the object is not smooth or has a discontinuity due to the nature of the 3D scanner. The point groups which are at a distance from the main point cloud based on some given

Fig. 3. Area of interest of the object while scanning with the measuring system.

threshold value were considered as noise. The resulting noise was filtered based on the Hausdorff distance criterion. After the removal of the noise point groups, the point clouds were put into the best-fit plane to reconstruct the 3D model based on iterative closest point algorithm [20] both for the before and after the cleaning process.

The registration of the before state point clouds based on the ICP algorithm resulted in a total of 13 204 857 points after reconstruction. The registration was done in Frames which is developed in Warsaw University of Technology using C++. For the registration and to fit the point clouds in the best plane, it took a couple of minutes resulting in an RMS of 0.32 mm. In Fig. 4 the 3D model obtained after registration is shown.

Fig. 4. Reconstruction of the object before cleaning and the density of the point cloud

Following the same algorithm the 3D model of the object after the state was obtained as well. The reconstruction of the scans after cleaning obtained a total of 13 298 867 points with an RMS of 0.31 mm. The 3D model obtained after stitching the point clouds were presented in Fig. 5.

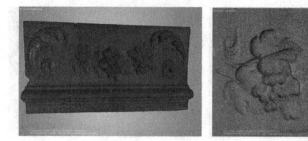

Fig. 5. Reconstruction of the object after cleaning and the density of the point cloud

5.3 Analysis of the Obtained Data

The analysis of the 3D models obtained after reconstruction was done. During the analysis, the change that appeared on the object surface was measured by calculation according to the chosen parameter of the distance between the point cloud before cleaning and the cleaning plane of the object. First of all, both the reconstructions obtained after registering the point clouds were tried to register to base on the best-fit plane. The reconstruction was stored in two different nodes as original_cloud and changed_cloud. This registration stored the global transformation of the changed_cloud concerning the original_cloud. The registration obtained with an RMS error equal to 0.15 mm.

5.4 Analysis of Global Distance

The registration of both before and after state 3D models was done to computationally analyze the global distance between these two reconstructions. In this section, the 3D distance from each point from the before state point cloud was calculated (Fig. 6) to the plane of the after state point cloud plane.

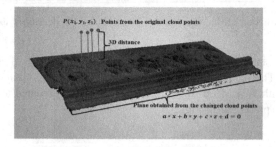

Fig. 6. General representation of the global distance analysis

Pseudo code1:

Input data: Two integrated models from before and after object state in form of XYZ, RGB, NormalsXYZ. They are stored in two nodes: original_cloud and changed_cloud. Radius.

    1) Get the 3D details (x,y,z) for each point in the point cloud of original_cloud

    2) Get the 3D plane $(ax+by+cz+d=0)$ from the point cloud of changed_cloud with the input radius.

    3) For(int i=0;i<size_of_original_cloud_points;i++)

        a. Calculate the perpendicular distance to the plane obtained in 2.

        b. Store the value of maximum and minimum distance obtained (3a).

Output: global distance histogram and the visualization of the surface with a color map.

In this analysis, a constant value of radius 3 mm is considered to calculate the plane for the points in changed_cloud. And results obtained ranging from a minimum distance 0 to a maximum of 2.73 mm is shown in Fig. 7 as 'global distance histogram'. The color map visualization from the calculated distance is also presented.

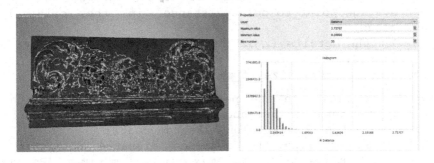

Fig. 7. Results analysis on parameter distance and global distance histogram

5.5 Local Neighbourhood Analysis

In In this work, the goal is to propose a new local measure for change assessment and discrimination on different types. The general idea behind this proposed algorithm is to calculate the neighborhood points for a focus point where we want to assess the change occurrence up to a user-specified range of neighborhood size. And based on this local neighborhood analysis our goal is to visualize the results only for that focus point to present the results like in Fig. 8. The general logic behind this approach is to categorize

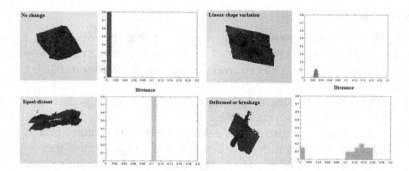

Fig. 8. General representation of the histogram behavior based on the geometry change.

the change occurrences on the surface before and after conservation, based on the behavior of the local-neighborhood distance histogram.

Pseudo code2:

Input: original_cloud, changed_cloud, radius, focus_point, size_of_neighborhood.
Execute Pseudo code1
  1) Get the details (x,y,z) for the chosen focus_point.
  2) Calculate the average_point_to_point distance for the original_cloud.
  3) Find the neighbourhood points for the selected focus point in the range of size_of_neighborhood * average_point_to_point_distance in the original cloud.
  4) Get the details (x,y,z) of the obtained neighbourhood points.
  5) For(int i=0;i<size_of_neighborhood_points,i++)

      a. Calculate the perpendicular distance for the obtained neighbourhood points to the plane of the changed cloud (Pseudo code1: 2).
      b. Normalize the obtained distance within the range of global maximum and minimum (Pseudo code1: 3(b)).
      c. Visualize the calculated distance for the neighbourhood points for the focus point to the original_cloud.
Output: local_neighborhood_histogram for different selection based focus points.

The work considered a relatively small area to analyze based on a selected focus point with the size of the neighborhood as 10. The *Pseudo code2* was performed taking different focus points from different parts of the object's surface as in Fig. 9.

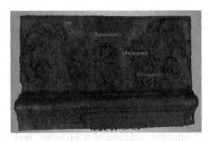
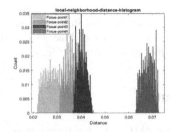

Fig. 9. Selected focus points and the local neighborhood distance histogram

All the distances obtained are almost linear for each focus point since the cleaning was the removal of a layer from the surface. The maximum cleaning part as in focus-point4 reached the maximum value compared to the other parts, focus-point2 and focus-point3 with lesser value in focus-point1 which is presented in local-neighborhood-distance-histogram Fig. 9. The results obtained from both the global distance and local distance analyses are fair enough to justify the amount of material removal from the object's surface.

6   Conclusion and Discussion

There are various techniques and thousands of research is going on to develop the 3D imaging technology both in terms of image capturing set up and based on computational analysis of the captured images. This work introduces the analysis of neighborhood points for a selected point in terms of local histogram calculation and assesses the change of surface geometry for that selected point. This study can solve the real-world bunch of problems in various application areas of 3D imaging.

The challenging part of this analysis was the critical global registration of the obtained scans from the measuring device. This analysis is hard to claim its correctness if the point clouds before and after are not fitted in the same coordinate system. The registration of the scans has been done with minimum RMS ICP to calculate the distance accurately between before and after the state of the object which is very crucial stage to follow the proposed analysis. The cleaning of the object was done on the entire object's surface thus we have no reference for global registration. The results obtained for the analysis is sensitive to the surface curvature.

This object while choosing to work on it due to the unawareness of its change causes the result analysis only linear change type. This analysis will be carried out to analyze the various other CH objects to validate its strength considering all types of categorized change as in Fig. 8. The analysis faced a critical stage of showing inaccurate change while calculating the local neighborhood distance histogram due to its global registration. The future work also includes in solving that critical phase.

Acknowledgement. This work is carried out at the beneficiary partner of CHANGE: *Cultural Heritage Analysis for New Generation,* Warsaw University of Technology, Poland, received funding from the European Union's Horizon 2020 research and innovation program under the

Marie Skłodowska-Curie grant agreement No. 813789. Also, we would like to acknowledge the Studio of Conservation and Restoration of Ceramic, Faculty of Conservation and Restoration of Works of Art, Academy of Fine Arts in Warsaw, Poland for collaborating in this work by providing the cultural heritage objects.

References

1. Karatas, O.H., Toy, E.: Three-dimensional imaging techniques: a literature review. Eur. J. Dent. **8**(1), 132–140 (2014). https://doi.org/10.4103/1305-7456.126269
2. Mączkowski, G., Krzesłowski, J., Bunsch, E.: How to capture aesthetic features of complex cultural heritage objects – active illumination data fusion. In: IS&T International Symposium on Electronic Imaging 2018 3D Image Processing, Measurement (3DIPM), and Applications (2018)
3. Chane, C.S., Mansouri, A., Marzani, F., Boochs, F.: Integration of 3D and multispectral data for cultural heritage applications: survey and perspectives. Image Vis. Comput. **31**(1), 91–102 (2013). <hal-00783985>
4. Kamińska, D., et al.: Virtual reality and its applications in education: survey. Information **10**, 318 (2019)
5. Cohen-Or, D., et al.: A Sampler of Useful Computational Tools for Applied Geometry, Computer Graphics, and Image Processing, 1st edn. A K Peters/CRC Press, Natick (2015)
6. Schurmans, U.A., et al.: Advances in geometric modeling and feature extraction on pots (2001)
7. Volonakis, P.: Use of various surveying technologies to 3D digital mapping and modelling of cultural heritage structures for maintenance and restoration purposes: the Tholos in Delphi, Greece. Mediterr. Archaeol. Archaeom. **17**(3), 311–336 (2017). https://doi.org/10.5281/zenodo.1048937
8. Grabowski, B., Masarczyk, W., Głomb, P., Mendys, A.: Automatic pigment identification from hyperspectral data. J. Cult. Herit. **31**, 1–12 (2018). https://doi.org/10.1016/j.culher.2018.01.003. ISSN 1296-2074
9. Sitnik, R., Lech, K., Bunsch, E., Michoński, J.: Monitoring surface degradation process by 3D structured light scanning. In: Proceedings of the SPIE, Optics for Arts, Architecture, and Archaeology VII, vol. 11058, p. 1105811, 12 July 2019. https://doi.org/10.1117/12.2525668
10. Pintus, R., Pal, K., Yang, Y., Weyrich, T., Gobbetti, E., Rushmeier, H.: Geometric analysis in cultural heritage, pp. 1–17 (2014)
11. Manfredi, M., Williamson, G., Kronkright, D., Doehne, E., Jacobs, M., Bearman, G.: Measuring changes in cultural heritage objects with reflectance transform imaging (2013). https://doi.org/10.1109/digitalheritage.2013.6743730
12. Marengo, E., et al.: Development of a technique based on multi-spectral imaging for monitoring the conservation of cultural heritage objects. Analytica Chimica Acta **706**(2), 229–237 (2011). https://doi.org/10.1016/j.aca.2011.08.045
13. Geng, J.: Structured-light 3D surface imaging: a tutorial. Adv. Opt. Photonics **3**, 128–160 (2011)
14. Adamczyk, M., Kamiński, M., Sitnik, R., Bogdan, A., Karaszewski, M.: Effect of temperature on calibration quality of structured-light three-dimensional scanners. Appl. Opt. **53**(23), 5154–5162 (2014)
15. Sitnik, R.: New method of structure light measurement system calibration based on adaptive and effective evaluation of 3D-phase distribution. Proc. SPIE **5856**, 109–117 (2005)

16. DLP LightCrafter 4500 evaluation module. User's Guide, July 2013. http://www.ti.com/lit/ug/dlpu011e/dlpu011e.pdf. Accessed Sept 2015
17. UI-3180CP-C-HQ Rev. 2 datasheet. Datasheet, July 2017. https://en.ids-imaging.com/IDS/datasheet_pdf.php?sku=AB00686
18. Arduino uno rev3. https://store.arduino.cc/arduino-uno-rev3
19. Adamczyk, M., Sieniło, M., Sitnik, R., Woźniak, A.: Hierarchical, three-dimensional measurement system for crime scene scanning. J. Forensic Sci. **62**(4), 889–899 (2017)
20. Mavridis, P., Andreadis, A., Papaioannou, G.: Efficient sparse ICP. Comput. Aided Geom. Des. **35–36**, 16–26 (2015). https://doi.org/10.1016/j.cagd.2015.03.022

Towards the Tactile Discovery of Cultural Heritage with Multi-approach Segmentation

Ali Souradi[1], Christele Lecomte[1], Katerine Romeo[1(✉)] [ID],
Simon Gay[1] [ID], Marc-Aurele Riviere[1] [ID], Abderrahim El Moataz[2],
and Edwige Pissaloux[1] [ID]

[1] Rouen Normandy University, Saint Etienne du Rouvray, France
`katerine.romeo@univ-rouen.fr`
[2] Caen Normandy University, Caen, France

Abstract. This paper presents a new way to access visual information in museums through tactile exploration, and related techniques to efficiently transform visual data into tactile objects. Accessibility to cultural heritage and artworks for people with visual impairments requires the segmentation of images and paintings to extract and classify their contents into meaningful elements which can then be presented through a tactile medium. In this paper, we investigate the feasibility and how to optimize the tactile discovery of an image. First, we study the emergence of image comprehension through tactile discovery, using 3D-printed objects extracted from paintings. Later, we present a dynamic Force Feedback Tablet (F2T) used to convey the 2D shape and texture information of objects through haptic feedback. We then explore several image segmentation methods to automate the extraction of meaningful objects from selected artworks, to be presented to visually impaired people through the F2T. Finally, we evaluate how to best combine the F2T's haptic effects in order to convey the extracted objects and features to the users, with the aim of facilitating the comprehension of the represented objects and their affordances.

Keywords: Accessibility to artworks · Cultural heritage · Visual impairment · Image segmentation · Haptic interface

1 Introduction

Access to art and culture for visually impaired people (VIP) is often complex, with most of the artworks exhibited in museums relying on visual (2D) content. Currently, the most common solutions to overcome this problem are audio descriptions and 3D models for tactile exploration. However, these solutions convey limited information and have several drawbacks: audio descriptions are sequential, passive, and they monopolize the attention and the listening of the user. An active exploration (gaze or finger-guided) is paramount to form a holistic and coherent mental picture of the explored content (and thus appreciate its beauty), which is often not compatible with the linearity and passivity of audio descriptions. On the other hand, 3D-printed or thermoformed objects are usually expensive to manufacture (due to the transposition of

© Springer Nature Switzerland AG 2020
A. El Moataz et al. (Eds.): ICISP 2020, LNCS 12119, pp. 14–23, 2020.
https://doi.org/10.1007/978-3-030-51935-3_2

the artwork to its tactile representation often done manually by artists), and usually provides too much detail for an efficient tactile exploration.

Allowing VIP to explore tactile representations of artworks autonomously is a challenge that requires both the automated extraction of the meaningful content from an image and its adaptation to the specificities of haptic exploration. Each artwork being specific, classical automatic methods fail to provide a universal solution.

In order to improve accessibility to art and culture for VIP, we want to design and prototype a solution allowing to display transposed artworks, combining tactile and kinesthetic perceptions, which can be actively explored by the user. This solution will cover the two main problematics linked with artwork accessibility: (1) the reliable automatic extraction of meaningful objects from artworks and (2) their efficient transposition into haptic representations allowing VIP to understand their meaning and appreciate their beauty.

This paper first presents how a tactile representation of artworks is perceived on a printed object and the difficulties of understanding the shapes (Sect. 2). In Sect. 3, we introduce the semantic segmentation methods we explored in order to extract meaningful content from an image, and their results on the Bayeux Tapestry. The 4th section presents the F2T, a multimodal (audio-tactile) interface for the active exploration of graphical content, while the 5th section explores the possible applications of our system to cultural heritage of Bayeux Tapestry. Finally, we discuss the results and suggest possible future improvements.

2 Tactile Representation of Artwork

Tactile representations of artworks must be tailored to the specificities of haptic perception in order to be easily accessible to VIP [1, 2]. It must have, above all, outlines highlighted to detect the represented objects, and some indication should be given on their number and their location relative to the neighboring objects [3, 4]. This will allow users to better understand the explored object and its meaning in relation to the rest of the scene [5]. Furthermore, transposing information from one modality to another (i.e. visual to tactile) is an extremely complex process [6], comprised of two main problematics: selecting the most useful characteristics to convey (here, the spatial and visual features of an object), and finding the optimal way to encode those features in the space of the output modality (i.e. the tactile stimulation parameters of the haptic device used as an interface). Although it is an inclusive approach it involves a creation process towards tactile representation in order to communicate messages to users.

To experience how variations of the chosen tactile parameters might influence the recognition of the selected scene elements, we segmented (using inkscape) salient objects from a painting (Fig. 1a), such as a spinning wheel (Fig. 1b), and then 3D-printed it as a slightly elevated relief on a 2D plate, as shown in Fig. 1c. Visually, it is easily recognizable, but blindfolded participants were not able to locate or identify any part of the object without external help. We then printed a 3D model of this object (as shown in Fig. 1d) and presented both prints (2.5D and 3D) to blindfolded participants, explaining they were two representations of the same object. Again, the correspondence between the two models was not perceived through touch alone. As mentioned by Hatwell in [3], tactile recognition can be acquired with training, but is usually an arduous process due to the interference of vision-specific elements, such as perspective cues, which will hinder recognition [7]. A drawn object with added relief information (2.5D, see Fig. 1c) has very little actual correspondences with the tactile experience of the actual object (in 3D, see Fig. 1d). Indeed, the projection of an object on a surface relies on perspective cues which are specific to vision. VIP thus have to learn to interpret those perspective cues in order to make sense of this type of images. A tactile representation must be simplified compared to the visual object it is representing, keeping only the essential information to allow its recognition. It must "preserve the overall meaning" (or gist) of the represented object [8, 9].

Fig. 1. a) Painting of L. Minet, Chateau de Martainville; **b)** The spinning wheel in the painting. **c)** printed with a small relief (2.5D); **d)** printed as a 3D model.

3   Semantic Segmentation of an Artwork

In this section, we present several segmentation methods we selected in order to extract meaningful elements from the cultural heritage of Bayeux Tapestry. The particularity of this piece of art is its lack of perspective or shadows. It is comprised of a multitude of characters with their accessories, but also animals and some simplified buildings, depicting important scenes of the joint history of England and Normandy.

Semantic segmentation allows extracting and grouping elements of an image into meaningful categories. Several approaches exist, which could be categorized as: (1) segmentation based on edge detection; (2) segmentation based on the perceived similarities of spatially closed pixels (i.e. clustering), and (3) the cooperative segmentation of contours and regions.

These classes of methods all aim to segment the image using low-level information such as pixel color and its gradients along the image's dimensions. Selecting the appropriate methods will depend on the target images characteristics and on the goal of the segmentation. Our final objective is to segment the image elements in way that is relevant to the tactile transposition (and later, comprehension) of its content.

3.1 Contour Detection

Edge detection consists in the identification of delimitations between meaningful elements in an image, and often relies on the identification of strong oriented gradients of chrominance or luminance. Conventional methods such as high-pass or Canny filters often produce discontinuous contours, which does not allow users to follow contours in order to identify an object. They also tend to over-segment the image by producing contours and noise where no meaningful borders are present due to their sensitivity to lighting and exposure variations. We applied the following methods to the Bayeux tapestry (scene N°16 in Fig. 2a):

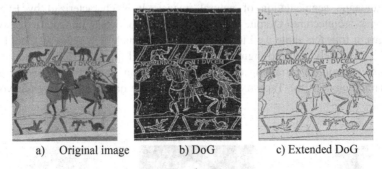

a)   Original image   b) DoG   c) Extended DoG

Fig. 2. Bayeux Tapestry contour segmentation with Difference of Gaussians method.

1) *The Difference of Gaussians (DoG)* is an edge extraction algorithm that relies on computing the difference between two images blurred with Gaussian kernels of different spread, producing an image where the areas most affected by the Gaussian blurring (i.e. high-frequency spatial information) are over-represented. The DoG algorithm produced continuous edges suitable for tactile transposition (Fig. 2b). Its main inconvenient is the exacerbation of background noise which will result in a grainy tactile transposition if kept. The Extended Difference of Gaussian (XDoG) [10] gives cleaner results (Fig. 2c).

2) ***The HED (Holistically-Nested Edge Detection)*** [11] is an end-to-end deep neural
network architecture inspired from Fully Convolutional Networks. Based on human
perception in the search for the contours of objects, it uses different levels of
perception, structural information and context (Fig. 3b). Nested multi-scale feature
learning is taken from deeply supervised nets. The output is combined from multiple scales. The details of the figures on the tapestry are correctly segmented by this
method.

a)   b)

Fig. 3. **a:** A detail of the Bayeux Tapestry (scene n°16), **b:** Contour image obtained with HED,
Holistically Nested Edge Detection method applied to the image (right).

3) ***DeepLab V3*** (Fig. 4b) [12]: deep convolutional neural network architecture used
for semantic segmentation of natural images. It is characterized by Encoder-
Decoder architecture. Despite being pre-trained to detect human in real-world
images, it transfers quite well to painting datasets (such as IconArtV2), but less so
to the extremely stylized figures of the Bayeux Tapestry. This method would
require a complete retraining or a fine-tuning on an image dataset closer to the
domain content of the Tapestry, which could be the object of future endeavors.

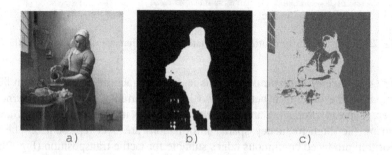

a)   b)   c)

Fig. 4. **a** The painting milkwoman by Jan Vermeer, **b** DeepLab gives the region where the
woman is detected in the image. **c** GMM with 2 regions (dark/light)

For the extraction of contours, HED seems to provide the most relevant contours, with the least discontinuities and thus the easiest to follow.

3.2   Clustering

We also considered several clustering-based approaches to segment the Tapestry's content in an unsupervised manner:

1) **K-means clustering algorithm** aims to partition pixels into a preset number of regions based on their perceived similarity. For the Tapestry, the number of clusters giving the best segmentation results is between 4 to 6 clusters. But some segmentation errors occur where adjacent pixels of different objects have low contrast.

2) **Gaussian mixture model** (GMM) [13] is a latent probabilistic clustering method which is used to estimate the parameters of distributions representing features potentially involved in the generation of the observed data (i.e. the content of the image), by modeling them as a finite mixture of Gaussians. The results obtained with this method are presented in Fig. 4c.

3) **Slic Superpixels** [14]: is a clustering method that divides the image into groups of connected pixels based on a weighted combination of their chrominance similarity and their spatial proximity. These superpixels can later be merged in bigger regions based on their semantic meaning. This solution could make a useful interactive tool to help exhibition curators when manually segmenting paintings.

Between the aforementioned methods, K-means and GMM are better suited for the segmentation of the Bayeux Tapestry than DeepLab and Superpixels. Indeed, DeepLab requires further domain specialization for the specific painting style of the tapestry.

K-Means and GMM will therefore be used in our approach for preliminary segmentations. The proposed image processing pipeline is based on several of the methods previously evaluated:

1. Preprocessing is done to choose the object semi-automatically:
   1.1. Mean-Shift method for minimizing the number of color clusters and eliminating the noisy background (fabric),
   1.2. Grabcut to extract relevant objects and reduce errors due to the influence of spatial proximity during clustering.
2. Integrating extracted contours and clusters:
   2.1. We apply a GMM on the extracted image (since K-means resulted in more noisy regions) to obtain better clusters. In Fig. 5b, the methods of K-Means with the cluster number of 6, is applied to the extracted Horseman of Fig. 5a. GMM gives a better result with 4 clusters; the horse and the horseman are better separated (Fig. 5c).
   2.2. To produce a more intuitive tactile image with relevant details and relief, we then apply HED to the GMM output. Then contours obtained that way are cleaner and the overall output less noisy than after applying HED to the original image, (Fig. 6).

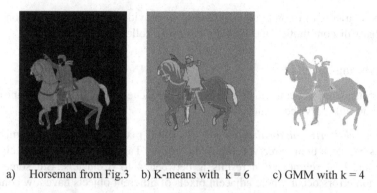

a)   Horseman from Fig.3   b) K-means with k = 6   c) GMM with k = 4

Fig. 5. K-Means and GMM are applied on a Horseman in Bayeux Tapestry.

Fig. 6. A horseman with HED applied to the GMM image from Fig. 5c (left), HED contours are highlighted with the integration of region information from GMM image.

4   Audio and Tactile Interfaces

Various interfaces have been designed to allow haptic-based communication, such as taxel matrices and vibrating surfaces. However, most of those devices do not allow the simultaneous display of contours and textures, and are often expensive. After reviewing and evaluating different technologies, we developed our own device: a Force Feedback Tablet (F2T) shown in Fig. 7.

When using the F2T, the exploration is controlled by a micro joystick on a motorized mobile support, which provides information on the underlying image through force-feedback variations, making it possible to create different passive (in response to user movements) and active (guidance) effects. Furthermore, audio information (such as audio-descriptions or ambient sounds) can be linked to the movements of the user to provide additional semantic information. Preliminary evaluations of the F2T were carried out on the recognition of simple geometric shapes, directions, perceived angles and spatial structures (arrangement of rooms) [15].

Fig. 7. F2T Prototype (left), concept of a tactile tablet (right).

5   Application to Cultural Heritage

After obtaining the final segmented image (See Fig. 6), we attribute different tactile effects to each cluster based on its meaning (human, horse, etc.). Figure 8 shows the regions and contours of the horseman linked to different tactile textures (i.e. different friction effects of the F2T), making it possible to feel the difference between the two regions. The contours in the image (green overlay) are simulated as domed surface with a small slope which is perceived as an edge during exploration with the F2T.

Fig. 8. Application on the Force Feedback Tablet with the contours in green producing a slope effect under the moving touch of a finger, the texture shown in red is simulated with a liquid friction effect and the texture shown with blue with a solid friction effect. (Color figure online)

Future evaluations will be conducted to assess the efficiency of each tactile effect (and its intensity) on the overall understanding of the displayed image.

6   Conclusion and Perspectives

This paper introduced preliminary results on a system facilitating the transposition of graphical artworks into haptic representations, using a specific interface, the F2T, which could improve the accessibility of Museums to VIP and facilitate the discovery of cultural heritage. This innovative audio-haptic interface allows VIP to actively and autonomously explore simplified version of artworks, where meaningful objects are extracted from the overall scene by a combination of edge detection and semantic segmentation algorithms.

Further research and development will be conducted to improve the segmentation pipeline, and to further optimize the conversion of segmented objects into haptic representations. We plan to develop solutions to facilitate a collaborative and iterative segmentation between the selected algorithms and domain specialists such as museum curators, in order to better capture the intentions of artworks' authors.

Acknowledgment. This research work is the result of an ongoing collaboration between the University of Rouen-Normandy, the University of Caen-Normandy, the Region of Normandy, the Department of Seine Maritime, the city of Bayeux, the Museum of Bayeux Tapestry, the Museum "Château of Martainville", the Museum of Quai Branly and several Associations of Visually Impaired People (FAF, ANL, AVH).

References

1. Chen, Y.: Analyse et interpretation d'images à l'usage des personnes non-voyantes, Application à la generation automatique d'images en relief à partir d'équipements banalisés. Thesis, University of Paris 8 (2015)
2. Chen, Y., Haddad, Z., Lopez Krahe, J.: Contribution to the automation of the tactile images transcription process. In: Miesenberger, K., Fels, D., Archambault, D., Peňáz, P., Zagler, W. (eds.) ICCHP 2014. LNCS, vol. 8547, pp. 642–649. Springer, Cham (2014). https://doi.org/10.1007/978-3-319-08596-8_99
3. Hatwell, Y.: Le développement perceptivo-moteur de l'enfant aveugle. Enfance 1 **55**, 88–94 (2003)
4. Hatwell, Y., Streri, A., Gentaz, E.: Toucher pour Connaître-Psychologie cognitive de la perception tactile manuelle. Psychologie et science de la pensée, Presse Universitaire de France, p 193 (2000)
5. Eriksson, Y.: The Swedish Library of Talking Books and Braille. How to make tactile pictures understandable to the blind reader, Disability Information Resources, JSRPD. https://www.dinf.ne.jp/doc/english/Us_Eu/conf/z19/z19001/z1900116.html. Accessed 29 Mar 2020
6. Strickfaden, M., Vildieu, A.: Questioning communication in tactile representation. In: International Conference on Inclusive Design, Include 2011, London (2011)
7. Ansaldi, B.: Perspective and the blind. In: Cocchiarella, L. (ed.) ICGG 2018. AISC, vol. 809, pp. 541–552. Springer, Cham (2019). https://doi.org/10.1007/978-3-319-95588-9_44
8. Romeo, K., Chottin, M., Ancet, P., Pissaloux, E.: Access to artworks and its mediation by and for visually impaired persons. In: Miesenberger, K., Kouroupetroglou, G. (eds.) ICCHP 2018. LNCS, vol. 10897, pp. 233–236. Springer, Cham (2018). https://doi.org/10.1007/978-3-319-94274-2_32

9. Romeo, K., Chottin, M., Ancet, P., Lecomte, C., Pissaloux, E.: Simplification of painting images for tactile perception by visually impaired persons. In: Miesenberger, K., Kouroupetroglou, G. (eds.) ICCHP 2018. LNCS, vol. 10897, pp. 250–257. Springer, Cham (2018). https://doi.org/10.1007/978-3-319-94274-2_35

10. Winnemöller, H., Kyprianidis, J.E., Olsen, S.C.: XFoG: an eXtended difference-of-Gaussians compendium including advanced image stylization. Comput. Graph. **36**(6), 720–753 (2012)

11. Xie, S., Zhuowen, T.: Holistically-nested edge detection. Int. J. Comput. Vis. **125**(1–3), 3–18 (2017). https://doi.org/10.1007/s11263-017-1004-z

12. Chen, L.-C., et al.: DeepLab: semantic image segmentation with deep convolutional nets, atrous convolution, and fully connected CRFs. IEEE Trans. Pattern Anal. Mach. Intell. **40** (4), 834–848 (2017)

13. Reynolds, D.A.: Gaussian mixture models. In: Encyclopedia of Biometrics (2009)

14. Achanta, R., Smith, K., Lucchi, A., Fua, P., Susstrunk, S.: SLIC superpixels, Technical report, EPFL, 149300 (2010)

15. Pissaloux, E., Romeo, K., Chottin, M., Ancet, P.: TETMOST-Toucher et Etre Touché, Les vertus inclusives du Mouvement et de la Sensibilité Tactiles. HANDICAP 2018, Paris (2018)

Use of Imaging Techniques as a Support for the Preventive Conservation Strategy of Wall Paintings: Application to the Medieval Decors of the Château de Germolles

Christian Degrigny[1,3]($\boxtimes$) (iD), Frank Boochs[2], Laura Raddatz[2],
Jonas Veller[2], Carina Justus[2], and Matthieu Pinette[3] (iD)

[1] Haute Ecole Arc Conservation-restauration HES-SO University of Applied Sciences and Arts Western Switzerland, Espace de l'Europe 11, 2000 Neuchâtel, Switzerland
christian.degrigny@he-arc.ch

[2] Institute for Spatial Information and Surveying Technology, University of Applied Sciences - i3Mainz, Lucy-Hillebrand-Straße 2, 55128 Mainz, Germany
frank.boochs@geoinform.hs-mainz.de,
{laura.raddatz, jonas.veller, carina.justus}@hs-mainz.de

[3] Château de Germolles, 100 place du 5 Septembre 1944, 71640 Mellecey, France
matthieu.pinette@gmail.com

Abstract. Imaging techniques were used to document and monitor physical damage to the unique wall paintings at the Château de Germolles, Burgundy, France. Photogrammetry combined with scanning are the most appropriate techniques to monitor the evolution of microcrack networks in the cornice overhanging the paintings and preserved as a witness to 19th century additions.

However, the application of these techniques was challenged due to the given constraints of a working height of 4 m and the required accuracy of a tenth of a millimetre. A special effort was therefore necessary to ensure sufficient stability of the acquisition protocol and to make it relevant during the four measurement campaigns planned over the two years of the project.

The analysis of photogrammetric data has made it possible to document certain macro-deformations of the cornice according to the seasons of the year. The microcracks could be visualized and monitored from transformed 3D models of each segment for the different campaigns. The results obtained show only local movements, mainly on the walls that are most exposed to the specific climatic conditions of each season.

Keywords: Medieval decorations · Physical damage · Imaging techniques · Photogrammetry · Scanning · Monitoring · Preventive conservation

© Springer Nature Switzerland AG 2020
A. El Moataz et al. (Eds.): ICISP 2020, LNCS 12119, pp. 24–34, 2020.
https://doi.org/10.1007/978-3-030-51935-3_3

1   Introduction

The château de Germolles, the only princely residence of the Dukes of Burgundy Valois to be so well preserved, has been benefiting for about ten years from a major campaign to document its 14th century mural paintings [1].

This work has made it possible to specify the original character of the residual decorations, to study the nature of the constituent materials and the pictorial techniques used, to evaluate their conservation condition and to propose hypotheses for restitution to the public visiting the château.

Cracks are visible in the upper parts of the decorations, seeming to indicate certain physical damages. These have been confirmed by IR thermography: they appear to be detachments between the painted plaster and the walls.

The risk of detachment of the painted decorations and loss of this unique heritage associated with the negative impact on the tourist activities of the site has led the managers of the château to set up a campaign to monitor the damages observed. This preventive conservation approach is the subject of this article.

In the following, we describe the site and its painted decorations, the knowledge recently acquired about them, including the physical damages observed. We then present the imaging techniques used to understand the risks of evolution of these damages. Finally, we present some results which are then discussed in the general context of safeguarding this unique Burgundian heritage.

2   Wall Paintings in the Château de Germolles: An Overview

2.1   Historical Background, Rediscovery and Restoration of the Paintings

The Château de Germolles is located in Southern Burgundy, 10 km west of Chalon-sur-Saône, France. Built in the second half of the 14th century, this residence is the only example of its type still well preserved in France. Philip the Bold, Duke of Burgundy and brother of King Charles V, acquired the estate of Germolles in 1380. The following year, he offered it to his wife, Margaret of Flanders. Important works are immediately undertaken; they will last 20 years. The aim was to transform the austere and archaic 13th century fortress into a country palace. To do this, the Duchess summoned the artists who were attached to the ducal couple: the architect Drouet de Dammartin, the sculptor Claus Sluter and the painter Jean de Beaumetz [2].

The main building is undoubtedly the most exceptional part of the place. It has retained its three-storey structure, where the first floor is the suite of the hold. This is where the paintings of Jean de Beaumetz and his workshop are located. Medieval account books provide information on the artists involved and the list of materials used. We learn that important metal decorations were used [3].

Part of the decorations were accidentally rediscovered in a room during World War II under the layer of plaster applied after the wall paintings had been keyed at the beginning of the 19th century. However, the clearing of the walls was not completed until the 1970s. White letters P and M, initials of the first names of the princes, repeated diagonally, and thistles, also white, appear between these letters. According to

Fig. 1. Photograph of the west wall of Margaret of Bavaria's dressing-room in visible light, showing the initials of the Dukes of Burgundy separated by thistles, © Papiashvili.

medieval records, we were able to identify the room as the dressing-room of Margaret of Bavaria, daughter-in-law of Margaret of Flanders. The château was listed as historical monument in 1989 and a major restoration campaign carried out from 1989 to 1995 led to filling in the holes due to the keying process and re-homogenizing the decoration using the *trattegio* technique. Figure 1 shows the preserved decorations of the west wall of the room. The lower panelling, as well as the cornice and ceiling, still bear witness to the transformations undergone by the château following the French Revolution.

2.2   Re-documentation of the Paintings

Apart from a few rare analyses, the restoration work was not documented. Moreover, they do not reflect the extreme richness of the decorations, especially the presence of the metallic ornaments mentioned above.

The involvement of the main author of this article both in the activities of the château as co-manager and in those of the European COSCH action (Colour and Space in Cultural Heritage - www.cosch.info) as Swiss national delegate for the Haute Ecole Arc Conservation-restoration, Neuchâtel, led him to carry out some preliminary work on these paintings through short term scientific missions of the COSCH action. This work, which revealed the original character of these decorations and thus their interest, made it possible to make these wall paintings a case study of the COSCH Action. It then benefited from greater technical and scientific support from the action, as well as from the Regional Direction of Cultural Affairs (DRAC) Burgundy Franche-Comté, which provided additional funding.

The "Germolles" project took place over 4 years, from 2012 to 2016. Although other decorations had been discovered since then, it focused on the dressing-room of Margaret of Bavaria and made it possible to:

- Collect all existing historical records on the paintings and their restoration;
- Appreciate the extent of the original decoration and confirm the respectful character of the restoration intervention of 1989–1995;

- Study with the most innovative and least possible invasive imaging and analytical techniques the constituent materials, in particular the residues of the metallic decorations found on the thistles;
- Specify the pictorial techniques used: the letters M and P were created with stencils and painted in reserve on the green background while the thistles, made of a green tin foil covered with fine gold leaf enhanced with paint, were glued on the latter. The letters P were also embellished with arabesques.
- Suggest the initial rendering using augmented reality (Fig. 2).

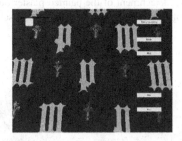

Fig. 2. Proposal for rendering the wall decoration of Margaret of Bavaria's dressing-room using augmented reality, © Germolles. (Colour figure online)

The information collected have been partly integrated into the guided tour and the public enjoying the augmented reality experience can better perceive the refinement of the interior decoration which, in the Middle Ages, was supposed to reflect the interest of the princes in the subtleties of the surrounding nature. Walking through the princely floor of the château, one wanders through green meadows dotted with symbolic floral decorations (the thistle is the symbol of fidelity and protection) separating the letters of the Dukes, including the P of Philip courteously embellished.

2.3   Physical Damage to Be Documented

The documentation work carried out on the paintings also revealed significant physical damage to their upper part under the 19th century cornice. Figure 3 shows the areas of detachment between the painted plaster and the west wall of the Margaret of Bavaria's dressing-room quantitatively documented through IR thermography. The same damage was observed on the other three walls, of which only the east wall is interior.

The 19th century cornice is also dotted with networks of cracks (Fig. 4a) resulting in other damage, including local detachment of the filling material used during the conservation intervention of 1989–1995 (Fig. 4b).

The mural decorations of the Château de Germolles are one of the highlights of the tour of the site. The physical detachment of the upper part of the painted plaster, induced or not by movements of the cornice, would not only be an irreversible damage but would also be detrimental to the appreciation of the painted decoration by the visitors. The site managers therefore wished to be reassured about the overall stability.

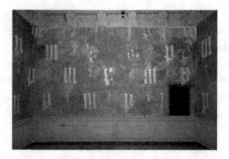

Fig. 3. Distribution of areas of detachment of the painted plaster detected by IR thermography on the west wall of Margaret of Bavaria's dressing-room, © Tedeschi & Cucchi. (Colour figure online)

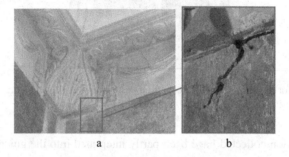

a   b

Fig. 4. Cracks in the 19th-century cornice and ceiling in the southwest corner of Margaret of Bavaria's dressing-room (a) and a detail showing a plaster filling detachment (b), © Germolles.

To this end, they requested the technical support of the Institute for Spatial Information and Surveying Technology of the University of Applied Sciences (i3mainz) in Mainz, Germany.

3 Imaging Techniques Used

Since the 19th century cornice is believed to play a key role in the observed physical damage, we decided to document and monitor its network of microcracks over time. Because of the geometric accuracy and micro-resolution required, only high-resolution structured light techniques, embedded in a frame of close-range photogrammetry, are able to provide the necessary quality. Although these techniques are well known and well tested [4–6], their application has been made difficult by the position of the object about 4 m above the ground and the limited possibilities to use installations on or around the object. The analysis of changes over time therefore also had to be done in a relative manner.

3.1 Preparation of the Cornice

Fortunately, a few targets could be placed over the entire surface of the cornice providing a minimum of geometric information for data acquisition. As these have to be discreet and removed non-invasively, circular uncoded black and white targets of 1 cm diameter were used, glued with Paraloid B72 diluted in acetone. In order to prevent the latter from irreparably marking the plaster of the cornice when the targets are removed, a layer of cyclododecane, which sublimates in the atmosphere, was previously applied between the plaster and the adhesive. The targets are not referenced in a superordinate system (Fig. 5a).

a   b

Fig. 5. Target distribution on a cornice segment of Margaret of Bavaria's dressing-room (a) and photogrammetric campaign on the south wall (triangulated area in red) (b), © Germolles. (Colour figure online)

3.2 Workflow of the Photogrammetric and Scanning Campaigns

In order to obtain the actual geometry of the targets, we applied a photogrammetric triangulation prior to each campaign. Due to the location of the cornice, we established a temporary point field allowing to connect the images and to position scale bars. During the different campaigns, it was visible that the movement of the scaffolding used by the operator (Fig. 5b) induced vibrations resulting in movements of the temporary points. The points frame was therefore reinforced in order to improve the stability of the network. As camera, we used a Nikon D800 with a 35 mm lens that was pre-calibrated on-site.

We used GOM TRITOP Professional 2016 as a tool for triangulation and achieved an average rms quality of 0.1 pixel in the images, 0.033 mm at the object points and 0.05 mm at the scale bars. This gave us the basis for the geometrical analyses of the object in the requested order of a tenth of millimetre. The target points give the frame for the following scanning process performed with a GOM ATOS Core 500, providing individual scans with a resolution of 0.19 mm and a coverage of 360 * 380 mm^2 per scan. This resulted in 20 scans per segment. During the scanning process, the scanner was mounted on a tripod, which was securely attached to the scaffolding platform (Fig. 6). This non-standard build-up resulted in an increased vulnerability to scanner movements during the measurement. Although the scanning software checks the data in order to detect measurement inaccuracies due to scanner movements, some perturbations were created in the 3D mesh which were eliminated through post processing. GOM ATOS Professional 2016 software was used for data acquisition and processing.

Fig. 6. Scanning campaign with the scanner GOM ATOS Core on the east wall, © Germolles. (Colour figure online)

Four measurement campaigns were carried out: two in late spring (campaigns 1 and 3 in May–June 2018 and 2019), a few weeks after the château's heating system was shut down, and two in autumn (campaigns 2 and 4 in late September-early October 2018 and 2019), just before the heating system was restarted. Each campaign provided individual 3D models of the cornice for each of the walls together with the corresponding field of target points.

4   Results

4.1   Deformation of the Cornice Over Time

Both data sets (3D models, 3D target points) allow us to analyse changes of the geometry over time. In this context, two questions need to be addressed: (1) does an individual wall show a deformation over time, and (2) are there microscopic changes (such as cracks) that can be identified and express possible local deterioration? Since we do not have an absolute reference serving as base for an analysis, we compared all individual data sets by means of geometric transformations. In the case of the targets points we could apply regular coordinate transformations, while for the 3D models we applied best fit transformations which are part of the GOM software package. One example for transformation results is shown in Fig. 7, comparing the 3D models of the summer and early autumn campaign (west wall). It shows clear global changes where the centre of the wall behaves opposite to the corners.

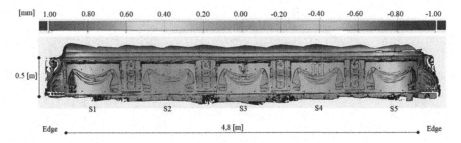

Fig. 7. Result of best fit transformation for two 3D models of the cornice of the west wall (campaigns 1 & 2), © i3Mainz. (Colour figure online)

Further studies of these global changes use the field of target points transformed by means of a least square 6 parameter transformation. Figures 8a & 8b compare the results of campaigns 1 & 2 and 2 & 3. As in Fig. 7, we identify global changes from spring to autumn 2018 (1 & 2) and also from autumn 2018 to spring 2019 (2 & 3). As local trends are similar but in opposite directions, it appears that cornice movements are documented as small oscillations (less than 0.5 mm) where areas move back and forth.

Similar phenomena were observed on the south wall between the 1st and 3rd campaigns, although more pronounced and to a much lesser extent between the 3rd and 4th campaigns. The north wall and the east inner wall were each measured in two parts. It is therefore not possible to comment on the movement of the whole wall as for the other walls. The cornice moved much less in both parts of the east wall. The smallest movement was observed in the parts of the north wall.

4.2 Visualization of Micro-deformations

Global deformations of the walls make it difficult to analyse deformations on the millimetre scale (at the level of the cracks). In order to answer the second question (are there local deformations), we analysed the individual wall segments (S1–S5 on Fig. 7 for west wall) separately. For this purpose, we applied a best fit transformation between two entire 3D models of an individual segment and visualized the results. As a visualization tool, we used 3DHOP and added functionality to display all difference models (from 1 & 2, 1 & 3,… up to 3 & 4).

Figure 9 gives examples for the west wall (left side: S1, right side: S4). Figures 9a and 9c show that between scan campaigns 2 and 3, segment 1 of the cornice of the west wall of Margaret of Bavaria's dressing-room moved little, except in the area indicated by a red circle. In this wide crack, it appears that one of the fragments moved from the

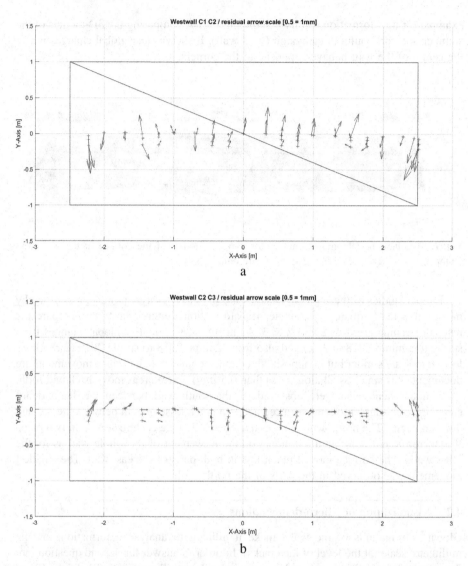

Fig. 8. Visualization of the relative displacement of the targets of the cornice of the west wall of Margaret of Bavaria's dressing-room between the 1st and 2nd (a) and between the 2nd and 3rd campaigns (b), © i3Mainz. (Colour figure online)

inside to the outside. This movement is confirmed between scan campaigns 3 and 4 (Fig. 9e) but to a lesser extent and extends to the lower part of the cornice segment. Colour/uncoloured edges of the difference model (red arrows Fig. 9e) result from difficulties to access the area directly below the ceiling, resulting in less overlap of the scans. These side-effects are not documenting physical movements. A larger object movement is observed on the side of one of the volutes between campaigns 3 and 4 (Fig. 9f).

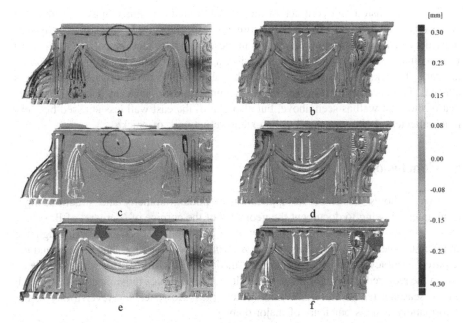

Fig. 9. 3D models of segments 1 and 4 of the cornice of the west wall of Margaret of Bavaria's dressing-room and comparison between scans 1/2 (a), 2/3 (c) and 3/4 (e) for segment 1 and 1/2 (b), 1/3 (d) and 1/4 (f) for segment 4, © i3Mainz. (Colour figure online)

These few damages are the only ones that could be clearly observed on the four walls during the four campaigns.

4.3 Discussion

It should be noted that in the absence of an absolute reference, we only analysed the relative deformations of the different elements or segments of the 19th century cornice. This choice of approach is linked to the prestigious nature of the mural paintings, which did not allow targets to be fixed on them.

The movements of the cornice seem more related to its position on a wall exposed to more or less sun and western rain than to the change of seasons. This is the case for the south and west walls and appears to be reflected in an outward expansion movement of the cornice during the summer and early autumn of 2018, while it appears to have contracted from late autumn 2018 to spring 2019. This phenomenon seems to be repeated for the west wall, but to a lesser extent, the following year. It is less acute in the second year for the south wall.

The cornice of the north wall, which is only sunny at the end of the day in summer and receives some rain, is much less affected. The fact that it is interrupted in its central part by a window may also play a role. The cornice of the inner east wall seems to be subject to some movement in the southeast corner.

These expansion and contraction movements remain minimal (max. 0.5 mm) but cause local damage of some cornice segments, especially those that are already partially detached or near major cracks. It is interesting to note that the cornice of the west wall seems to be the most affected as shown in Fig. 9 but also with its important network of cracks. The cornice of the south wall, which is under great stresses, is less affected. Its movements, however, have an impact on the adjacent cornice elements, that of the west wall as seen above, but also that of the east wall less stressed but very much affected as shown again by the extensive network of cracks.

5   Conclusion

This work shows results of the application of photogrammetry and structured light techniques allowing to document the geometry of the cornice of Margaret of Bavaria's dressing-room at Château de Germolles over a period of two years. Although the constraints given by the object as such (size, extend, position under the ceiling) and the required accuracy (tenth of a millimetre) made the task difficult, it was possible to obtain the required quality for the analysis. It revealed cornice movements, as expected, due to seasonal temperature and humidity changes. However, the reversibility of the deformations dispels our fears of major damage.

Acknowledgments. This project would not have been possible without the financial support by Regional Direction of Cultural Affairs of Burgundy – Franche-Comté which the authors wish to sincerely thank.

References

1. Degrigny, C., Piqué, F., et al.: Wall paintings in the château de Germolles: an interdisciplinary project for the rediscovery of a unique fourteenth century decoration. In: Bentkowska-Kafel, A., MacDonald, L. (eds.) Digital Techniques for Documenting and Preserving Cultural Heritage, pp. 67–86. Arc Humanities Press, Kalamazoo (2017)
2. Beck, P.: Vie de cour en Bourgogne à la fin du Moyen Âge. Alan Sutton, Saint-Cyr-sur-Loire (2002)
3. ADCO B 4434-1. Archives départementales de Côte d'Or: Baillage de Dijon, compte ordinaire – 1389–1390. f. 22v° - 24
4. Hanke, K., Böhler, W.: Recording and visualizing the cenotaph of German Emperor Maximilian I. Int. Arch. Photogram. Remote Sens. **35**(5), 413–418 (2004)
5. Wiemann, A.-K., Boochs, F., Karmacharya, A., Wefers, S.: Characterisation of spatial techniques for optimised use in cultural heritage documentation. In: Ioannides, M., Magnenat-Thalmann, N., Fink, E., Žarnić, R., Yen, A.-Y., Quak, E. (eds.) EuroMed 2014. LNCS, vol. 8740, pp. 374–386. Springer, Cham (2014). https://doi.org/10.1007/978-3-319-13695-0_36
6. Luhmann, Th.: Nahbereichsphotogrammetrie, Grundlagen – Methoden – Beispiele. Wichmann, Berlin (2018). ISBN 978-3-87907-640-6

Multispectral Dynamic Codebook and Fusion Strategy for Moving Objects Detection

Rongrong Liu[(✉)], Yassine Ruichek, and Mohammed El Bagdouri

Connaissance et Intelligence Artificielle Distribuées (CIAD),
University Bourgogne Franche-Comté, UTBM, 90010 Belfort, France
{rongrong.liu,yassine.ruichek,mohammed.el-bagdouri}@utbm.fr

Abstract. The Codebook model is one of the popular real-time models for background subtraction to detect moving objects. In this paper, we propose two techniques to adapt the original Codebook algorithm to multispectral images: dynamic mechanism and fusion strategy. For each channel, only absolute spectral value is used to calculate the spectral similarity between the current frame pixel and reference average value in the matching process, which can simplify the matching equations. Besides, the deciding boundaries are obtained based on statistical information extracted from the data and always adjusting themselves to the scene changes. Results demonstrate that with the proposed techniques, we can acquire a comparable accuracy with other methods using the same multispectral dataset for background subtraction.

Keywords: Moving objects detection · Background subtraction · Multispectral images · Dynamic Codebook · Fusion strategy

1 Introduction

Moving object detection is one of the most commonly encountered low-level tasks in computer vision and a prerequisite of many intelligent video processing applications, such as automated video surveillance [1], object tracking [2,3] and parking lot management [4], to just name a few.

A common and widely used approach for detecting moving objects from stationary camera is background subtraction. As the name suggests, it is the process to automatically generate a binary mask which classifies the set of pixels into foreground and background, during which, the moving objects are always called the foreground and the static information is called the background. Thus, background subtraction is sometimes also known as foreground detection and foreground-background segmentation [5].

The last decade witnessed very significant publications on background subtraction. A quick search for "background subtraction" on IEEE Xplore returns over 2370 publications in the last ten years (2009–2019). Among these, Gaussian

A. El Moataz et al. (Eds.): ICISP 2020, LNCS 12119, pp. 35–43, 2020.
https://doi.org/10.1007/978-3-030-51935-3_4

Mixture Model (GMM) [6] modeled every pixel with a mixture of k Gaussian to handle multiple backgrounds. To deal with cluttered and fast variation scenes, a non-parametric technique known as Kernel Density Estimation (KDE) [7] was also proposed. There are also other methods that process images not only at the level of pixel, but also at region or even at frame levels [8]. Moreover, Kim et al. [9], with Codebook algorithm, summarized each background pixel by one or more codewords to cope with illumination changes. In our work, Codebook approach has been chosen as the base modeling frame for its simplicity and efficiency.

According to [10], most of the methods are based on color images, namely Red-Green-Blue (RGB). In recent years, thanks to the technological advances in video capture, multispectral imaging is becoming increasingly accepted and used by the computer vision and robotic communities, in particular, Benezeth et al. [11] present a publicly available collection of multispectral video sequences called multispectral video sequences (MVS) dataset. This is the first dataset on MVS available for research community in background subtraction.

In this paper, we try to utilize the different channels of multispectral images from a new perspective and propose a novel background subtraction framework motivated by Codebook algorithm. The goal is to simplify the similarity calculation between the current frame pixel and reference average value in the matching process.

The remainder of this paper is organized as follows. The proposed methods is presented in detail in Sect. 2. Section 3 demonstrates the experiments and comparisons with other methods on the MVS dataset. Final conclusions and future works are given in Sect. 4.

2   Proposed Multispectral Codebook Algorithms

The original Codebook background modeling algorithm was proposed in 2005 by Kim [9] to construct a background model from long observation sequences. It is quite simply a non-statistical clustering scheme with several important additional elements to make it robust against moving background. The motivation for adopting such a model is that it is fast to run, because it is deterministic; efficient for requiring little memory; adaptive; and able to handle complex backgrounds with sensitivity [12]. It has been proved to be very efficient in dealing with dynamic backgrounds.

During the last decade, many works have been dedicated to improve the original Codebook model [13,14]. For example, [15] has adopted multi-scale multi-feature codebook-based background subtraction targeting for challenging environments. In the work of [16], the background model is constructed by encoding each pixel into a codebook consisting of codewords based on a box model and it is also appropriate in the Hue-Saturation-Value (HSV) color space. Besides, [17] has rewritten the model parameters and then processed the three channels separately to simplify the matching equations. In [18], a dynamic boundary of codebook under the Lab color space has been developed.

2.1 Multispectral Dynamic Codebook

As a parametric method, the original Codebook needs parameter tuning to find the appropriate values for every scene. In this paper, we utilize a self-adaptive method motivated by [19], to select the optimal parameters automatically. In order of achieve this, some statistical information need to be calculated iteratively and recorded for each codeword during the whole process.

Multispectral Dynamic Codebook Model Initialization. At the beginning, the codebook is an empty set and the number of the codewords is set to 0. When the first multispectral frame comes, the Codebook model is initialized by constructing an associated codeword for each pixel. The corresponding vector $\mathbf{v}_m$ is set to be the average spectral values for all the channels and it is initialized as below:

$$\mathbf{v} = \mathbf{x} = (X_1, X_2, \ldots, X_n) \tag{1}$$

where n is the number of channels. The auxiliary information will be defined as a four-tuple $\mathbf{aux}_m = (f_m, \lambda_m, p_m, q_m)$, where f_m is the frequency of access or the number of times that the codeword is matched, λ_m is maximum length of time between consecutive accesses, p_m and q_m are the first and the last accesses times of the codeword respectively. They are initialized as:

$$\mathbf{aux} = (1, 0, 1, 1) \tag{2}$$

In spite of the vector $\mathbf{v}_m$ and the auxiliary tuple $\mathbf{aux}_m$, we also record a third vector named $\mathbf{S}_m$, which represents the set of the variance of the separate spectrum σ_i^2 and is initialized as:

$$\mathbf{S} = (X_1^2, X_2^2, \ldots, X_n^2) \tag{3}$$

What's more, another two vectors $\mathbf{B\_min}$ and $\mathbf{B\_max}$ need to be used to record the minimum and maximum values for each channel and they are initialized with the values of the spectral values of the first frame.

Multispectral Dynamic Codebook Matching Process. For an input pixel at time instant t, with the current value $\mathbf{x}_t = (X_1, X_2, \ldots, X_n)$, the matching codeword is found if

$$B\_low_i \leq X_i \leq B\_high_i \tag{4}$$

is satisfied for each channel, where B_low and B_high denote the lower and upper boundaries, respectively and are obtained by:

$$B\_low_i = max(B\_min_i - \sigma_i, \; 0) \tag{5}$$

$$B\_high_i = max(B\_max_i + \sigma_i, \; 255) \tag{6}$$

where σ_i is the standard deviation of the ith band value in the current codeword, whose square is the corresponding i^{th} element of $\mathbf{S}_m$. Thus, during this whole process, the boundaries are obtained from the data themselves and no manual parameters tuning is required, which makes the proposed model more practical.

Multispectral Dynamic Codebook Model Updating. When a new frame arrives, the matching criteria is first evaluated to see whether it is satisfied. If a match is found, the corresponding codeword is updated with the information of the current pixel, as illustrated in Algorithm 1, where N is the number of frames.

Algorithm 1. Multispectral dynamic Codebook construction

1: $L \leftarrow 0$, $\mathbf{C} \leftarrow \phi$
2: **for** $t = 1 \rightarrow N$ **do**
3:     $x_t = (X_1, X_1, ..., X_n)$
4:     Find the matching codeword to x_t in C if the following condition is satisfied for each channel.
5:         $B\_low_i \leq X_i \leq B\_high_i$
6:     **if** $\mathbf{C} = \phi$ or there is no match **then**
7:         $L \leftarrow L + 1$, create a new codeword $\mathbf{c}_L$
8:         $\mathbf{v}_0 = \mathbf{x}_t$
9:         $\mathbf{aux}_0 = (1, t - 1, t, t)$
10:         $\mathbf{S}_0 = (X_1^2, X_2^2, \ldots, X_n^2)$
11:         $\mathbf{B\_min} = (X_1, X_2, \ldots, X_n)$
12:         $\mathbf{B\_max} = (X_1, X_2, \ldots, X_n)$
13:     **else**
14:         update the matched codeword
15:         $\mathbf{v}_m \leftarrow \left(\frac{f_m \overline{X}_{m1} + X_1}{f_m + 1}, \frac{f_m \overline{X}_{m2} + X_2}{f_m + 1}, \ldots, \frac{f_m \overline{X}_{mn} + X_n}{f_m + 1} \right)$
16:         $\mathbf{aux}_m \leftarrow (f_m + 1, max\lambda_m, t - q_m, p_m, t)$
17:         $\mathbf{S}_m \leftarrow \left(\frac{f_m \overline{\sigma}_{m1} + (X_1) - \overline{X}_{m1}^2}{f_m + 1}, \frac{f_m \overline{\sigma}_{m2} + (X_2) - \overline{X}_{m2}^2}{f_m + 1}, \ldots, \frac{f_m \overline{\sigma}_{mn} + (X_n) - \overline{X}_{mn}^2}{f_m + 1} \right)$
18:         $B\_min_i \leftarrow min(B\_min_i, X_i)$
19:         $B\_max_i \leftarrow max(B\_max_i, X_i)$
20:     **end if**
21: **end for**

Multispectral Dynamic Codebook Model Foreground Detection. After constructing the Codebook model, the moving objects are detected by conducting the matching process. The pixel is detected as foreground if no acceptable matching codeword exists. Otherwise, it is classified as background and the corresponding codeword is updated according to the updating strategy.

2.2   Multispectral Fusion Strategy

Another step forward to exploit benefits of each spectral channel of multispectral images is to fuse the detection results of the monochromatic channels. The idea is very straight forward. We first employ the multispectral dynamic Codebook which has been discussed in detail in the last subsection to every channel separately and obtain seven binary foreground-background masks independently. Then the detection results of the monochromatic channels (3 bands, 4 bands, 5 bands, 6 bands and 7 bands) are fused via union, vote or intersection to get the final foreground background segmentation result. The workflow for multispectral self-adaptive fusion strategy is shown in Fig. 1.

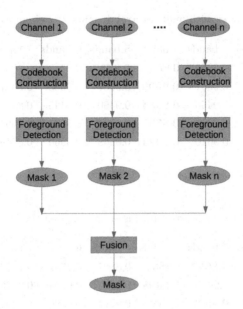

Fig. 1. Workflow of fusion strategy for multispectral self-adaptive Codebook

3 Experimental Results

To evaluate the performance of the proposed approaches for background subtraction, the MVS dataset presented by Benezeth et al. [11] is adopted for testing in this section. The proposed approaches are also compared with other methods using the same dataset [5]. Both visual and numerical results are displayed.

The MVS dataset contains a set of five challenging video sequences with seven multispectral channels or bands (six visible spectra and one near-infrared spectrum) captured simultaneously. These sequences are all publicly available, and the ground truth images are obtained manually. Note that the first scene is indoor, while the other four are outdoor.

The measure of accuracy employed in these experiments is F-measure, also known as balanced F-score or F1 score, which reaches its best value at 1 and worst score at 0. It is a harmonic mean of the precision and recall.

Firstly, the experiments for multispectral dynamic Codebook are conducted on the thirty-five different three-band-based combinations, thirty-five different four-band-based combinations, twenty-one different five-band-based combinations, seven different six-band-based combinations, and total seven-band, for the five videos composing the MVS dataset. Then the largest F-measures are selected and listed in Table 1. Accordingly, Table 2 shows the best F-measures with the fusion strategy. The largest F-measure for each video is in bold.

Table 1. Best F-measures with multispectral dynamic Codebook on the MVS dataset

Video	3 bands	4 bands	5 bands	6 bands	7 bands
1	0.8669	0.8750	**0.8757**	0.8749	0.7759
2	0.9607	**0.9609**	0.9607	0.9603	0.9307
3	0.9456	**0.9464**	0.9460	0.9445	0.8321
4	0.8693	**0.8731**	0.8734	0.8728	0.8661
5	0.9010	**0.9024**	0.8969	0.8966	0.8719

Table 2. Best F-measures with fusion strategy on the MVS dataset

Video	3 bands	4 bands	5 bands	6 bands	7 bands
1	**0.8967**	0.8836	0.8805	0.8712	0.7796
2	0.9607	**0.9608**	0.9607	0.9604	0.9575
3	**0.9372**	0.9346	0.9277	0.9020	0.8059
4	0.8685	0.8732	0.8741	**0.8742**	0.8709
5	0.8847	**0.8913**	0.8820	0.8646	0.8429

Figure 2 shows the visual results on the five video sequences. The top row is the original frames and the second row is the corresponding ground truth frames provided together with the dataset. The third and forth rows are the results obtained by the proposed multispectral dynamic Codebook and fusion strategy, respectively.

From Tables 1 and 2, the largest F-measure never appear when all seven channels are used. The combination with four channels has the largest possibility to achieve the best performance. This agrees with the assertion deduced with the Pooling method proposed by Yannick et al. [11] that only few channels actually define better the moving objects.

The results of the proposed methods are further compared in Table 3 with other methods [5] using the same dataset, in which, the brightness (B), spectral distortion (SD) and spectral information divergence (SID) are adopted to calculate the spectral distance in the matching process. The average F-measures for

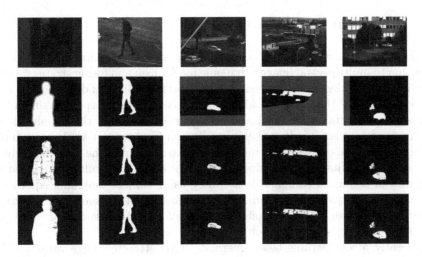

Fig. 2. Visual results of the proposed methods on MVS dataset

all the five videos and outdoor sequences are calculated and listed in the last two rows.

From average F-measures for the whole dataset, our approaches produce comparable results to the best method that utilizes all the three criteria in [5] (4th column in Table 3), which proves the effectiveness of our methods. As we can see for the method adopting B and SD, the accuracy for outdoor scenes outperforms in average all the other mechanisms, but it achieves less satisfactory result for the indoor video. The two techniques proposed in this paper can be compromising solutions. Another advantage of the proposed methods is the low complexity of the matching equations, as the multispectral channels are processed separately only utilizing the intensity value of each channel and no correlation between channels need to be considered and calculated in the matching process. Besides, like other compared methods, it is also quite easy to adapt the proposed two algorithms to situation with any number of channels for background subtraction.

Table 3. F-measures with different methods on MVS dataset

Video	B+SD	B+SID	B+SD+SID	Proposed dynamic	Proposed fusion
1	0.8060	0.9219	0.9147	0.8757	0.8967
2	0.9643	0.9538	0.9642	0.9609	0.9608
3	0.9248	0.8939	0.9213	0.9464	0.9372
4	0.9001	0.8807	0.8979	0.8734	0.8742
5	0.9198	0.8842	0.8948	0.9024	0.8913
Mean	0.9030	0.9069	0.9186	0.9118	0.9121
Outdoor mean	0.9273	0.9032	0.9196	0.9208	0.9160

4   Conclusion and Future Works

In this paper, we have proposed two techniques to adapt the original Codebook algorithm to multispectral images: dynamic mechanism and fusion strategy, both of which process the seven channels of multispectral images independently. For each channel, only the intensity value is used to calculate the spectral similarity between the current frame pixel and reference one. Besides, the thresholds to determine are not set in advance empirically and fixed for the whole procedure, but obtained based on statistical information extracted from the data themselves and can always adjusting themselves to the scene changes. Results demonstrated that we can acquire a comparable accuracy using simpler matching equations than other techniques, when conducting experiments on the same multispectral public dataset. Our work may offer a new way for future works for applying multispectral images in moving objects detection.

Recently, deep learning-based methods have attracted huge attention in research community for its impressive performance for classification, semantic segmentation, localization, object detection and instance detection. Motivated by the recent success of deep neural networks for foreground segmentation [20], our next work is to explore deep learning besides traditional machine learning methods to investigate the benefits of multispectral images to improve the performance of background subtraction.

References

1. Khan, A., Janwe, N.J.: Review on moving object detection in video surveillance. Int. J. Adv. Res. Comput. Commun. Eng. **6**, 664–670 (2017)
2. Yang, T., Cappelle, C., Ruichek, Y., El Bagdouri, M.: Online multi-object tracking combining optical flow and compressive tracking in Markov decision process. J. Vis. Commun. Image Represent. **58**, 178–186 (2019)
3. Yang, T., Cappelle, C., Ruichek, Y., El Bagdouri, M.: Multi-object tracking with discriminant correlation filter based deep learning tracker. Integr. Comput.-Aided Eng. **26**, 1–12 (2019)
4. Yusnita, R., Norbaya, F., Basharuddin, N.: Intelligent parking space detection system based on image processing. Int. J. Innov. Manag. Technol. **3**(3), 232–235 (2012)
5. Liu, R., Ruichek, Y., El-Bagdouri, M.: Extended codebook with multispectral sequences for background subtraction. Sensors **19**(3), 703 (2019)
6. Bouwmans, T., El Baf, F., Vachon, B.: Background modeling using mixture of gaussians for foreground detection-a survey. Recent Pat. Comput.Sci. **1**(3), 219–237 (2008)
7. Elgammal, A., Harwood, D., Davis, L.: Non-parametric model for background subtraction. In: Vernon, D. (ed.) ECCV 2000. LNCS, vol. 1843, pp. 751–767. Springer, Heidelberg (2000). https://doi.org/10.1007/3-540-45053-X_48
8. Toyama, K., Krumm, J., Brumitt, B., Meyers, B.: Wallflower: principles and practice of background maintenance. In: Proceedings of the Seventh IEEE International Conference on Computer Vision, vol. 1, pp. 255–261. IEEE, Kerkyra (1999). https://doi.org/10.1109/ICCV.1999.791228

9. Kim, K., Chalidabhongse, T.H., Harwood, D., Davis, L.: Real-time foreground-background segmentation using codebook model. Real-Time Imaging **11**(3), 172–185 (2005)
10. Bouwmans, T.: Traditional and recent approaches in background modeling for foreground detection: an overview. Comput. Sci. Rev. **11**, 31–66 (2014)
11. Benezeth, Y., Sidibé, D., Thomas, J.B.: Background subtraction with multispectral video sequences. In: IEEE International Conference on Robotics and Automation Workshop on Non-Classical Cameras, Camera Networks and Omnidirectional Vision (OMNIVIS), Hong Kong, China (2014)
12. Doshi, A., Trivedi, M.: "Hybrid Cone-Cylinder" codebook model for foreground detection with shadow and highlight suppression. In: 2006 IEEE International Conference on Video and Signal Based Surveillance. IEEE, Sydney (2006). https://doi.org/10.1109/AVSS.2006.1
13. Liyun, G., Miao, Y., Timothy, G.: Online codebook modeling based background subtraction with a moving camera. In: 3rd International Conference on Frontiers of Signal Processing (ICFSP), pp. 136–140. IEEE, Paris (2017)
14. Mousse, M. A., Ezin, E.C., Motamed, C.: Foreground-background segmentation based on codebook and edge detector. In: Tenth International Conference on Signal-Image Technology and Internet-Based Systems, pp. 119–124. IEEE, Marrakech (2014)
15. Zaharescu, A., Jamieson, M.: Multi-scale multi-feature codebook-based background subtraction. In: 2011 IEEE International Conference on Computer Vision Workshops (ICCV Workshops), pp. 1753–1760. IEEE, Barcelona (2011). https://doi.org/10.1109/ICCVW.2011.6130461
16. Tu, Q., Xu, Y., Zhou, M.: Box-based codebook model for real-time objects detection. In: 2008 7th World Congress on Intelligent Control and Automation, pp. 7621–7625. IEEE, Chongqing (2008). https://doi.org/10.1109/WCICA.2008.4594112
17. Ruidong, G.: Moving object detection based on improved codebook model. In: 2nd International Conference on Modelling, Identification and Control. Atlantis, Paris (2015). https://doi.org/10.2991/mic-15.2015.4
18. Kusakunniran, W., Krungkaew, R.: Dynamic codebook for foreground segmentation in a video. Trans. Comput. Inf. Technol. (ECTI-CIT) **10**(2), 144–155 (2016)
19. Shah, M., Deng, J.D., Woodford, B.J.: A Self-adaptive CodeBook (SACB) model for real-time background subtraction. Image Vis. Comput. **38**, 52–64 (2015)
20. Lim, L.A., Keles, H.Y., Woodford, B.J.: Learning multi-scale features for foreground segmentation. Pattern Anal. Appl. **23**, 1369–1380 (2019). https://doi.org/10.1007/s10044-019-00845-9

A Spectral Hazy Image Database

Jessica El Khoury[1]($\boxtimes$), Jean-Baptiste Thomas[2], and Alamin Mansouri[1]

[1] ImViA Laboratory, Université Bourgogne Franche-Comté, Dijon, France
`jessica.el-khoury@u-bourgogne.fr`
[2] The Norwegian Colour and Visual Computing Laboratory,
Norwegian University of Science and Technology (NTNU),
2815 Gjøvik, Norway

Abstract. We introduce a new database to promote visibility enhancement techniques intended for spectral image dehazing. SHIA (Spectral Hazy Image database for Assessment) is composed of two real indoor scenes M1 and M2 of 10 levels of fog each and their corresponding fog-free (ground-truth) images, taken in the visible and the near infrared ranges every 10 nm starting from 450 to 1000 nm. The number of images that form SHIA is 1540 with a size of 1312 × 1082 pixels. All images are captured under the same illumination conditions. Three of the well-known dehazing image methods based on different approaches were adjusted and applied on the spectral foggy images. This study confirms once again a strong dependency between dehazing methods and fog densities. It urges the design of spectral-based image dehazing able to handle simultaneously the accurate estimation of the parameters of the visibility degradation model and the limitation of artifacts and post-dehazing noise. The database can be downloaded freely at http://chic.u-bourgogne.fr.

Keywords: Hazy image database · Dehazing · Image quality

1 Introduction

In computer vision applications, dehazing is applied to enhance the visibility of outdoor images by reducing the undesirable effects due to scattering and absorption caused by atmospheric particles.

Dehazing is needed for human activities and in many algorithms like objects recognition, objects tracking, remote sensing and sometimes in computational photography. Applications that are of interest in this scope include fully autonomous vehicles typically that use computer vision for land or air navigation, monitored driving or outdoor security systems. In bad visibility environments, such applications require dehazed images for a proper performance.

Image dehazing is a transdisciplinary challenge, as it requires knowledge from different fields: meteorology to model the haze, optical physics to understand how light is affected through haze and computer vision as well as image and signal processing to recover the parameters of the scene. Researchers have been always

A. El Moataz et al. (Eds.): ICISP 2020, LNCS 12119, pp. 44–53, 2020.
https://doi.org/10.1007/978-3-030-51935-3_5

searching for an optimal method to get rid of degradation by light scattering along aerosols. Many methods have been proposed and compared to each other. However, available methods still do not meet efficient recovery standards and show a varying performance depending on the density of haze [7].

Earlier approaches involve multiple inputs to break down the mathematical ill-posed problem. Narasimhan *et al.* [18] calculate the haze model parameters by considering the variation of the color of pixels under different weather conditions. Feng *et al.* [9] take advantage of the deep penetration of near-infrared wavelength to unveil the details that could be completely lost in the visible band. Other ways consist in employing depth data [13] or images differently polarized [20]. Later techniques mainly focus on single image dehazing approach, which is more challenging but more suitable for the purpose of real time and costless computer vision applications. Single image dehazing was promoted through the work of He *et al.* [11], the well-known Dark Channel Prior, which gained its popularity thanks to its simple and robust real assumption based on a statistical characteristic of outdoor natural images. Therefore, numerous versions were released later, some of them propose an improvement in estimating one or more of the model's parameters and others extend the approach to other fields of application [14]. This approach, like others such as filtering based method [21], estimates explicitly transmission and airlight. Other methods overlook the physical model and improve contrast through multi-scale fusion [3], variational [10] or histogram equalization approaches [27]. Recently, like many research domains, several machine learning approaches for image dehazing have come to light [5,15]. These models are trained on synthetic images built upon a simplistic model comparing to reality [24]. Hence the importance to build a large number of real hazy images.

To evaluate various dehazing methods, some databases of hazy images are available. Tarel *et al.* [22,23] used FRIDA and FRIDA2 as two dehazing evaluating databases dedicated to driving assistance applications. They are formed of synthetic images of urban road scenes with uniform and heterogenous layers of fog. There exist also databases of real outdoor haze-free images, for each, different weather conditions are simulated [28].

Given the significant research that has been conducted through the last decade, it turns out that synthetic images formed upon a simplified optical model do not simulate faithfully the real foggy images [6]. Therefore, several databases of real images and real fog with the groundtruth images have emerged. The real fog was produced using a fog machine. This was first used in our previous work presented in [8] and it was used later by Ancuti *et al.* [2,4] to construct a good number of outdoor and indoor real hazy images covering a large variation of surfaces and textures. Lately, they introduced a similar database containing 33 pairs of dense haze images and their corresponding haze-free outdoor images [1].

The main contribution of this paper is SHIA, which is inspired from our previous color image database CHIC [8]. To the best of our knowledge, SHIA is the first database that presents for a given fog-free image, a set of spectral images with various densities of real fog. We believe that such database will promote visibility enhancement techniques for drone and remote sensing images. It will represent also a useful tool to valid future methods of spectral image dehazing. Although it contains only 2 scenes, it stands for an example to consider and to integrate efforts on a larger scale to increase the number of such complex databases.

After the description of the used material and the acquisition process of the scenes, we provide a spectral dimension analysis of the data. Then, we evaluate the three color dehazing methods that have been applied to single spectral images. Our experimental results underline a strong dependency between the performance of dehazing methods and the density of fog. Comparing to color images, the difference between dehazing methods is minor, since color shifting, which is usually caused by dehazing methods is not present here. The difference is mainly due to the low intensity, especially induced by the physical based methods and the increase of noise after dehazing.

2   Data Recording

2.1   Used Material

The hyperspectral data was obtained using the Photon focus MV1-D1280I-120-CL camera based on e2v EV76C661 CMOS image sensor with 1280×1024 pixel resolution.

In order to acquire data in visible and Near-infrared (VNIR) ranges, two models of VariSpec Liquid Crystal Tunable Filters (LCTF) were used: VIS, visible-wavelength filters with a wavelength range going from 450 to 720 nm. NIR, near-infrared wavelength filter with a wavelength range going from 730 to 1100 nm. Every 10 nm in the VIS range and in the NIR range, we captured a picture with a single integration time of 530 ms, which allows a sufficient light to limit the noise without producing saturated pixels over channels. This reduces as well the complexity of the preprocessing spectral calibration step (cf. Sect. 2.3).

In order to provide the image depth of the captured scenes, which could be a relevant data to assess approaches, a Kinect device was used. The Kinect device can detect objects up to 10 m but it induces some inaccuracies beyond 5 m [16]. Therefore, the camera was standing at 4.5 m from the furthest point at the center of the scene.

To generate fog, we used the fog machine FOGBURST 1500 with the flow rate 566 m³/min and a spraying distance of 12 m, which emits a dense vapor that appears similar to fog. The particles of the ejected fog are water droplets whose radius is close to the radius size of the atmospheric fog (1–10 μm) [19].

2.2   Scenes

Scenes were set up in a closed rectangular room (length = 6.35 m, width = 6.29 m, height = 3.20 m, diagonal = 8.93 m) with a window (length = 5,54 m, height = 1.5 m), which is large enough to light up the room with daylight. The acquisition session was performed on a cloudy day to ensure an unfluctuating illumination. The objects forming the scenes were placed in front of the window, by which the sensors were placed. This layout guarantees a uniform diffusion of daylight over the objects of the scenes.

After the set up of each scene and before introducing fog, a depth image was first captured using the Kinect device, and it was then replaced by the Photon focus camera, which kept the same position through the capture of images at various fog densities of the same scene. The different densities of fog were generated by spreading first an opaque layer of fog, which was then evacuated progressively through the window. The same procedure was adopted for the acquisition of visible and near infrared images.

Hence, the dataset consists of two scenes, M1 and M2. The images of the scene M1 are only acquired over the visible range (450–720 nm) for technical reasons. M2's images are captured in visible and NIR (730–1000 nm) ranges (Figs. 1 and 2). In the first set the lamp in the middle of the scene is turned off and turned on in the second. For each acquisition set, 10 levels of fog were generated besides the fog-free scene. As result, there are 308 images for M1: 11 levels (10 levels of fog + fog-free level), in each there are 28 spectral images taken at every 10 nm from 450 to 720 nm). On the other hand, there are 1232 images for M2: on the basis of M1's images calculation, M2_VIS and M2_NIR's images are 616 each (308 for lamp on scene and 308 for lamp off scene).

2.3 Data Processing

We performed a dark correction to get rid of the offset noise that appears all over the image, and a spectral calibration to deal with the spectral sensitivities of the sensor and the used filters. The dark correction consists in taking several images in the dark with the same integration time. For each pixel, we calculate the median value over these images. Therefore, we obtain the dark image. We then subtract the dark image from the spectral images taken with the same integration time. The negative values are set equal to zero [17].

For the spectral calibration, we considered the relative spectral response of the camera and the filter provided in the user manuals. For each captured image at each wavelength band with an integration time of 530 ms, we divided by the maximum peak value of the spectral response of the sensor and the corresponding filter.

3 Spectral Dimension Analysis

In order to investigate the effective spectral dimension of the spectral images, we used the Principal Component Analysis (PCA) technique [12]. For this analysis, we computed the minimum number of dimensions required to preserve 95% and 99% of the total variance of the spectral images of a given scene. Table 1 shows the minimum number of dimensions computed on the scenes M2 Lamp on and M2 Lamp off considering visible and NIR components at the three levels of fog,

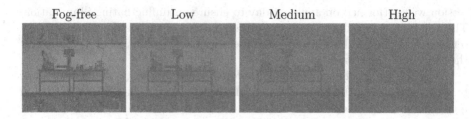

Fig. 1. M2_VIS Lamp off. Visible image at 550 nm with its corresponding images taken under low, medium and high levels of fog.

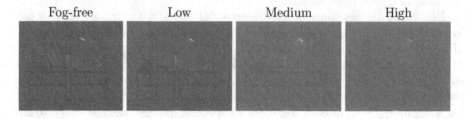

Fig. 2. M2_NIR Lamp on. NIR image at 850 nm with its corresponding images taken under low, medium and high levels of fog.

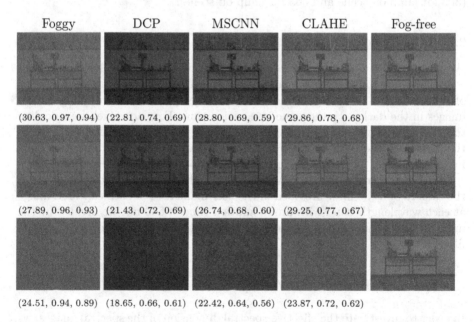

Fig. 3. M2_VIS Lamp off taken at 550 nm. Dehazed images processed by DCP, MSCNN and CLAHE methods, with the foggy images presented in the first column and the corresponding fog-free images in the last column. The first, second and third rows correspond to the low, medium and high levels of fog, respectively. Under each image, its corresponding scores are showed as follows: (PSNR, SSIM, MS-SSIM).

Table 1. The minimum number of dimensions required to preserve 95% and 99% of the total variance of M2 Lamp on and M2 Lamp off scenes.

	VIS_on		VIS_off		NIR_on		NIR_off	
	95%	99%	95%	99%	95%	99%	95%	99%
Fog-free	2	3	2	5	1	2	2	21
Low	2	3	2	5	1	2	6	23
Medium	1	2	1	5	16	25	21	26
High	1	2	1	8	21	26	20	26

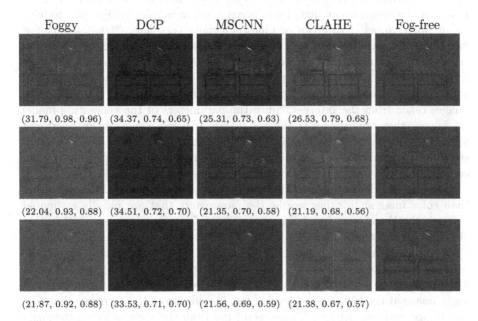

Foggy   DCP   MSCNN   CLAHE   Fog-free

(31.79, 0.98, 0.96) (34.37, 0.74, 0.65) (25.31, 0.73, 0.63) (26.53, 0.79, 0.68)

(22.04, 0.93, 0.88) (34.51, 0.72, 0.70) (21.35, 0.70, 0.58) (21.19, 0.68, 0.56)

(21.87, 0.92, 0.88) (33.53, 0.71, 0.70) (21.56, 0.69, 0.59) (21.38, 0.67, 0.57)

Fig. 4. M2_NIR Lamp on taken at 850 nm. Dehazed images processed by DCP, MSCNN and CLAHE methods, with the foggy images presented in the first column and the corresponding fog-free images in the last column. The first, second and third rows correspond to the low, medium and high levels of fog, respectively. Under each image, its corresponding scores are showed as follows: (PSNR, SSIM, MS-SSIM).

low, medium and high. We can observe that for M2 Lamp on, fog-free images and images with very light fog, show an effective dimensions of one or two to preserve 95% of the total variance and from two to four dimensions to preserve 99% of the total variance. Images with a denser fog, require almost the same dimensions as light-fog images in the visible range and more dimensions in the NIR range. This is very likely to be caused by the sensor noise, which is accentuated on dark images. This can be observed from the number of dimensions required at NIR range of M2 lamp off (darker than M2 Lamp on), which is significantly high at fog-free and low levels. Considering the particles size of the fog used to construct this database, the spectral properties at NIR wavelengths is very close to visible wavelengths. To take advantage of its spectral particularities and get better visibility, images at higher infrared wavelengths will be required [6].

4   Evaluation of Dehazing Techniques

The images of SHIA database have been used to evaluate three of the most representative categories of single image dehazing approaches: DCP [11], MSCNN [5], and CLAHE [27]. DCP and MSCNN are both physical-based methods. DCP relies on the assumption that, for a given pixel in a color image of a natural scene, one channel (red, green or blue) is usually very dark (it has a very low intensity). The atmospheric light tends to brighten these dark pixels. Thus, it is estimated over the darkest pixels in the scene. MSCNN introduces DehazeNet, a deep learning method for single image haze removal. DehazeNet is trained with thousands of hazy image patches, which are synthesized from haze-free images taken from the Internet. Since the parameters of the generating model are known, they are used for training. CLAHE, which is a contrast enhancement approach, consists of converting an RGB image into HSI color space. The intensity component of the image is processed by contrast limited adaptive histogram equalization. Hue and saturation remain unchanged.

These methods have been adjusted to be applied on spectral images rather than color images. In other words, the parameters that are usually estimated through the three color bands, were estimated from the single spectral image. For the sake of readability, we have selected three levels of fog, which are denoted by low, medium and high levels (Figs. 1 and 2). In this article, we only display the dehazed images of the scene M2_VIS Lamp off at 550 nm (Fig. 3) and the scene M2_NIR lamp on at 850 nm (Fig. 4). The first row in these figures represent the foggy image at the low selected level of fog, in addition to the corresponding fog-free image and the dehazed images resulting from the three selected dehazing methods. Similarly, the second and the third rows represent the foggy image at the medium and low levels, respectively. We have calculated the scores of the classical metrics used to evaluate spectral images: PSNR, which calculate the absolute error between images; SSIM [25], which consider the neighborhood dependencies while measuring contrast and structure similarity; and MS-SSIM, which is a multiscale SSIM [26], and performs particularly well in assessing sharpness [7]. A higher quality is indicated by a higher PSNR and closer SSIM and MS-SSIM to 1. The corresponding values are written under the images in Figs. 3 and 4. The average values calculated over a few selected wavelengths in the VNIR range are given in Table 2.

Through the visual assessment of the dehazed images presented in Figs. 3 and 4, we can observe that all methods, regardless their approach and hypotheses, perform better at low fog densities, either at visible or near infrared range. CLAHE, which does not consider the physical model of image degradation, eliminates well the fog. However, it induces an important amount of noise that increases with the density of fog. DCP, which is a physical-based approach, fails to estimate accurately the unknown parameters of the image degradation model, the airlight and the transmission of light [11]. This bad estimate produces dim dehazed images, especially at high densities of fog, where the dark channel hypothesis fails. This accords with the observation made on color images presented in our previous work [7]. MSCNN performs also an inversion of the physical model of visibility degradation. However, it estimates better the unknown parameters comparing to DCP since it is trained on a large number of hazy images. This can be deduced through its dehazed images, which are not as dark as the DCP's dehazed images are.

The metrics values provided in Figs. 3 and 4 have the same trends for color dehazed images across fog densities [7]. They show an increase in quality when the density of fog decreases. However, they underline a global low performance of dehazing methods. This means that haze removal is associated with secondary effects that restrains quality enhancement. This is likely to be handicapped by the noise and the artifacts induced in the image and the dark effect resulted from wrong estimation of visibility model parameters. These effects seem to have a more severe impact on image quality than the fog itself.

From Table 2, we can conclude that foggy images are quantitatively of better quality comparing to the dehazed images, which suffer from noise, low intensity and structural artifacts; the scores resulting from different dehazing methods are very close to each other across wavelengths; the metrics values demonstrate a correlated performance between MSCNN and CLAHE over wavelengths. Although DCP has relatively higher scores, this does not mean it is the best performing method. The dimness of its resulting images seems to minimize the effect of the artifacts.

Table 2. The average values of PSNR, SSIM and MS-SSIM (MS) metrics calculated on the images taken under 10 densities of fog at 550, 650, 750 and 850 nm

	Foggy			DCP			MSCNN			CLAHE		
	PSNR	SSIM	MS	PSNR	SSIM	MS	PSNR	SSIM	MS	PSNR	SSIM	MS
550 nm	26.98	0.94	0.91	22.34	0.69	0.63	24.51	0.65	0.56	25.03	0.73	0.63
650 nm	21.63	0.88	0.78	23.13	0.77	0.64	21.06	0.61	0.45	20.72	0.63	0.48
750 nm	33.23	0.97	0.95	24.03	0.69	0.66	26.07	0.70	0.63	28.10	0.70	0.66
850 nm	29.53	0.93	0.89	26.71	0.71	0.67	26.92	0.69	0.61	25.90	0.70	0.59

5   Conclusions

We introduce a new database to promote visibility enhancement techniques intended for spectral image dehazing. For two indoor scenes, this hard built database SHIA, contains 1540 images taken at 10 levels of fog, starting from a very light to a very opaque layer, with the corresponding fog-free images. The applied methods introduce the same effects induced in color images, such as structural artifacts and noise. This is underlined by pixelwise quality metrics when they are compared to foggy images. Accordingly, future works should focus on reducing these effects while considering the particularities of spectral foggy images that need to be further investigated.

References

1. Ancuti, C.O., Ancuti, C., Sbert, M., Timofte, R.: Dense haze: a benchmark for image dehazing with dense-haze and haze-free images. arXiv preprint arXiv:1904.02904 (2019)
2. Ancuti, C.O., Ancuti, C., Timofte, R., De Vleeschouwer, C.: O-haze: a dehazing benchmark with real hazy and haze-free outdoor images. In: Proceedings of the IEEE Conference on Computer Vision and Pattern Recognition Workshops, pp. 754–762 (2018)
3. Ancuti, C.O., Ancuti, C.: Single image dehazing by multi-scale fusion. IEEE Trans. Image Process. **22**(8), 3271–3282 (2013)
4. Ancuti, C., Ancuti, C.O., Timofte, R., De Vleeschouwer, C.: I-HAZE: a dehazing benchmark with real hazy and haze-free indoor images. In: Blanc-Talon, J., Helbert, D., Philips, W., Popescu, D., Scheunders, P. (eds.) ACIVS 2018. LNCS, vol. 11182, pp. 620–631. Springer, Cham (2018). https://doi.org/10.1007/978-3-030-01449-0_52
5. Cai, B., Xu, X., Jia, K., Qing, C., Tao, D.: DehazeNet: an end-to-end system for single image haze removal. IEEE Trans. Image Process. **25**(11), 5187–5198 (2016)
6. El Khoury, J.: Model and quality assessment of single image dehazing (2016). http://www.theses.fr/s98153
7. El Khoury, J., Le Moan, S., Thomas, J.-B., Mansouri, A.: Color and sharpness assessment of single image dehazing. Multimed. Tools Appl. **77**(12), 15409–15430 (2017). https://doi.org/10.1007/s11042-017-5122-y
8. El Khoury, J., Thomas, J.-B., Mansouri, A.: A database with reference for image dehazing evaluation. J. Imaging Sci. Technol. **62**(1), 10503-1 (2018)
9. Feng, C., Zhuo, S., Zhang, X., Shen, L., Süsstrunk, S.: Near-infrared guided color image dehazing. In: 2013 IEEE International Conference on Image Processing, pp. 2363–2367. IEEE (2013)
10. Galdran, A., Vazquez-Corral, J., Pardo, D., Bertalmio, M.: Enhanced variational image dehazing. SIAM J. Imaging Sci. **8**(3), 1519–1546 (2015)
11. He, K., Sun, J., Tang, X.: Single image haze removal using dark channel prior. IEEE Trans. Pattern Anal. Mach. Intell. **33**(12), 2341–2353 (2010)
12. Hotelling, H.: Analysis of a complex of statistical variables into principal components. J. Educ. Psychol. **24**(6), 417 (1933)
13. Kopf, J., et al.: Deep photo: model-based photograph enhancement and viewing. ACM Trans. Graph. (TOG) **27**(5), 1–10 (2008)
14. Lee, S., Yun, S., Nam, J.-H., Won, C.S., Jung, S.-W.: A review on dark channel prior based image dehazing algorithms. EURASIP J. Image Video Process. **2016**(1), 4 (2016). https://doi.org/10.1186/s13640-016-0104-y
15. Liu, R., Ma, L., Wang, Y., Zhang, L.: Learning converged propagations with deep prior ensemble for image enhancement. IEEE Trans. Image Process. **28**(3), 1528–1543 (2018)

16. Mankoff, K.D., Russo, T.A.: The Kinect: a low-cost, high-resolution, short-range 3D camera. Earth Surf. Proc. Land. **38**(9), 926–936 (2013)
17. Mansouri, A., Marzani, F.S., Gouton, P.: Development of a protocol for CCD calibration: application to a multispectral imaging system. Int. J. Robot. Autom. **20**(2), 94–100 (2005)
18. Narasimhan, S.G., Nayar, S.K.: Chromatic framework for vision in bad weather. In: Proceedings IEEE Conference on Computer Vision and Pattern Recognition, CVPR 2000 (Cat. No. PR00662), vol. 1, pp. 598–605. IEEE (2000)
19. Nayar, S.K., Narasimhan, S.G.: Vision in bad weather. In: Proceedings of the Seventh IEEE International Conference on Computer Vision, vol. 2, pp. 820–827. IEEE (1999)
20. Schechner, Y.Y., Narasimhan, S.G., Nayar, S.K.: Instant dehazing of images using polarization. In: CVPR, vol. 1, pp. 325–332 (2001)
21. Tarel, J.-P., Hautiere, N.: Fast visibility restoration from a single color or gray level image. In: 2009 IEEE 12th International Conference on Computer Vision, pp. 2201–2208. IEEE (2009)
22. Tarel, J.-P., Hautiere, N., Caraffa, L., Cord, A., Halmaoui, H., Gruyer, D.: Vision enhancement in homogeneous and heterogeneous fog. IEEE Intell. Transp. Syst. Mag. **4**(2), 6–20 (2012)
23. Tarel, J.-P., Hautiere, N., Cord, A., Gruyer, D., Halmaoui, H.: Improved visibility of road scene images under heterogeneous fog. In: 2010 IEEE Intelligent Vehicles Symposium, pp. 478–485. IEEE (2010)
24. Valeriano, L.C., Thomas, J.-B., Benoit, A.: Deep learning for dehazing: comparison and analysis. In: 2018 Colour and Visual Computing Symposium (CVCS), pp. 1–6. IEEE (2018)
25. Wang, Z., Bovik, A.C., Sheikh, H.R., Simoncelli, E.P.: Image quality assessment: from error visibility to structural similarity. IEEE Trans. Image Process. **13**(4), 600–612 (2004)
26. Wang, Z., Simoncelli, E.P., Bovik, A.C.: Multiscale structural similarity for image quality assessment. In: The Thirty-Seventh Asilomar Conference on Signals, Systems & Computers, vol. 2, pp. 1398–1402. IEEE (2003)
27. Xu, Z., Liu, X., Ji, N.: Fog removal from color images using contrast limited adaptive histogram equalization. In: 2009 2nd International Congress on Image and Signal Processing, pp. 1–5. IEEE (2009)
28. Zhang, Y., Ding, L., Sharma, G.: HazeRD: an outdoor scene dataset and benchmark for single image dehazing. In: 2017 IEEE International Conference on Image Processing (ICIP), pp. 3205–3209. IEEE (2017)

A Bottom-Up Approach for Pig Skeleton Extraction Using RGB Data

Akif Quddus Khan, Salman Khan, Mohib Ullah[✉], and Faouzi Alaya Cheikh

Norwegian University of Science and Technology, 2815 Gjøvik, Norway
mohib.ullah@ntnu.no

Abstract. Animal behavior analysis is a crucial task for the industrial farming. In an indoor farm setting, extracting Key joints of animals is essential for tracking the animal for a longer period of time. In this paper, we proposed a deep network that exploits transfer learning to train the network for the pig skeleton extraction in an end to end fashion. The backbone of the architecture is based on an hourglass stacked dense-net. In order to train the network, keyframes are selected from the test data using K-mean sampler. In total, 9 Keypoints are annotated that gives a brief detailed behavior analysis in the farm setting. Extensive experiments are conducted and the quantitative results show that the network has the potential of increasing the tracking performance by a substantial margin.

Keywords: Pig · Behavior analysis · Hourglass · Stacked dense-net · K-mean sampler

1 Introduction

Automatic behavior analysis of different animal species is one of the most important tasks in computer vision. Due to variety of applications in the human social world like sports player analysis [1], anomaly detection [2], action recognition [3], crowd counting [4], and crowd behavior [5,6], humans have been the main focus of research. However, due to the growing demands of food supplies, vision-based behavior analysis tools are pervasive in the farming industry and demands for cheaper and systematic solutions are on the rise. From the algorithmic point of view, other than the characterization of the problem, algorithm design for humans and the farm animals are similar. Essentially, behavior analysis is a high-level computer vision task and consists of feature extraction, 3D geometry analysis, and recognition, to name a few. As far as the input data is concerned, it could be obtained through smart sensors (Radio-frequency identification [7], gyroscope [8], GPS [9]). Depending on the precision of measurements, such sensors give acceptable results but using such sensors has many drawbacks. For example, in most cases, it is required to remove the sensor from the animal to collect the data. Such a process is exhausting for the animals and laborious for the human operator. Compared to this, a video-based automated behavior

© Springer Nature Switzerland AG 2020
A. El Moataz et al. (Eds.): ICISP 2020, LNCS 12119, pp. 54–61, 2020.
https://doi.org/10.1007/978-3-030-51935-3_6

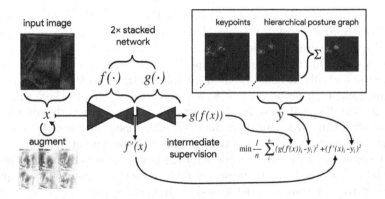

Fig. 1. Detailed illustration of model training process

analysis offers a non-invasive solution. Due to cheaper hardware, it is not only convenient for the animals but also cost-effective for the industry. Automatic behavior analysis and visual surveillance [10–12] has been used for the security of public places (airports, shopping malls, subways, etc.) and turned into a mature field of computer vision.

In this regard, Hu et al. [13] proposed a recurrent neural network named MASK-RNN for the instance level video segmentation. The network exploits the temporal information from the long video segment in the form of optimal flow and perform binary segmentation for each object class. Ullah et al. [14] extracted low level global and Keypoint features from video segments to train a neural network in a supervised fashion. The trained network classifies different human actions like walking, jogging, running, boxing, waving and clapping. Inspired from the human social interaction, a hybrid social influence model has been proposed in [15] that mainly focused on the motion segmentation of the moving entities in the scene. Lin et al. [16] proposed a features pyramid network that extract features at different level of a hierarchical pyramid and could potentially benefit several segmentation [13,17,18], detection [19,20], and classification [21,22] frameworks. Additionally, in the field of cybersecurity [23,24], such techniques are very beneficial. By addressing the problem of scale variability for object detection, Khan et al. [20] proposed a dimension invariant convolution neural network (DCNN) that compliment the performance of RCNN [19] but many other state-of-the-art object detectors [4,25] could take advantage of it. Inspired by the success of deep features, [26] proposed a two-stream deep convolutional network where the first stream focused on the spatial features while the second stream exploit the temporal feature for the video classification. The opensource deep framework named OpenPose proposed by Cao et al. [27] focuses on the detection of Keypoints of the human body rather than the detection of the whole body. Detection of Keypoints has potential applications in the pose estimation and consequently behavior analysis. Their architecture consists of two convolutional neural networks were the first network extract features and gives the location of the main joints of the human body in the form of a heat map. While the second network is responsible for associating the corresponding body joints. For the feature extraction, they used the classical VGG architecture. The frameworks like OpenPose are very helpful in skeleton extraction of the human body and potentially, it could be used in tracking framework. For example, the Bayesian framework proposed in [28] works as a Keypoint tracker, where any Keypoints like the position of head [29], or neck or any other body organ can be used to do tracking for longer time. Such Keypoints can be obtained from a variety of human pose estimation algorithms. For example, Sun et al. [30] proposed a parallel multi-resolution subnetworks for the human pose estimation. The idea of a parallel network helps to preserver high resolution and yield high quality features maps that results in better spatial Keypoints locations. Essentially, in such a setting, the detection module is replaced by [27,30]. In this regard, a global optimization approach like [31] could be helpful for accurate tracking in an offline setting. By focusing only on pose estimation of humans, Fang et al. [32] proposed a top-down approach where first the humans are detected in the form of bounding boxes and later, the joints and Keypoints are extracted through a regional multi-person pose estimation framework. Such a framework is helpful in not only in the localization and tracking of tracking in the scene, but also getting the pose information of all the targets sequentially. For a robust appearance model that could differentiate between different targets, a sparse coded deep features framework was proposed in [33] that accelerate the extraction of deep features from different layers of a pre-trained convolution neural network. The framework helps handle the bottleneck phenomenon of appearance modeling in the tracking pipeline. Alexander et al. [34] used transfer learning [35] and fine-tuned ResNet to detect 22 joints of horse for the pose estimation. They used the data collected from 30 horses for the within domain and out of domain testing. The work by Mathis et al. [36] analyzed the behavior of mice during the experimental tasks of tail-tracking, and reach & pull of joystick tasks. They also analyze the behavior of drosophila while it lays eggs. The classical way of inferring behavior is to perform segmentation and tracking [37] first, and based on the temporal evolution of the trajectories, perform behavior analysis. However, approaches like [38] can be used to directly infer predefined actions and behaviors from the visual data. In addition to the visual data, physiological signals [39,40], and acoustic signals [41] can be used to identify different emotional states and behavioral traits in farm animals.

Compared to the existing methods, our proposed framework is focused on the extraction of the key joints of the pig in an indoor setting. The visual data is obtained from a head-mounted Microsoft Kinect sensors. Our proposed framework is inspired by [42] where a fully-convolutional stacked hourglass-shaped network is proposed that converts the image into a 16-channel space representing detection maps. For the part detection, the thresholds are set from 0.10 to 0.90. These thresholds are used while evaluating the recall, precision, and F-measure metrics for both the vector matching and euclidean matching results. Such an analysis provides a detailed overview of the trade-offs between precision and recall while maintaining an optimal detection threshold. The loss function, the optimizer, and the training details are also given in Sect. 3. The qualitative results are mentioned in Sect. 4 and the remarks are given in Sect. 5 which concludes the paper.

The rest of the paper is organized in the following order. In Sect. 2 the proposed method is briefly explained including the Keypoints used in the experiment, the data filtration and annotation, and the augmentation. Model architecture along with the loss function, the optimizer, and the training details are elaborated in Sect. 3. The qualitative results are given in Sect. 4 and the remarks are given in Sect. 5 which concludes the paper.

2   Proposed Approach

The block diagram of the network is given in Fig. 1. It mainly consists of two encoder-decoder stacked back to back. The convolution neural network used in each encoder-decoder network is based on the dense net. The network takes the input as the visual frame. To train the model, we annotate the data by first converting videos into individual frames, and then annotating each frame separately by specifying the important key points on the animal's body. After sufficient training, the model returns a 9×3 matrix for each frame, where each row corresponds to one keeping, the first two columns specify the x and y coordinates of the detected point, and the third column contains the confidence score of the model. After obtaining the x and y coordinates for each frame, we visualize these key points on each frame and stitch all the individual frames into a single video. A total of nine (Nose, Head, Neck, Right Foreleg, Left Foreleg, Right Hind leg, Left Hind Leg, Tail base, and tail tip) key points is being focused for each pig.

2.1   Data Filtration

The given RGB data consists of three sets with three pigs, six pigs, and ten pigs. Each dataset has 2880 images. To get better and more accurate results, a larger dataset was required. However, K-Mean clustering is applied to each dataset for selecting the most informative frames. As a result, only 280 images are extracted from the larger dataset. The count is approximately 10% of the size of the original dataset. Data annotator developed by Jake Graving is used to annotate the dataset. It provides a simple graphical user interface that reads key points data from the CSV file and saves the data in .h5 format once the annotation is completed. DeepPoseKit works with augmenters from the imgaug package. We used spatial augmentations with axis flipping and affine transforms.

3   Model Architecture

The proposed framework is based on an hourglass densenet which is an efficient multi-scale deep-learning model. The architecture consists of an encoder and decoder where dense nets are stacked in sequential order. Densenet is a densely Connected Convolutional Networks [43]. DenseNet can be seen as the next generation of convolutional neural networks that are capable of increasing the depth of the model with every decreasing the number of parameters.

3.1 Loss Function and Optimizer

Mathematically, the loss function is:

$$L(x, y) = \frac{1}{n} \sum_{i}^{n} ((g(f(x))_i - y_i)^2 + (f'(x) - y_i)^2) \qquad (1)$$

where x is the input sample and y corresponds to the network prediction. We used the callback function using *ReduceLROnPlateau*. ReduceLROnPlateau automatically reduces the learning rate of the optimizer when the validation loss stops improving. This helps the model to reach a better optimum at the end of training. While training a model on three pigs' data, first a test was run for data generators. Creating a TrainingGenerator from the DataGenerator for training the model with annotated data is an important factor. The TrainingGenerator uses the DataGenerator to load image-keypoints pairs and then applies the augmentation and draws the confidence maps for training the model. The validationSplit argument defines how many training examples to use for validation during training. If a dataset is small (such as initial annotations for active learning), we can set this to validationSplit = 0, which will just use the training set for model fitting. However, when using callbacks, we made sure to set monitor = "loss" instead of monitor = "valloss". To make sure the Training Reduce learning rate parameters saves useless resource utilization and model overfeeding. For this particular reason, the parameter is set to reduce the learning rate by 0.2 if the loss does not improve after 20 iterations. Another parameter that is used to prevent resource exploitation is Early Stopping. Patience is set 100 iterations which means training would stop automatically if the loss does not improve after 100 iterations. Training started at a loss of 220, after running 400 iterations, the loss stopped showing improvement at 4.5. In the test case, when given the same video from which dataset was generated, very accurate results are produced.

4 Experiments

The proposed framework is implemented in Python with the support of Keras backend by Tensorflow. The processing is performed on Nvidia P-100 with 32 GB RAM. The qualitative results are shown in Fig. 2. It can be seen that the keypoints are successfully extracted from the pig joints. However, sometimes, when the pigs are very close to each other, the extracted keypoints are associated with the wrong animal. It is simply because the network is trained on very limited data. Training the network with more data to help improve the association of keypoints.

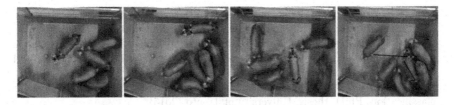

Fig. 2. Qualitative results of the proposed method.

5   Conclusion

We proposed a deep network for animal skeletonization based on only RGB data. The network exploits use varies data augmentation and transfer learning to fine-tune the parameters. The backbone of the network is based on an hourglass stacked dense-net. In order to train the network, keyframes are selected from the test data using K-mean sampler. In total, 9 Keypoints are annotated that gives a brief detailed behavior analysis in the farm setting. Experiments are conducted on pig data and the quantitative results that training the network with only 280 frames yields promising results.

References

1. Khan, S.D., et al. Disam: density independent and scale aware model for crowd counting and localization. In: 2019 IEEE International Conference on Image Processing (ICIP), pp. 4474–4478. IEEE (2019)
2. Ullah, H., Altamimi, A.B., Uzair, M., Ullah, M.: Anomalous entities detection and localization in pedestrian flows. Neurocomputing **290**, 74–86 (2018)
3. Yang, J., Shi, Z., Ziyan, W.: Vision-based action recognition of construction workers using dense trajectories. Adv. Eng. Inform. **30**(3), 327–336 (2016)
4. Khan, S.D., et al.: Person head detection based deep model for people counting in sports videos. In: 2019 16th IEEE International Conference on Advanced Video and Signal Based Surveillance (AVSS), pp. 1–8. IEEE (2019)
5. Ullah, M., Ullah, H., Conci, N., De Natale, F.G.B.: Crowd behavior identification. In: 2016 IEEE International Conference on Image Processing (ICIP), pp. 1195–1199. IEEE (2016)
6. Ullah, H., Ullah, M., Conci, N.: Dominant motion analysis in regular and irregular crowd scenes. In: Park, H.S., Salah, A.A., Lee, Y.J., Morency, L.-P., Sheikh, Y., Cucchiara, R. (eds.) HBU 2014. LNCS, vol. 8749, pp. 62–72. Springer, Cham (2014). https://doi.org/10.1007/978-3-319-11839-0_6
7. Maselyne, J., et al.: Measuring the drinking behaviour of individual pigs housed in group using radio frequency identification (RFID). Animal **10**(9), 1557–1566 (2016)
8. Ullah, M., Ullah, H., Khan, S.D., Cheikh, F.A.: Stacked LSTM network for human activity recognition using smartphone data. In: 2019 8th European Workshop on Visual Information Processing (EUVIP), pp. 175–180. IEEE (2019)
9. Pray, I.W., et al.: GPS tracking of free-ranging pigs to evaluate ring strategies for the control of cysticercosis/taeniasis in Peru. PLoS Negl. Trop. Dis. **10**(4), e0004591 (2016)

10. Alreshidi, A., Ullah, M.: Facial emotion recognition using hybrid features. In: Informatics, vol. 7, p. 6. Multidisciplinary Digital Publishing Institute (2020)
11. Chen, J., Li, K., Deng, Q., Li, K., Philip, S.Y.: Distributed deep learning model for intelligent video surveillance systems with edge computing. IEEE Trans. Ind. Inform. (2019)
12. Ullah, H.: Crowd motion analysis: segmentation, anomaly detection, and behavior classification. Ph.D. thesis, University of Trento (2015)
13. Hu, Y.-T., Huang, J.-B., Schwing, A.: MaskRNN: instance level video object segmentation. In: Advances in Neural Information Processing Systems, pp. 325–334 (2017)
14. Ullah, M., Ullah, H., Alseadonn, I.M.: Human action recognition in videos using stable features (2017)
15. Ullah, H., Ullah, M., Uzair, M.: A hybrid social influence model for pedestrian motion segmentation. Neural Comput. Appl. **31**, 7317–7333 (2018). https://doi.org/10.1007/s00521-018-3527-9
16. Lin, T.-Y., et al. Feature pyramid networks for object detection. In: Proceedings of the IEEE Conference on Computer Vision and Pattern Recognition, pp. 2117–2125 (2017)
17. Ullah, H., Uzair, M., Ullah, M., Khan, A., Ahmad, A., Khan, W.: Density independent hydrodynamics model for crowd coherency detection. Neurocomputing **242**, 28–39 (2017)
18. Ullah, M., Mohammed, A., Alaya Cheikh, F.: PedNet: a spatio-temporal deep convolutional neural network for pedestrian segmentation. J. Imaging **4**(9), 107 (2018)
19. Girshick, R.: Fast R-CNN. In: Proceedings of the IEEE International Conference on Computer Vision, pp. 1440–1448 (2015)
20. Khan, S., et al.: Dimension invariant model for human head detection. In: 2019 8th European Workshop on Visual Information Processing (EUVIP), pp. 99–104. IEEE (2019)
21. Wei, Y., Sun, X., Yang, K., Rui, Y., Yao, H.: Hierarchical semantic image matching using cnn feature pyramid. Comput. Vis. Image Underst. **169**, 40–51 (2018)
22. Ullah, M., Ullah, H., Cheikh, F.A.: Single shot appearance model (SSAM) for multi-target tracking. Electron. Imaging **2019**(7), 466-1 (2019)
23. Yamin, M.M., Katt, B.: Modeling attack and defense scenarios for cyber security exercises. In: 5th Interdisciplinary Cyber Research Conference 2019, p. 7 (2019)
24. Yamiun, M.M., Katt, B., Gkioulos, V.: Detecting windows based exploit chains by means of event correlation and process monitoring. In: Arai, K., Bhatia, R. (eds.) FICC 2019. LNNS, vol. 70, pp. 1079–1094. Springer, Cham (2020). https://doi.org/10.1007/978-3-030-12385-7_73
25. Ren, S., He, K., Girshick, R., Sun, J.: Faster R-CNN: towards real-time object. IEEE Trans. Pattern Anal. Mach. Intell. **39**(6), 1137–1149 (2017)
26. Ullah, H., et al.: Two stream model for crowd video classification. In: 2019 8th European Workshop on Visual Information Processing (EUVIP), pp. 93–98. IEEE (2019)
27. Cao, Z., Hidalgo, G., Simon, T., Wei, S.-E., Sheikh, Y.: OpenPose: real-time multi-person 2D pose estimation using part affinity fields. arXiv preprint arXiv:1812.08008 (2018)
28. Ullah, M., Cheikh, F.A., Imran, A.S.: Hog based real-time multi-target tracking in Bayesian framework. In: 2016 13th IEEE International Conference on Advanced Video and Signal Based Surveillance (AVSS), pp. 416–422. IEEE (2016)

29. Ullah, M., Mahmud, M., Ullah, H., Ahmad, K., Imran, A.S., Cheikh, F.A.: Head-based tracking. In: IS&T International Symposium on Electronic Imaging 2020: Intelligent Robotics and Industrial Applications using Computer Vision proceedings, San Francisco, USA 2020. Society for Imaging Science and Technology. https://doi.org/10.2352/ISSN.2470-1173.2020.6.IRIACV-074

30. Sun, K., Xiao, B., Liu, D., Wang, J.: Deep high-resolution representation learning for human pose estimation. arXiv preprint arXiv:1902.09212 (2019)

31. Ullah, M., Cheikh, F.A.: A directed sparse graphical model for multi-target tracking. In: IEEE Conference on Computer Vision and Pattern Recognition Workshops, pp. 1816–1823 (2018)

32. Fang, H.-S., Xie, S., Tai, Y.-W., Lu, C.: RMPE: regional multi-person pose estimation. In: Proceedings of the IEEE International Conference on Computer Vision, pp. 2334–2343 (2017)

33. Ullah, M., Mohammed, A.K., Cheikh, F.A., Wang, Z.: A hierarchical feature model for multi-target tracking. In: 2017 IEEE International Conference on Image Processing (ICIP), pp. 2612–2616. IEEE (2017)

34. Mathis, A., Yüksekgönül, M., Rogers, B., Bethge, M., Mathis, M.W.: Pretraining boosts out-of-domain robustness for pose estimation (2019)

35. Ullah, M., Kedir, M.A., Cheikh, F.A.: Hand-crafted vs deep features: a quantitative study of pedestrian appearance model. In: 2018 Colour and Visual Computing Symposium (CVCS), pp. 1–6. IEEE (2018)

36. Mathis, A., et al.: Markerless tracking of user-defined features with deep learning. arXiv preprint arXiv:1804.03142 (2018)

37. Ullah, M., Cheikh, F.A.: Deep feature based end-to-end transportation network for multi-target tracking. In: IEEE International Conference on Image Processing (ICIP), pp. 3738–3742 (2018)

38. Nasirahmadi, A., Edwards, S.A., Sturm, B.: Implementation of machine vision for detecting behaviour of cattle and pigs. Livestock Sci. **202**, 25–38 (2017)

39. Kanwal, S., et al.: An image based prediction model for sleep stage identification. In: 2019 IEEE International Conference on Image Processing (ICIP), pp. 1366–1370. IEEE (2019)

40. Atlan, L.S., Margulies, S.S.: Frequency-dependent changes in resting state electroencephalogram functional networks after traumatic brain injury in piglets. J. Neurotrauma **36**, 2558–2578 (2019)

41. da Cordeiro, A.F.S., et al.: Use of vocalisation to identify sex, age, and distress in pig production. Biosyst. Eng. **173**, 57–63 (2018)

42. Psota, E.T., Mittek, M., Pérez, L.C., Schmidt, T., Mote, B.: Multi-pig part detection and association with a fully-convolutional network. Sensors **19**(4), 852 (2019)

43. Huang, G., Liu, Z., Van Der Maaten, L., Weinberger, K.Q.: Densely connected convolutional networks. In: Proceedings of the IEEE Conference on Computer Vision and Pattern Recognition, pp. 4700–4708 (2017)

Data and Image Processing for Precision Agriculture

Deep Transfer Learning Models for Tomato Disease Detection

Maryam Ouhami[1(✉)], Youssef Es-Saady[1], Mohamed El Hajji[1],
Adel Hafiane[2], Raphael Canals[2], and Mostafa El Yassa[1]

[1] IRF-SIC Laboratory, Ibn Zohr University, Agadir, Morocco
maryam.ouhami@edu.uiz.ac.ma
[2] PRISME Laboratory, INSA CVL University of Orléans, Orléans, France

Abstract. Vegetable crops in Morocco and especially in the Sous-Massa region are exposed to parasitic diseases and pest attacks which affect the quantity and the quality of agricultural production. Precision farming is introduced as one of the biggest revolutions in agriculture, which is committed to improving crop protection by identifying, analyzing and managing variability delivering effective treatment in the right place, at the right time, and with the right rate.

The main purpose of this study is to find the most suitable machine learning model to detect tomato crop diseases in standard RGB images. To deal with this problem we consider the deep learning models DensNet, 161 and 121 layers and VGG16 with transfer learning. Our study is based on images of infected plant leaves divided into 6 types of infections pest attacks and plant diseases. The results were promising with an accuracy up to 95.65% for DensNet161, 94.93% for DensNet121 and 90.58% for VGG16.

Keywords: Precision agriculture · Plant disease detection · Transfer learning · DensNet

1 Introduction

With a surface area of nearly 8.7 million hectares, the Moroccan department of agriculture assumes that agricultural field produces a very wide range of products and generates 13% of the gross domestic product (GDP) [1]. This sector has experienced a significant evolution of the GDP due to the exploitation of fertilization and plant protection systems [1]. Despite the efforts, it still faces important challenges, such as diseases. Pathogen is the factor that causes disease in the plant; it is induced either by physical factors such as sudden climate changes or chemical/biological factors like viruses and fungi [2].

Market gardening and especially tomato crops in the Sous-Massa region are ones of the crops that are exposed to several risks which increase the quantity and quality of the agriculture products. The most important damages are caused by pests' attacks (leaf-miner flies, Tuta absoluta and Thrips) in addition to cryptogamic pathogens infections (early blight, late blight and Powdery Mildew).

Since diagnosis can be performed on plant leaves, our study is conducted as a task of classification of symptoms and damages on those leaves. Ground imaging with RGB

© Springer Nature Switzerland AG 2020
A. El Moataz et al. (Eds.): ICISP 2020, LNCS 12119, pp. 65–73, 2020.
https://doi.org/10.1007/978-3-030-51935-3_7

camera presents an interesting way for this diagnosis. However, robust algorithms are required to deal with different acquisition conditions: light changes, color calibration, etc. For several years, great efforts have been devoted to the study of plant disease detection. Indeed, feature engineering models [3–6] on one side with Convolutional Neural Networks (CNN) [7–10] on the other side; are carried out to solve this task.

In [6], the study is based on a database of 120 images of infected rice leaves divided into three classes bacterial leaf blight, brown spot, and leaf smut (40 images for each class), Authors have converted the RGB images to an HSV color space to identify lesions, with a segmentation accuracy up to 96.71% using k-means. The experiments were carried out to classify the images based on multiple combinations of the extracted characteristics (texture, color and shape) using Support Vector Machine (SVM). The weakness of this method consists in a moderate accuracy obtained of 73.33%. In fact, the image quality was decreased during the segmentation phase, during which some holes were generated within disease portion, which could be a reason for the low classification accuracy. In addition, leaf smut is misclassified with an accuracy of 40%, which requires other types of features to improve le results.

In the same context, in [4], the authors have proposed an approach for diseases recognition on plant leaves. This approach is based on combining multiple SVM classifiers (sequential, parallel and hybrid) using color, texture and shape characteristics. Different preprocessing have been performed, including normalization, noise reduction and segmentation by the k-means clustering method. The database of infected plant leaves contains six classes including three types of insect pest damages and three forms of pathogen symptoms. The hybrid approach has outperformed the other approaches, achieving a rate of 93.90%. The analysis of the confusion matrix for these three methods has highlighted the causes of misclassification, which are essentially due to the complexity of certain diseases for which it is difficult to differentiate their symptoms during the different stages of development, with a high degree of confusion between powdery mildew and thrips classes in all the combination approaches.

In another study, the authors have used a maize database acquired by a drone flying at a height of 6 m [11]. They have selected patches of 500 by 500 pixels of each original image of 4000 by 6000, and labelled them according to the fact that the most central 224 by 224 area contained a lesion. For the classification step between healthy and diseased, a sliding window of 500 by 500 on the image is used and was introduced in the convolutional neural network (CNN) ResNet model [8]. With a test precision of 97.85%, the method remains non generalizable since the chosen disease has important and distinct symptoms compared to other diseases and the acquisitions are made on a single field. For that reason, it is not clear how the model would perform to classify different diseases with similar symptoms.

Another work that uses aerial images, with the aim of detecting disease symptoms in grape leaves [9]. Authors have used CNN approach by performing a relevant combination of image features and color spaces. Indeed, after the acquisitions of RGB images using UAV at 15 m height. The images were converted into different colorimetric spaces to separate the intensity information from the chrominance. The color spaces used in this study were HSV, LAB and YUV, in addition to the extracted vegetation indices (Excessive Green (ExG), Excessive Red (ExR), Excessive Green-Red (ExGR), Green-Red Vegetation Index (GRVI), Normalized Difference Index

(NDI) and Red-Green Index (RGI)). For classification, they have used the CNN model Net-5 with 4 output classes: soil, healthy, infected and susceptible to infection. The model has been tested on multiple combinations of input data and three patch sizes. The best result was obtained by combining ExG, ExR & GRVI with an accuracy of 95.86% on 64 × 64 patches.

In [10], the authors tested several existing state-of-the-art CNN architectures for plant disease classification. The public PlantVillage database [12] was used in this study. The database consists of 55,038 images of 14 plant types, divided into 39 classes of healthy and infected leaves, including a background class. The best results were obtained with the transfer learning ResNet34 model, achieving an accuracy of 99.67%.

Several works have been carried out to deal with the problem of plant diseases detection using images provided by remote sensing materials (smartphones, drones…). Nevertheless, CNN have demonstrated high performances to solve this problem compared to models based on the classic feature extracting methods.

In the present study we took advantages from the deep learning and transfer learning approaches to address the problem of the most important damages caused by pests' attacks and cryptogamic pathogens infections in tomato crops. The rest of the paper is organized as follows. Section 2 presents a comparative study and discusses our preliminary results. The conclusion and perspectives are presented in Sect. 3.

2 Materials and Methods

2.1 Data Description

The study was conducted on a database of images of infected leaves, developed and used in [3–5].The images were taken with a digital camera, Canon 600D, in several farms in the area of Sous Massa, Morocco. Additional images are collected from the Internet in order to increase the size of the database. The dataset is composed of six classes, three of damage caused by insect pests (leafminer flies, thrips and tuta abso-luta), and three classes of cryptogamic pathogens symptoms (Early blight, late blight and powdery mildew). The dataset is validated with the help of an agricultural experts. Figure 1 depicts the types of symptoms on tomato leaves, Table 1 presents the composition of the database and the symptoms of each class. The images were resized in order to put the leaves in the center of the images.

2.2 Architecture Model

The motivation behind using deep learning for computer vision is the direct exploitation of image without any hand-crafted features. In plant disease detection field, many researchers have chosen deep models DensNets and VGGs for their high performance in standard computer vision tasks.

DensNets.
The idea behind the DensNet architecture is to avoid creating short paths from the early layers to the later layers and to ensure maximum information flow between the layers of the network. Therefore, DensNet connects all its layers (with corresponding feature

Fig. 1. Images from the dataset where: (a) Early blight, (b) late blight, (c) Powdery mildew, (d) leafminer fly, (e) Thrips and (f) Tuta absoluta.

Table 1. Dataset distribution.

Name	Early blight	Late blight	Powdery mildew	Leaf miner flies	Thrips	Tuta absoluta	Total
Size	111	111	111	111	111	111	666

map sizes) directly to each other. In addition, each layer obtains additional inputs from all previous layers and transmits its own characteristic maps to all subsequent layers [13]. Indeed, according to [13] DensNets require substantially fewer parameters and less computation to achieve state-of-the-art performances. Figure 2(a) gives an example of a five dense layers convolutional model. In this study DensNet was used with 121 layers and 161 layers.

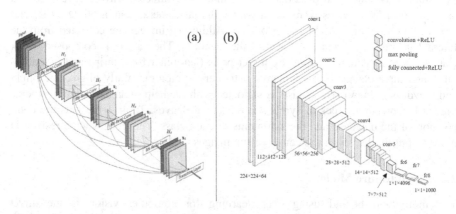

Fig. 2. (a) Architecture of a five dense convolutional layers model [13], (b) Architecture of VGG16 model.

VGG.
Very deep convolutional networks or VGG ranked second in ILSVRC-2014 challenge [14]. The model is widely used for image recognition task especially in the crop field [15, 16]. Consequently, we used VGG with 16 layers. The architecture has 16

convolutional and 5 pooling layers, followed by 3 fully connected layers. The filters are of size 3 × 3 × m where m is the number of feature maps. Figure 3(b) illustrates the architecture of VGG16.

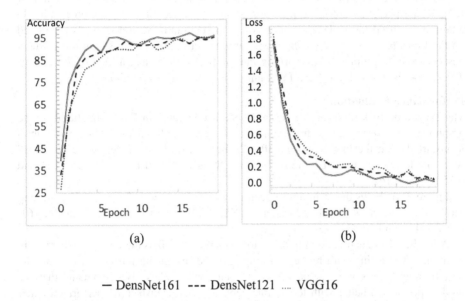

(a)   (b)

— DensNet161 --- DensNet121 .... VGG16

Fig. 3. (a) Evolution of the loss during training for DensNet161, DensNet121 and VGG16, (b) Evolution of accuracy during training for DensNet161, DensNet121 and VGG16.

Fine-Tuning the Models.

Fine-tuning is a transfer learning method that allows to take advantage of models trained on another computer vision task where a large number of labelled images is available. Moreover, it reduces the needs on having a large dataset and computation power to train the model from scratch [16]. Fine-tuned learning experiments are much faster and more accurate compared to models trained from scratch [10, 15, 17, 18]. Hence, we fine-tuned all network layers of the 3 models based on learned features on the ImageNet dataset [19]. The idea is to take pre-trained weights from the VGG16, Densnet121 and Densnet161 trained on the ImageNet dataset, use those weights as first step of our learning process, then we keep them for every convolutional layer in all the iterations and update only the weights of the linear layers.

2.3 Results and Discussion

Experimental Setup

Experiments were run on a Google Compute Engine instance named Google Colaboratory (Colab) [20] as well as a local machine LenovoY560 with 16 GB of RAM. Colab notebooks are based on Jupyter and work as a Google Docs object. In addition to that, the notebooks are pre-configured with the essential machine learning and artificial intelligence libraries, such as TensorFlow, Matplotlib, and Keras. Colab operates under Ubuntu 17.10 64 bits and it is composed of an Intel Xeon processor and 13 GB RAM. It is equipped with a NVIDIA Tesla K80 (GK210 chipset), 12 GB RAM, 2496 CUDA.

We implemented and executed the experiments in Python, using PyTorch library [21], which performs automatic differentiation over dynamic computation graphs. In addition, we used the PyTorch model Zoo, which contains various models pretrained on the ImageNet dataset [19]. The model architecture is train with stochastic gradient descent (SGD) optimizer with learning rate of 1e-3 (0.005) and a total of 20 epochs. The dataset is divided into 80% for training and 20% for evaluation.

Performance Evaluation.
The evaluation of loss during the training phase illustrated in Fig. 3(a). Based on the graph we can observe that the training loss converged for all models. A big reduction of loss started from the first 5 epochs, after 20 iterations all the models were optimized with low losses reaching a score of 0.12 for DensNet161, 0.14 for DensNet121 and 0.15 for VGG16.

Figure 3(a) shows the training set accuracy score for epoch of 1 to 20. The training set accuracy score at epoch 20 reach 96.4%, 95.27% and 94.7% for DensNet161, DensNet121 and VGG16 respectively.

After the 14th epoch the training start to converge as well as the training accuracy. In addition, after testing with higher learning rate and increasing number of epochs, the best training scores were achieved using DensNets models. Which means that the models performed better with less parameters. Besides, DensNet161 performed better than Densnet121 due to the deeper architecture of the model.

We can observe in Table 2 that DensNets performed better than VGG model during the test even if their losses reached a score around 0.14 during. Note that DensNet with deeper layers had better test score. In the test phase DensNet161 outperformed DensNet121 and VGG16 with an accuracy 95.65%, 94.93% and 90.58% respectively. We can clearly see from the Table 2 that DensNet161 outperformed the other models in classifying leafminer fly, thrips and powdery mildew with an accuracy up to 100%, 95.65% and 100% respectively. Furthermore, DensNet121 had the best classification rate for early blight, late blight and tuta absoluta with an accuracy of 100%, 95.65% and 95.65 respectively.

Table 2. Accuracy on test set for each class and average accuracy overall classes for DensNet161, DensNet121 and VGG6

Accuracy	DensNet161	DensNet121	VGG16
Leafminer fly	**100**	91.30	78.26
Thrips	**95.65**	91.30	91.30
Powdery mildew	**100**	95.65	95.65
Early blight	95.65	**100**	**100**
Late blight	91.30	**95.65**	86.96
Tuta absoluta	91.30	**95.65**	91.30
Over all	**95.65**	**94.93**	90.58

In order to compare the two models having the best accuracies, we calculated the confusion matrix of the testing dataset for those models. Figure 4 represents the confusion matrix for the DensNet classification models with 161 and 121 layers. More images were misclassified for DensNet121 compared to DensNet161. Moreover, the most confused classes for the DensNet121 are leafminer fly and thrips with two thrips images classified as leafminer fly and one leaf miner classified as thrips. In DensNet161 model, the confusion is more likely between early blight and late blight with one early blight image classified as late blight and one image late blight classified as early blight.

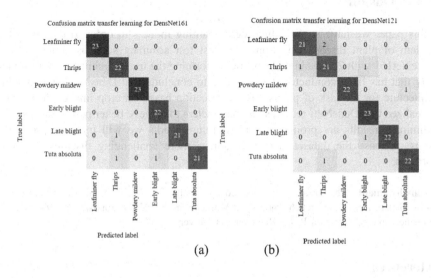

(a) (b)

Fig. 4. Confusion matrix of the DensNet architectures. (a) DensNet161 (b) DensNet121

Thrips and early blight are the most misclassified classes for both models, which is due to the similarity between the symptoms making it difficult to differentiate between these classes.

Table 3. State of art comparison.

Authors	Crop	Data size	Method	Accuracy
Brahimi [10]	PlantVillage	55,038 images	Inception_v3	99.76%
Hanks [8]	Maize	8766 UAV Images	ResNet	97.85%
Kerkech [9]	Vine	70,560 patch, UAV images	Le Net-5 on YUV and ExGR	95.92%
El Massi [4]	Tomato	600 images	SVM combination	93.90%
Harshadkumar [6]	Rice	120 images	k-means	73.33%
Our model	**Tomato**	**666 images**	**DensNet161**	**95.65%**

State of Art Comparison

Table 3 describes the studies we cited earlier in Sect. 1, aligned with our model results. Each approach is using different dataset. Nevertheless, according to the accuracies listed, the approaches based on deep learning models outperformed the approaches based on feature engineering. The results of our model are promising starting with a dataset with the size of 666 images – see Sect. 2.1 and achieving an accuracy of 95.65% using DensNet161 model.

3   Conclusion

In this paper we have studied three deep learning models in order to deal with the problem of plant disease detection. The best test accuracy score is achieved with DenseNet161 with 20 training epochs, outperforming the tested architectures. From the study that has been conducted it is possible to conclude that DensNet has a suitable architecture for the task of plants disease detection based on crop images. Moreover, we realized that DensNets requires less parameters to achieve better performances. The preliminary results are promising. In our future works we will try to improve the results, increase the dataset size and address more challenging diseases detection problems.

Acknowledgment. This study was supported by Campus France, Cultural Action and Cooperation Department of the French embassy in Morocco, in the call of proposals, 2019 campaign, under the name of "Appel à Projet Recherche et Universitaire".

References

1. MAPM du développement rural et des eaux et forêts, L'agriculture en chiffre_Plan Maroc vert, L'agriculture en chiffre (2018)
2. Jullien, A., Huet, P.: Agriculture de précision. In: Agricultures et territoires, Editions L, pp. 1–15 (2005)
3. Es-Saady, Y., El Massi, I., El Yassa, M., Mammass, D., Benazoun, A.: Automatic recognition of plant leaves diseases based on serial combination of two SVM classifiers. In: Proceedings of the International Conference on Electrical and Information Technologies (ICEIT 2016), pp. 561–566 (2016)
4. El Massi, I., Es-Saady, Y., El Yassa, M., Mammass, D., Benazoun, A.: A hybrid combination of multiple SVM classifiers for automatic recognition of the damages and symptoms on plant leaves. In: Mansouri, A., Nouboud, F., Chalifour, A., Mammass, D., Meunier, J., ElMoataz, A. (eds.) ICISP 2016. LNCS, vol. 9680, pp. 40–50. Springer, Cham (2016). https://doi.org/10.1007/978-3-319-33618-3_5
5. El Massi, A., Es-Saady, I., El Yassa, Y., Mammass, M., Benazoun, D.: Automatic recognition of the damages and symptoms on plant leaves using parallel combination of two classifiers. In: 13th International Conference on Computer Graphics, Imaging and Visualization (CGiV), Morocco, pp. 131–136. IEEE (2016)
6. Prajapati, H.B., Shah, J.P., Dabhi, V.K.: Detection and classification of rice plant diseases. Intell. Decis. Technol. **11**(3), 357–373 (2017)

7. El, I., Es-Saady, Y., El, M., Mammass, D., Benazoun, A.: Automatic recognition of vegetable crops diseases based on neural network classifier. Int. J. Comput. Appl. **158**(4), 48–51 (2017)

8. Wu, H., et al.: Autonomous detection of plant disease symptoms directly from aerial imagery. Plant Phenom. J. **2**(1), 1–9 (2019)

9. Kerkech, M., Hafiane, A., Canals, R.: Deep leaning approach with colorimetric spaces and vegetation indices for vine diseases detection in UAV images. Comput. Electron. Agric. **155** (July), 237–243 (2018)

10. Brahimi, M., Arsenovic, M., Laraba, S., Sladojevic, S., Boukhalfa, K., Moussaoui, A.: Deep learning for plant diseases: detection and saliency map visualisation. In: Zhou, J., Chen, F. (eds.) Human and Machine Learning. HIS, pp. 93–117. Springer, Cham (2018). https://doi.org/10.1007/978-3-319-90403-0_6

11. Wiesner-Hanks, T., et al.: Image set for deep learning: field images of maize annotated with disease symptoms. BMC Res. Notes **11**(1), 10–12 (2018)

12. Hughes, D., Salathe, M.: An open access repository of images on plant health to enable the development of mobile disease diagnostics (2015). arXiv:1511.0806

13. Huang, G., Weinberger, Q.K.: Densely connected convolutional networks. In: The IEEE Conference on Computer Vision and Pattern Recognition (CVPR), pp. 4700–4708 (2018). arXiv:1608.06993

14. Simonyan, K., Zisserman, A.: Very deep convolutional networks for large-scale image recognition. In: 3rd International Conference on Learning Representations, ICLR 2015, pp. 1–14 (2015)

15. Chebet, E., Yujian, L., Njuki, S., Yingchun, L.: A comparative study of fine-tuning deep learning models for plant disease identification. Comput. Electron. Agric. **161**, 272–279 (2019)

16. Abas, M.A.H., Ismail, N., Yassin, A.I.M., Taib, M.N.: VGG16 for plant image classification with transfer learning and data augmentation. Int. J. Eng. Technol. (UAE) **7**, 90–94 (2018)

17. Bah, M.D., Hafiane, A., Canals, R.: Deep learning with unsupervised data labeling for weed detection in line crops in UAV images. Remote Sens. **10**(11), 1–22 (2018)

18. Mohanty, S.P., Hughes, D.P., Salathé, M.: Using deep learning for image-based plant disease detection. Front. Plant Sci. **7**(9), 1–10 (2016)

19. Deng, J., Dong, W., Socher, R., Li, L.J., Li, K., Fei-Fei, L.: Imagenet: a large-scale hierarchical image database. In: 2009 IEEE Conference on Computer Vision and Pattern Recognition, pp. 248–255. IEEE (2009)

20. Carneiro, T., Da Nobrega, R.V.M., et al.: Performance analysis of Google Colaboratory as a tool for accelerating deep learning applications. IEEE Access **6**, 61677–61685 (2018)

21. Ketkar, N.: Introduction to PyTorch. Deep Learning with Python, pp. 195–208. Apress, Berkeley, CA (2017). https://doi.org/10.1007/978-1-4842-2766-4_12

Machine Learning-Based Classification of Powdery Mildew Severity on Melon Leaves

Mouad Zine El Abidine[1], Sabine Merdinoglu-Wiedemann[2], Pejman Rasti[1,3],
Helin Dutagaci[1], and David Rousseau[1(✉)]

[1] LARIS, UMR INRAE IRHS, Université d'Angers,
62 Avenue Notre Dame du Lac, 49000 Angers, France
david.rousseau@univ-angers.fr

[2] INRAE-Université de Strasbourg, 21 rue de Herrlisheim, 68000 Colmar, France

[3] Department of Big Data and Data Science, école d'ingénieur informatique
et environnement (ESAIP), Saint Barthelemy d'Anjou, France

Abstract. Precision agriculture faces challenges related to plant disease detection. Plant phenotyping assesses the appearance to select the best genotypes that resist to varying environmental conditions via plant variety testing. In this process, official plant variety tests are currently performed in vitro by visual inspection of samples placed in a culture media. In this communication, we demonstrate the potential of a computer vision approach to perform such tests in a much faster and reproducible way. We highlight the benefit of fusing contrasts coming from front and back light. To the best of our knowledge, this is illustrated for the first time on the classification of the severity of the presence of a fungi, powdery mildew, on melon leaves with 95% of accuracy.

Keywords: Machine learning · Classification · Plant disease

1 Introduction

During the last decades, precision agriculture benefited from advances in robotics [1,2], computer vision [3] and artificial intelligence [4,5] to automate the monitoring of crops [6] and harvesting [7]. However, some activities of major importance for agriculture are still to take benefit from these advances. One such activity is plant variety testing. To register and protect a new variety in a country, a plant breeding company has to follow a process managed by a national examination office within an official framework. The national examination offices run tests to register new varieties in the official catalogue, protect them with «plant variety rights» and post control of certified seed lots. Currently, most of these tests are based on manual measurements performed with visual inspection. This is an issue for the sake of efficiency due to the time consuming nature of these tests. In this context, we focus on one of these plant variety tests. We propose an automated algorithm to detect and quantify the presence of powdery mildew

© Springer Nature Switzerland AG 2020
A. El Moataz et al. (Eds.): ICISP 2020, LNCS 12119, pp. 74–81, 2020.
https://doi.org/10.1007/978-3-030-51935-3_8

on melon leaves via in vitro imaging to assess the resistance capability of the tested varieties. Powdery mildew is a fungal disease infecting melon leaves and causing a major reduction of yield. The typical symptoms of powdery mildew are white colonies on the leaf surface consisting of mycelium and spores of the fungal pathogen. We first describe the current manual method and then explain the computer vision procedure that we propose, based on machine learning and fusion of front and back light images. The performance of this automated procedure is compared with the manual method and previous automated methods before conclusion.

2   Related Work

Plant disease detection using deep learning has attracted many attention in the recent past [8–10] due to the large variety of conditions in which diseases can be studied including conditions of plant-pathogen interactions (virus, bacteria, fungi), environmental conditions (field, controlled environment, in vitro, ...), imaging modalities (RGB, thermal, fluorescent, ...) and observation scales (tissue, leaf, canopy, ...). As closely related work, for powdery mildew detection observed on foliar disk in vitro, a spatio-spectral analysis based on hyperspectral images of wine grapes was developed to classify powdery mildew infection levels [11]. An accuracy of 87% was reported to classify "healthy", "infected" and "severely" diseased bunches. In another work, a machine vision based phenotyping system was developed to assess the severity of grapevine powdery mildew [12]. The system is based on a high-resolution camera and a long working distance macro-focusing lens. The system acquires an image of each foliar disk inside a Petri dish and requires an XY motorised stage to move above one foliar disk to another. A GoogleNet neural network architecture was trained on 9920 images of two classes "infected" and "not infected". The training lasted 3.4 h. The resulting CNN had a classification accuracy of 94.3%. By contrast with these methods, our computer vision system requires only a simple RGB camera with standard resolution and a lighting device. This simplicity and low-cost is important for dissemination of the method as the system is dedicated to pathology tests performed by biologists. Also, acquisition time is important as the global objective is to implement a high-throughput phenotyping system. Unlike microscope-based images of [12] that catches only a single foliar disk at a time, we acquire 9 foliar disks in the same snapshot. The previous works have improved disease detection accuracy by investing in the imaging system (hyper-spectral camera and microscope). On the side of optics, we propose to fuse front and back light images to improve classification accuracy and thus implement a high-throughput phenotyping system at a relatively low cost.

3   Current Manual Procedure

The current manual procedure to assess melon leaves resistance to powdery mildew is as follows. First, biologists extract foliar disks from melon leaves (as illustrated in Fig. 1(a)) and position them inside a Petri dish. A control foliar disk (very sensitive to powdery mildew), positioned at the center of the Petri dish, serves to validate the presence of powdery mildew. In the next step, foliar disks are inoculated with powdery mildew powder inside Petri dishes and left for an incubation period of 10 days. After that, biologists use a binocular loop to visualize leaves and assign an ordinal score according to powdery mildew density on the leaf surface as shown in Fig. 1(b). The encoding of the scoring is provided in Table 1.

Table 1. Annotation scale of powdery mildew propagation on melon leaves.

Score assigned to powdery mildew density	Observation
Resistant	One spore of powdery mildew
Moderate	50% of the leaf is infected
Severe	Leaf is totally infected

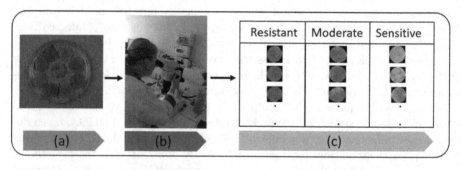

Fig. 1. Schematic visualisation of the current procedure. (a): Petri dish containing melon foliar disks inoculated with powdery mildew. (b): a biologist visualizes and annotates powdery mildew propagation using a binocular loop. (c): data generated after assessment following the annotation scale of Table 1. Each foliar disk is saved in a CSV file with its corresponding class.

4   Proposed Computer Vision Procedure

To automate the manual procedure described in the previous section, we propose to follow the pipeline given in Fig. 2. 70 Petri dish images are acquired with a digital color camera with resolution of 2448 by 2050 pixels. The size of each foliar disk is approximately 120 000 pixels. The camera is positioned vertically above the Petri dish as shown in Fig. 2(a). As illustrated in Fig. 2(b), Petri dish images are acquired under two lighting techniques, front and back light. After RGB to HSB conversion, the brightness channel of both images (front and back light) are fused in a linear blending to enhance the contrast between the lymb and the fungi.

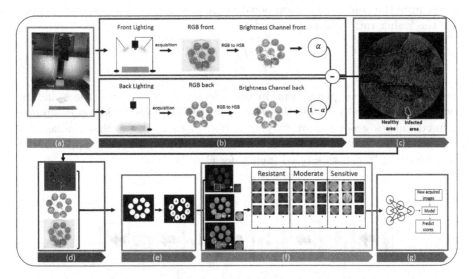

Fig. 2. Visual scheme of the proposed computer vision procedure. (a): imaging device. (b): fusing images following a linear blending. (c): generated fused image. (d): front, back and fused data sets. (e): The foliar disks are segmented from Petri dishes using the convolutional neural network architecture for semantic segmentation UNET [13] and cropped individually. (f): foliar disks are assigned to their corresponding ground truth, produced by an expert in the current manual procedure. (g): feeding training images to a supervised machine learning classification algorithm.

The partition of foliar disk images per class in front, back and fused data sets is as follows: 180 images for "Resistant" class, 62 images for "Moderate" class and 131 images for "Sensitive" class. This data set is rather small in this work (compared to standard large data sets in machine learning). This constitutes a possible limit to the use of an end-to-end deep learning method due to the tendency of overfitting. Instead, a shallow supervised learning scheme based on the concatenation of a deep-feature extraction [14] stage followed by a linear support vector machine classifier was used for the comparison of the performance with different images (fused, front and back light). Deep features trained on ImageNet from well-known architectures were tested in this study including VGG16 and Resnet50. In addition, a small end-to-end CNN model with the architecture shown in Fig. 3 was fine-tuned on a validation data set of 20% of training images. The accuracy of the classification of all tested models was computed per class from the confusion matrix. Due to the lack of enough data and the imbalance classes in our data set, a data augmentation was used to improve the classification accuracy and to be able to compare fairly with a deep learning architecture. Data were augmented to force invariance to rotation since the leaves are randomly positioned in the Petri dish invariance to shearing and zoom to allow for robustness to some plasticity of the leaf tissue. The mix parameter α in the linear blending to fuse front and back light images was chosen to maximize the contrast between powdery mildew and healthy lymb computed with the Fisher ratio which is defined as

$$FR = \frac{(\mu_2 - \mu_1)^2}{\sigma_2^2 + \sigma_1^2} \tag{1}$$

where μ is the mean pixel value in the selected area and σ is the standard deviation of the pixel values in the selected area. The Fisher ratio was computed on fused images generated by varying α from 0 to 1. The optimal value of α for which Fisher ratio is maximum, equals to 0.1. This value was applied in the linear blending to generate fused images for classification.

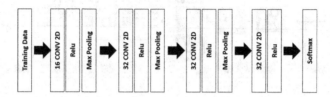

Fig. 3. CNN architecture proposed.

5   Results

Classification performances of the supervised machine learning algorithm described in Sect. 4, are given in Table 2 and Table 3. Best results, highlighted in blue were systematically obtained with the fused images for the three tested classifiers with or without data augmentation. Highest scores are obtained with association of deep features from Resnet50 coupled with a linear SVM with data augmentation. Other classical classifiers such as random forest or non linear SVM were also tested (not shown) but results were not significantly improved. The confusion matrix for this best classifier is illustrated in Fig. 4 which shows that most errors come from the confusion between moderate and sensitive classes. Finally, these two classes are merged by biologists when varieties are registered officially. The classification accuracy achieved in classification of resistant and sensitive levels are provided in Table 3 with best performances culminating at 95% accuracy.

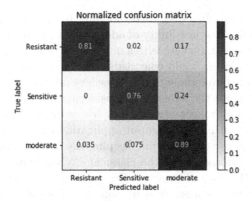

Fig. 4. Confusion Matrix of Resnet50 algorithm to classify powdery mildew on 3 infection levels: resistant & moderate & sensitive.

Table 2. Classification accuracy for 3 infection levels: resistant & moderate & sensitive.

Architecture	Training approach	Front	Back	Fused	Train/test
VGG16	Raw Data	0.61	0.53	0.7	135/88
	Data augmentation	0.69	0.49	0.75	1000/200
Resnet50	Raw Data	0.67	0.59	0.79	135/88
	Data augmentation	0.71	0.62	0.82	1000/200
Proposed CNN	Raw Data	0.56 ± 0.06	0.40 ± 0.04	0.67 ± 0.04	135/88
	Data augmentation	0.64 ± 0.05	0.48 ± 0.07	0.78 ± 0.02	1000/200

Table 3. Classification accuracy for 2 infection levels: resistant & sensitive.

Architecture	Training approach	Front	Back	Fused	Train/test
VGG16	Raw Data	0.82	0.72	0.86	280/100
	Data augmentation	0.81	0.74	0.83	1000/200
Resnet50	Raw Data	0.86	0.81	0.94	280/100
	Data augmentation	0.86	0.84	0.95	1000/200
Proposed CNN	Raw Data	0.71 ± 0.08	0.65 ± 0.02	0.86 ± 0.04	280/100
	Data augmentation	0.79 ± 0.11	0.68 ± 0.01	0.92 ± 0.01	1000/200

6 Discussion

The previous section presented successful results for the classification of the presence of powdery mildew in foliar disks containing melon leaves. The obtained performances are similar to the recently published work on the classification of powdery mildew in two [11] or three classes [12] as presented in the related work section. It is to be noticed that the closest related method of [12] is applied to another crop but in a similar in vitro imaging conditions protocol. While neural networks are also used as the main element of the image processing pipeline, [12] notably differs from our approach. The work of [12] focuses on a metrological measurement of the powdery mildew performed with a high resolution imaging system enabling to detect individual mycelium. By contrast, we propose an ordinal classification of the foliar disk corresponding to the final annotation of an expert. We investigated the possibility of addressing this less demanding task by considering foliar disks as a texture with a much lower spatial resolution. Working at such degraded resolution could constitute a risk of loosing accuracy specially at the low grade of the development of the powdery mildew. However, we demonstrated that this was not the case when considering the final score recorded in variety testing which only keeps two classes (resistant, sensitive). Our method is especially suitable for high-throughput application of variety testing to avoid an overwhelming increase of data while keeping the accuracy of the tests at the current level. The performance of the classical CNN architecture is promising and should exceed the 95% accuracy of Resnet50 in case more training images were provided. A comparison on the same samples of our classification approach with the metrological quantification of [12] would be an interesting perspective.

7   Conclusion and Future Work

In this paper, we presented a computer vision-based approach to automate a plant variety test performed to quantify the severity of powdery mildew infection levels on melon leaves. We demonstrated that fusing front light and back light images improved powdery mildew contrast. This fusion resulted an improvement of 10% accuracy with a very low-cost imaging system. Also, we highlighted the achievement of this performance level with a standard spatial resolution, while the state of the art on this problem reported the use of microscopic resolution to track individual mycellium. The use of deep features Resnet50 coupled with a standard SVM achieved an accuracy of 95%. This automated approach is expected to improve the speed and accuracy of disease detection and could be extended to other in vitro pathology tests. The fusion of front and back light was limited in this communication to a simple linear blending. In the future, we plan to explore various approaches of image fusion [15] to optimize the combination of front and back light images.

Acknowledgements. Authors thank INRAE for funding, V. Grimault, S. Perrot S. Houdault and H. Péteul from GEVES (French authority in variety testing) for acquisition trials.

References

1. Billingsley, J., Visala, A., Dunn, M.: Robotics in agriculture and forestry. In: Siciliano, B., Khatib, O. (eds.) Springer Handbook of Robotics, pp. 1065–1077. Springer, Heidelberg (2008). https://doi.org/10.1007/978-3-540-30301-5_47
2. Emmi, L., Gonzalez-de-Soto, M., Pajares, G., Gonzalez-de-Santos, P.: New trends in robotics for agriculture: integration and assessment of a real fleet of robots. Sci. World J. **2014**, 21 (2014)
3. Rousseau, D., et al.: Multiscale imaging of plants: current approaches and challenges. Plant Methods **11**(1), 6 (2015)

4. Kamilaris, A., Prenafeta-Boldú, F.X.: Deep learning in agriculture: a survey. Comput. Electron. Agric. **147**, 70–90 (2018)

5. Patrício, D.I., Rieder, R.: Computer vision and artificial intelligence in precision agriculture for grain crops: a systematic review. Comput. Electron. Agric. **153**, 69–81 (2018)

6. Negash, L., Kim, H.-Y., Choi, H.-L.: Emerging UAV applications in agriculture. In: 2019 7th International Conference on Robot Intelligence Technology and Applications (RiTA), pp. 254–257. IEEE (2019)

7. Zhuang, J., et al.: Computer vision-based localisation of picking points for automatic litchi harvesting applications towards natural scenarios. Biosyst. Eng. **187**, 1–20 (2019)

8. Mohanty, S.P., Hughes, D.P., Salathé, M.: Using deep learning for image-based plant disease detection. Front. Plant Sci. **7**, 1419 (2016)

9. Ferentinos, K.P.: Deep learning models for plant disease detection and diagnosis. Comput. Electron. Agric. **145**, 311–318 (2018)

10. Singh, A.K., Ganapathysubramanian, B., Sarkar, S., Singh, A.: Deep learning for plant stress phenotyping: trends and future perspectives. Trends Plant Sci. **23**(10), 883–898 (2018)

11. Knauer, U., Matros, A., Petrovic, T., Zanker, T., Scott, E.S., Seiffert, U.: Improved classification accuracy of powdery mildew infection levels of wine grapes by spatial-spectral analysis of hyperspectral images. Plant Methods **13**(1), 47 (2017). https://doi.org/10.1186/s13007-017-0198-y

12. Bierman, A., et al.: A high-throughput phenotyping system using machine vision to quantify severity of grapevine powdery mildew. Plant Phenom. **2019**, 9209727 (2019)

13. Ronneberger, O., Fischer, P., Brox, T.: U-Net: convolutional networks for biomedical image segmentation. In: Navab, N., Hornegger, J., Wells, W.M., Frangi, A.F. (eds.) MICCAI 2015. LNCS, vol. 9351, pp. 234–241. Springer, Cham (2015). https://doi.org/10.1007/978-3-319-24574-4_28

14. Vapnik, V.: The Nature of Statistical Learning Theory. Springer, New York (2013)

15. Sun, S.: A survey of multi-view machine learning. Neural Comput. Appl. **23**(7–8), 2031–2038 (2013). https://doi.org/10.1007/s00521-013-1362-6

Vine Disease Detection by Deep Learning Method Combined with 3D Depth Information

Mohamed Kerkech[1(✉)], Adel Hafiane[1], Raphael Canals[2], and Frederic Ros[2]

[1] INSA-CVL, Univ. Orléans, PRISME EA 4229, 18022 Bourges, France
{mohamed.kerkech,adel.hafiane}@insa-cvl.fr
[2] Univ. Orléans, INSA-CVL, PRISME EA 4229, 45072 Orléans, France
{raphael.canals,frederic.ros}@univ-orleans.fr

Abstract. Vine disease detection (VDD) is an important asset to predict a probable contagion of virus or fungi. Diseases that spreads through the vineyard has a huge economic impact, therefore it is considered as a challenge for viticulture. Automatic detection and mapping of vine disease in earlier stage can help to limit its impact and reduces the use of chemicals. This study deals with the problem of locating symptomatic areas in images from an unmanned aerial vehicle (UAV) using the visible and infrared domains. This paper, proposes a new method, based on segmentation by a convolutional neuron network SegNet and a depth map (DM), to delineate the asymptomatic regions in the vine canopy. The results obtained showed that SegNet combined with the depth information give better accuracy than a SegNet segmentation alone.

Keywords: Unmanned aerial vehicle · Deep learning · Depth map · 3D · Vine disease detection

1 Introduction

In recent years, remote sensing with unmanned aerial vehicles (UAV) for precision agriculture [1,2] has become a field of research in rapid progress, for different agricultural applications [3], using several types of data (visible, multi or hyper spectral) [4], and in several crop types notably in the viticulture [5].

Precision viticulture is an area of research that includes many applications, such as estimating growth [6], estimating evaporate-transpiration and harvest coefficients [7], vigor evaluation [8], water stress localization [9] or diseases detection [10–19].

Vine diseases detection (VDD) is a key issue for reducing the use of phytosanitary products and increasing the grapes production. So far, there is some researches on the different imaging systems for the VDD. Certain studies use images taken at the vine leaf level [10–14] which can be mounted in mobile robots. Other research is carried out on aerial images taken by drones at plot

© Springer Nature Switzerland AG 2020
A. El Moataz et al. (Eds.): ICISP 2020, LNCS 12119, pp. 82–90, 2020.
https://doi.org/10.1007/978-3-030-51935-3_9

scale targeting the vine canopies [15–19]. The VDD at the canopy level requires the vine isolation from the background. In some research works, the isolation of vine is carried out by using vegetation index (NDVI, ExG, ExGR ...), and machine learning methods. However, these methods are not always effective, especially when the vine inter-rows are covered with green grass, which can be confused with the green color of the vine and leads to misclassification of the vine and the soil. To solve this problem, the authors in [20–22] have used 3-dimension (3D) information to separate the soil and the vine by depth information, using the digital surface model (DSM). The results show the importance of the 3D information. However, the combination of the deep learning approach and 3D information is still less explored.

This paper presents a method based on deep learning network segmentation, combined with 3D depth information for VDD in UAV images of vineyards partially or totally covered with green or yellow grass. This method reduces confusion between different classes (healthy vine, diseased vine, green grass, yellow grass,...) and only keep the detections on the vine vegetation.

This paper is organized as follow, materials and methods are described in Sect. 2, experiences and results are presented in Sect. 3, discussion and interpretation in Sect. 4, and the conclusion in Sect. 5.

2 Materials and Methods

This section details material, and method used for data acquisition, the proposed system, the design of the vine and non-vine mask (depth map - DM), construction of the rasters, deep learning segmentation method, and finally, correction of the segmentation output.

2.1 Materials and Acquisition

The UAV which acquires image data is a Quad-copter type drone. It has a flight autonomy of 20 min, and it embeds a GPS module and two image sensors of MAPIR Survey2 model. The first sensor operates in visible spectrum (RGB), and the second one in near infrared spectrum, which records 3-band images (Near Infrared (NIR), Red and Normalized Difference Vegetation Index (NDVI)). Both sensors have a high resolution of 16 mega pixels (4608×3456).

The data acquisition method is realized by flying over the vineyard parcel at an altitude between 20 and 25 m and with an average speed of 10 km/h. The drone takes images of each area of the vine plot with a resolution of $1\,cm^2$/pixel and an overlap of 70% between each image taken. The acquired images are recorded with their GPS coordinates.

2.2 Method Overview

The system proposed in this study (Fig. 1) consists of three phases. The first one creates visible and infrared mosaic images for continuous view of the vineyard, and the DM (depth map). The second one segments the visible and infrared images by SegNet architecture, and merges the information of VDD. Finally, the last process corrects the result of the segmentation by using the DM.

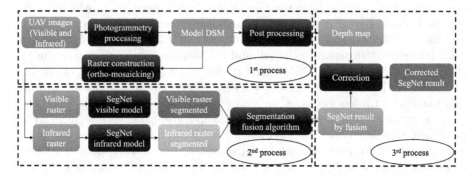

Fig. 1. Overview of the proposed method for VDD.

2.3   Depth Map

The design of the DM is carried out in two main stages, which are performed on images acquired by the drone in the two spectrum (visible and infrared). By using Agisoft Metashape software (version 1.5.5), the first step is to generate the DSM model, and the visible and infrared rasters (mosaic images), with the following steps; sparse point cloud (Tie points), dense point cloud (Dense cloud), digital surface model (DSM) and orthomosaic (Raster).

The second step is to extract the DM from the DSM model. It uses the same procedure in [21], which consists in three steps:

- DSM filtering: the DSM is filtered by a low pass filter of size 20×20, the filter has been chosen for smoothing the image and to keep only the digital terrain model (DTM).
- Subtraction of DTM from DSM: this step eliminates variations in the terrain and keep only the height of the vineyards.
- Contrast enhancement and thresholding: the result obtained by the subtraction has a weak dynamic of contrast. For this reason, a method for increasing the contrast based on the histogram was applied to improve the difference of the level between vine and soil. Then, an automatic thresholding (using Otsu's Method) is applied to obtain a binary DM.

2.4   Segmentation

In our previous study [16], a VDD method were proposed using a SegNet architecture. This study gave a good results on two parcels containing a non-grassy soil (the soil of these plots was brown). However, false diseases detections were observed in the segmentation results. The aim of this study is to introduce the depth information to separate the vine vegetation and the soil (whatever the type of soil). Therefore, it filters out the soil and thus reduces the detection errors.

2.5 Correction and Fusion

The aim of the correction phase is to reduce the errors of the SegNet, using the fusion. Indeed, the segmentation result often presents confusion between the green grassy soil, and the vine vegetation, or, confusion between the discolored grassy soil and the diseased vine. The correction phase proposed in this study takes as input the result of the SegNet segmentation by fusion, and the binary DM (vine and non-vine). Each pixel of the two images is compared and corrected (if necessary) by the rules described in Table 1.

Table 1. Operation of the correction.

Depth map	SegNet result by fusion	Corrected SegNet result
No-vine	Shadow	Shadow
	Ground	Ground
	Healthy	Ground
	Visible or infrared symptom	Ground
Vine	Shadow	Shadow
	Ground	Healthy
	Healthy	Healthy
	Visible or infrared symptom	Visible or infrared symptom

3 Experimentation and Results

This section presents experimental procedures, quantitative and qualitative results. The experiments were carried out with Python 2.7 language, using the Tensorflow 1.8.0 libraries for the SegNet architecture, and GDAL 3.0.3 for reading and writing the rasters (whole view of the plot) and their GPS information. The operating system used is Linux Ubuntu 16.04 LTS (64 bits). The programs were executed on a machine with characteristics: Intel Xeon 3.60 GHz × 8 processor, RAM of 32 GB, and a NVidia GTX 1080 Ti graphics card with an internal RAM of 11 GB.

3.1 Depth Map

To compute depth information, or the relief, we used depth from motion approach. The acquisition step is followed by the processing step, which consists in points matching between overlapped images. Matching points are represented by a sparse 3D point cloud, followed by a second processing to obtain a dense 3D point cloud. The DM is obtained by processing the DSM, which is created from the dense point cloud. The Fig. 2 represents an example of DM map, visible and infrared image.

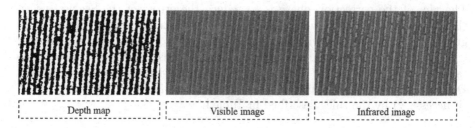

Depth map   Visible image   Infrared image

Fig. 2. Example of DM result.

3.2   Correction of SegNet Segmentation

SegNet segmentation is performed on a raster images with size of 12000×20000 pixels. A non overlapping sliding window of 360×480 pixels, is applied on the entire raster to segment each area of the parcel (visible and infrared spectra). It takes 45 min on average for each of them. Once the two rasters (visible and infrared) are segmented, they are merged using the segmentation fusion. Then, the DM is applied to the segmented image to isolate the background class. However, one can use background subtraction before the segmentation phase. But, we found that the SegNet is more precise when using soil classes. Table 2 shows the quantitative results of the SegNet segmentation by fusion, and its correction by the DM. Figures 3 and 4 are examples of qualitative results on healthy area with green grassy soil (Fig. 3), and another example of diseased area on discolored grassy soil (Fig. 4).

Table 2. Quantitative result on two temporal tests by measuring Recall (Rec.), Precision (Pre.), F1-Score/Dice (F1/D.) and Accuracy (Acc.) on the performance of SegNet by Fusion and corrected SegNet (values presented in percent).

Class name	Shadow			Ground			Healthy			Symptomatic			Total
Measure	Rec.	Pre.	F1/D.	Rec.	Pre.	F1/D.	Rec.	Pre.	F1/D.	Rec.	Pre.	F1/D.	Acc.
SegNet by fusion	71.56	82.63	76.69	55.78	96.58	70.71	90.25	50.32	64.61	89.01	80.47	84.52	75.31
Corrected SegNet	71.56	82.63	76.69	95.73	90.15	92.85	84.42	76.30	80.15	81.07	92.47	86.39	88.26

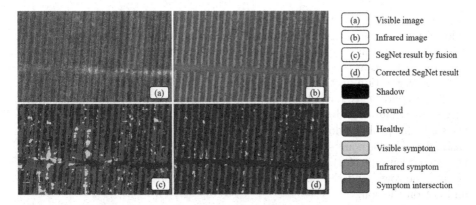

(a)	Visible image
(b)	Infrared image
(c)	SegNet result by fusion
(d)	Corrected SegNet result
■	Shadow
■	Ground
■	Healthy
■	Visible symptom
■	Infrared symptom
■	Symptom intersection

Fig. 3. Qualitative result of the healthy vine area using the depth map.

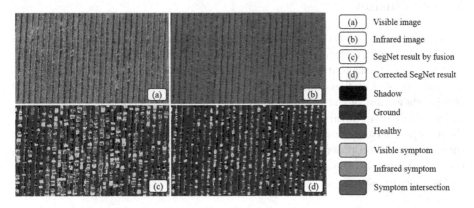

(a)	Visible image
(b)	Infrared image
(c)	SegNet result by fusion
(d)	Corrected SegNet result
■	Shadow
■	Ground
■	Healthy
■	Visible symptom
■	Infrared symptom
■	Symptom intersection

Fig. 4. Qualitative result of the diseased vine area using the depth map.

4 Discussion

This research work was set out with the aim of developing efficient methods for vine disease detection. Table 2 shows the quantitative results obtained for the SegNet segmentation experiments by fusion, and the corrected segmentation by DM. The results obtained are presented in terms of recall, precision, F1-score/Dice and accuracy, expressed in percentages. As shown in the accuracy column, the corrected method gives a better rate than the uncorrected method. This improvement is due to the correction of the soil areas and the vine vegetation. Also, the reduction of the over-detections of the disease areas, which can be observed on the individual results of each class.

Figures 3 and 4 represent respectively, the qualitative results, of the DM application on an area in good health and diseased. As can be seen in Fig. 3.c, the SegNet result by fusion gave several segmentation errors, in particular the vines detection and symptoms on the soil. These errors are mainly due to the presence of green grass mixed with a light brown color of the soil, which look like the color of vine disease. Figure 3.d shows a improvement of the segmentation result after correcting this result by the 3D information. Indeed, the correction brings a better distinction of the vine lines and reduces false detection of symptoms and vine vegetation on the grassy soil.

The second SegNet result on an area contaminated by Mildew disease (see Fig. 4.c) gave an over-detection of the symptomatic areas, which overflowed on the soil. This problem does not allow to evaluate the real diseased. Also, in some cases, it can cause confusion between the vine-rows that are contaminated. After the correction (Fig. 4.d), the result shows better interpretation and distinction of the vine-rows, and the detection of symptoms is observed only on the vine, and not on the soil.

5   Conclusion

This research work was set out with the aim of developing efficient methods for vine disease detection. We have developed a new method base on the deep learning segmentation approach and 3D depth map (DM). The method consists of three steps. The first one is mosaicking the visible and infrared pictures to obtain whole view of the vineyard, and their DM. The second step segments and merges visible and infrared rasters by using the SegNet architecture. Finally, the third step consists of correction of the SegNet result using the DM. This study showed that the proposed method reduces false detections of the vine vegetation, the vine symptoms, the soil, and therefore gives better precision and estimation on the disease map.

Acknowledgment. This work is part of the VINODRONE project supported by the Region Centre-Val de Loire (France). We gratefully acknowledge Region Centre-Val de Loire for its support.

References

1. Sona, G., et al.: UAV multispectral survey to map soil and crop for precision farming applications. ISPRS - Int. Arch. Photogram. Remote Sens. Spat. Inf. Sci. **XLI-B1**(June), 1023–1029 (2016)
2. Barbedo, J.G.A.: A review on the use of unmanned aerial vehicles and imaging sensors for monitoring and assessing plant stresses. Drones **3**(2), 40 (2019)
3. Otto, A., Agatz, N., Campbell, J., Golden, B., Pesch, E.: Optimization approaches for civil applications of unmanned aerial vehicles (UAVs) or aerial drones: a survey. Networks **72**(4), 411–458 (2018)
4. Teke, M., Deveci, H.S., Haliloglu, O., Gurbuz, S.Z., Sakarya, U.: A short survey of hyperspectral remote sensing applications in agriculture. In: RAST 2013 - Proceedings of 6th International Conference on Recent Advances in Space Technologies, pp. 171–176, June 2013

5. Santesteban, L.G., Di Gennaro, S.F., Herrero-Langreo, A., Miranda, C., Royo, J.B., Matese, A.: High-resolution UAV-based thermal imaging to estimate the instantaneous and seasonal variability of plant water status within a vineyard. Agric. Water Manag. **183**, 49–59 (2017)

6. Terrón, J.M., Blanco, J., Moral, F.J., Mancha, L.A., Uriarte, D., Marques Da Silva, J.R.: Evaluation of vineyard growth under four irrigation regimes using vegetation and soil on-the-go sensors. Soil **1**(1), 459–473 (2015)

7. Vanino, S., Pulighe, G., Nino, P., de Michele, C., Bolognesi, S.F., D'Urso, G.: Estimation of evapotranspiration and crop coefficients of tendone vineyards using multi-sensor remote sensing data in a mediterranean environment. Remote Sens. **7**(11), 14708–14730 (2015)

8. Mathews, A.J.: Object-based spatiotemporal analysis of vine canopy vigor using an inexpensive unmanned aerial vehicle remote sensing system. J. Appl. Remote Sens. **8**(1), 085199 (2014)

9. Bellvert, J., Zarco-Tejada, P.J., Girona, J., Fereres, E.: Mapping crop water stress index in a 'Pinot-noir' vineyard: comparing ground measurements with thermal remote sensing imagery from an unmanned aerial vehicle. Precis. Agric. **15**(4), 361–376 (2014). https://doi.org/10.1007/s11119-013-9334-5

10. Al-Saddik, H., Simon, J.C., Brousse, O., Cointault, F.: Multispectral band selection for imaging sensor design for vineyard disease detection: case of Flavescence Dorée. Adv. Anim. Biosci. **8**(2), 150–155 (2017)

11. Al-Saddik, H., Laybros, A., Billiot, B., Cointault, F.: Using image texture and spectral reflectance analysis to detect Yellowness and ESCA in grapevines at leaf-level. Remote Sens. **10**(4), 618 (2018)

12. Al-Saddik, H., Laybros, A., Simon, J.C., Cointault, F.: Protocol for the definition of a multi-spectral sensor for specific foliar disease detection: case of "Flavescence dorée". Methods Mol. Biol. **213–238**, 2019 (1875)

13. Rançon, F., Bombrun, L., Keresztes, B., Germain, C.: Comparison of SIFT encoded and deep learning features for the classification and detection of ESCA disease in Bordeaux vineyards. Remote Sens. **11**(1), 1–26 (2019)

14. MacDonald, S.L., Staid, M., Staid, M., Cooper, M.L.: Remote hyperspectral imaging of grapevine leafroll-associated virus 3 in cabernet sauvignon vineyards. Comput. Electron. Agric. **130**, 109–117 (2016)

15. Kerkech, M., Hafiane, A., Canals, R.: Deep leaning approach with colorimetric spaces and vegetation indices for vine diseases detection in UAV images. Comput. Electron. Agric. **155**(October), 237–243 (2018)

16. Kerkech, M., Hafiane, A., Canals, R.: Vine disease detection in UAV multispectral images using optimized image registration and deep learning segmentation approach. Comput. Electron. Agric. **174**(Apr), 105446 (2020). https://doi.org/10.1016/j.compag.2020.105446

17. Albetis, J., et al.: Detection of Flavescence dorée grapevine disease using Unmanned Aerial Vehicle (UAV) multispectral imagery. Remote Sens. **9**(4), 1–20 (2017)

18. Albetis, J., et al.: On the potentiality of UAV multispectral imagery to detect Flavescence dorée and Grapevine Trunk Diseases. Remote Sens. **11**(1), 0–26 (2019)

19. de Gennaro, S.F., et al.: Unmanned Aerial Vehicle (UAV)-based remote sensing to monitor grapevine leaf stripe disease within a vineyard affected by esca complex. Phytopathologia Mediterranea **55**(2), 262–275 (2016)

20. de Castro, A.I., Jiménez-Brenes, F.M., Torres-Sánchez, J., Peña, J.M., Borra-Serrano, I., López-Granados, F.: 3-D characterization of vineyards using a novel

UAV imagery-based OBIA procedure for precision viticulture applications. Remote Sens. **10**(4), 584 (2018)

21. Burgos, S., Mota, M., Noll, D., Cannelle, B.: Use of very high-resolution airborne images to analyse 3D canopy architecture of a vineyard. Int. Arch. Photogram. Remote Sens. Spat. Inf. Sci. - ISPRS Arch. **40**(3W3), 399–403 (2015)

22. Cinat, P., Gennaro, S.F.D., Berton, A., Matese, A.: Comparison of unsupervised algorithms for vineyard canopy segmentation from UAV multispectral images. Remote Sens. **11**(9), 1023 (2019)

A Random Forest-Cellular Automata Modeling Approach to Predict Future Forest Cover Change in Middle Atlas Morocco, Under Anthropic, Biotic and Abiotic Parameters

Anass Legdou[1(✉)], Hassan Chafik[2], Aouatif Amine[1], Said Lahssini[3], and Mohamed Berrada[2]

[1] LGS Laboratory, ENSA Kenitra, Ibn Tofail University, Kenitra, Morocco
anasslegdouseap@gmail.com, aouatif.amine@uit.ac.ma
[2] Department of Mathematics Informatics, Humanities and Sociology, University of Moulay Ismail, Meknes, Morocco
chafik1996hassan@gmail.com, berrada.mohamed@gmail.com
[3] National Forestry School of Engineers, Sale, Morocco
marghadi@gmail.com

Abstract. This study aims to predict forest species cover changes in the Sidi M'Guild Forest (Mid Atlas, Morocco). Used approach combines remote sensing and GIS and is based on training Cellular Automata and Random Forest (RF) regression model for predicting species cover transition. Five covariates that precludes such transition have been chosen according to Pearson's test. The model was trained and validated based on the use of forest cover stratum transition probabilities between 1990 and 2004 and then validated using 2018 forest species cover map. Validation of the predicted map with that of 2018 shows an overall agreement between the two maps (72%) for each number of RF's trees used. The 2032 projected forest species cover map indicate a strong regression of Cedar atlas and thuriferous juniper cover and a medium regression of mixture holm oak and thuriferous juniper, mixture of atlas cedar and thuriferous juniper, and sylvatic and asylvatic vacuums, a very strong progression of holm oak, and of mixture atlas cedar, holm oak and thuriferous juniper and medium progression of mixture of atlas cedar and holm oak. These findings provide important insights to planners, natural resource managers and policy-makers to reconsider their strategies to ensure the sustainability goals.

Keywords: Forest species cover change · Random forest regression · Cellular automata

1 Introduction

Forests as natural capital provides ecosystem services that contributes to human well being [1]. Sustainable management of forest resources try to maximize the

© Springer Nature Switzerland AG 2020
A. El Moataz et al. (Eds.): ICISP 2020, LNCS 12119, pp. 91–100, 2020.
https://doi.org/10.1007/978-3-030-51935-3_10

provided services without compromising their abilities to fulfil theses services in the future [2]. Nowadays due to social and human needs, forests resources are under high pressures. In many regions, the pressures are far beyond forests productive capacities. Moreover, climate change increase the fragility of the threatened natural ecosystems. As consequence, a diffuse and still progressing process of land use and forest cover changes are widely documented mainly in developing countries [4]. With the need to maintain forest contribution to global cycles and to protect this natural capital for the next generations, there is a need to understand their dynamic and to predict the tendencies in order to stress urgent actions and formulate appropriates policies to improve land use planing [5]. Such understanding relay on the Land Use Land Change Modeling LULC that try to explain human environment dynamics producing the changes [6]. LULC needs multi temporal land/forest cover maps as well as the driving forces conducting to that changes [7–9]. In addition, machine learning algorithms have been used extensively to explain LULCs. Several researches had combined Cellular automata (CA) with a plethora of modeling frameworks such as Markov chains [10], neural networks [11] support vector machines [12] and kernel-based methods [13] among others. More recently, CA have been successfully combined with Random Forest [14,15].

Moroccan forests hold a major part of its biological diversity. It covers 5.8 million hectares, including 132,000 ha of cedar, 1.36 million ha of holm oak, 830,000 ha of argan, 350,000 ha of cork oak, 600,000 ha of thuja and 1 million ha of Saharan acacia (HCEFLCD, 1992). The Atlas cedar occupies a prominent place among other species. Moroccan cedar forests, especially those in the Middle Atlas, show regressive trends. Cedar become limited to mountain tops. Among the factors generally blamed for the degradation: regeneration lack due to high grazing pressure [16]; high human pressure (overgrazing, cultivation, illegal cutting, fire, etc.); dieback phenomenon, which is becoming increasingly worrisome about the future of cedar stands, and damages caused by cedar's natural enemies (defoliating insects, wood-boring insects, fungi and the maggot monkey), which weakens stands stressed by climatic hazards.

Given the status of Moroccan cedar forests as threatened natural capital and in order to understand the driving forces toward its regressive tendencies and to predict its cover in the future, we focus within this work on cedar cover change modeling. The work concerns Sidi M'Guild Forest which belong to Middle Atlas National Park that holds a representative part of Cedar Ecosystem. Land Use change has been modeled using machine learning algorithm predicting future forest cover change as result of driving anthropic, biotic and abiotic factors. Our paper is organized in a Methods describing the used data and algorithms explanation and the achieved results which stress out the main results and the discussions and conclusion.

2   Materials and Methods

The study concerned Sidi M'Guild forest which is located in the Moroccan Middle Atlas. It covers about 29,000 hectares. Cedar covers about 51%, holm oak

34%, and Thuriferous juniper 3.6% of forest area. The overall methodology is shown in the Fig. 1. It consists on getting covariates, then processing and modeling.

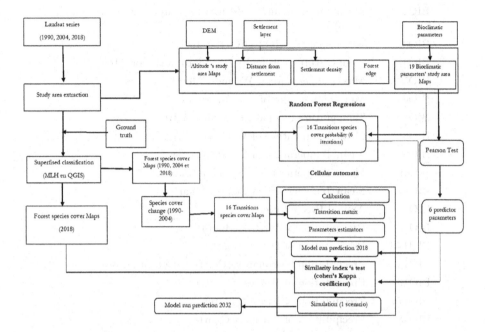

Fig. 1. Schematic of the used approach in this work.

2.1 Used Data

Landsat satellite images covering study used were downloaded from USGS Earth explorer platform. Landsat 4 TM (thematic Mappers) images with a 30 m spatial resolution and 7 spectral bands for 1990 to 2004. Landsat 8 OLI image (Operational Land Image) that contains 11 spectral bands at 30 m spatial resolution were used for 2018. Images were chosen based on the availability and being cloud free and captured during August to reduce atmospheric disturbance and confusion with herbaceous layers' spectral emission. The predictive variables used consists on bioclimatic variables, Digital elevation model (DEM) and human characteristics. Bioclimatic data were downloaded from worldclim's platform (see Table 1)[1]. Altitude Maps was extracted from Shuttle Radar Topography Mission (STRM) DEM[2]. The location of human settlement was linked to 2014 Morocco's 'general census' data. Distance from human settlement maps and distance from Forest edge map were generated using basic GIS functions.

Furthermore, a Pearson's r correlation coefficient has been calculated in order to identify the factors that are highly correlated.

---

[1] https://www.worldclim.org/.
[2] Data collected from http://dwtkns.com/srtm30m/.

Table 1. Used bioclimatic variables.

Symbology	Designation
BIO1	Annual Mean Temperature
BIO2	Mean Diurnal Range (Mean of monthly (max temp−min temp))
BIO3	Isothermality (BIO2/BIO7) (* 100)
BIO4	Temperature Seasonality (standard deviation *100)
BIO5	Max Temperature of Warmest Month
BIO6	Min Temperature of Coldest Month
BIO7	Temperature Annual Range (BIO5–BIO6)
BIO8	Mean Temperature of Wettest Quarter
BIO9	Mean Temperature of Driest Quarter
BIO10	Mean Temperature of Warmest Quarter
BIO11	Mean Temperature of Coldest Quarter
BIO12	Annual Precipitation
BIO13	Precipitation of Wettest Month
BIO14	Precipitation of Driest Month
BIO15	Precipitation Seasonality (Coefficient of Variation)
BIO16	Precipitation of Wettest Quarter
BIO17	Precipitation of Driest Quarter
BIO18	Precipitation of Warmest Quarter
BIO19	Precipitation of Coldest Quarter

2.2 Preprocessing and Image Classification

Satellite data were preprocessed QGIS and then classified using Maximum Likelihood algorithm. Images were standardized and accommodated to the same extent and spatial resolution.Image classification was based on training the classifier using, as ground truth, forest stand type maps released in 1990, 2004 and 2018. Six dominant forest cover stratum/classes were chosen while respecting the National forest inventory standards. The resulting classes are: atlas cedar (Ca), holm oak (Qr), thuriferous juniper (Jt), atlas cedar and thuriferous juniper mixture (CaJt), atlas cedar and holm oak mixture (CaQr), mixture of atlas cedar, holm oak and thuriferous juniper (CaQrJt), holm oak and thuriferous juniper mixture (QrJt) and sylvatic and asylvatic vacuums (V). In addition to Atlas cedar, we classified other species closed to or mixed with Atlas cedar, since when we talk about the cedar ecosystem, we refer to all the species that are closely linked to it. The accuracy assessment based on cross-validation of the classified images showed an overall accuracy of 88.86%, 90.21% and 92.33% respectively for the years 1990, 2004 and 2018.

2.3   Random Forest Regression and Calibration Model

Random Forest is a flexible and easy to use machine learning algorithm which performs both regression and classification tasks [17]. It combines multiple decision trees in determining the final output. It uses different technics for training the model. Bagging technique [18], which involves training each decision tree on a random set of data sampled, without replacement, from the training data set with a random split selection [19]. Random forest regression, which the performance has been proved in several studies [20, 21], was used to predict forest cover class transition probabilities. Transition maps for the period 1990–2004 were established by observing the behavior of the forest species. For each transition from one class to another, two possible values 1 denotes change from class to other class, and 0 denotes no change. Then, binary transition maps were produced 0 for stability and 1 for change occurrence and used for training 16 models. Following the approach adopted by Gounaridis et al. [15], the transition probability surfaces were generated through training Random Forest algorithm [18] using all variables and using the most independent covariates identified through the Pearson's correlation coefficient calculation (5 factors). The RF regression models were then implemented in python using the Random Forest regression. The model output is transition probability map. The model was run for each of the sixteen transitions defined before (Fig. 2) and for a set on number of trees (10, 20, 30, 40, 50, 100).

Cellular automata [22] were used to predict the future state of forest species distribution. Probability maps obtained with different number of trees were produced and for each time, the Kappa coefficient of Cohen was calculated through comparing predicted and observed results of the same year [23].

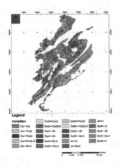

Fig. 2. Forest species cover change, during 1990–2000.

3   Results and Discussion

3.1   Forest Species Cover Classification

Accuracy assessment based on cross-validation showed an overall accuracy of 88.86%, 90.21% and 92.33% respectively for the years 1990, 2004 and 2018.

Thus, indicating the suitability of the derived classified maps for effective and reliable forest species cover change analysis and modeling. Post-classification analysis of the spatial metrics and their variations based on Table 2 showed that the area occupied by Ca, CaQr, CaQrJt and Jt classes had drastically decreased between 1990 and 2018 in the studied area. On the other hand, the area occupied by Qr, CaJt, QrJt and V had substantially increased during the same period. Such results seems to be coherent with literature describing Cedar as vulnerable species and holm oak as green cement with a high adaptive capacities.

Table 2. Temporal distribution of forest species cover classes (ha) and percentages of change.

Classes	Area size (ha)			Change in forest species cover (Δ%)		
	1990	2004	2018	2004–1990	2018–2004	2018–1990
Ca	8975,88	2171,34	2007,85	−313	−8	−347
CaQr	8482,05	9141,93	7138,62	7	−28	−19
Qr	44,37	551,88	233,1	92	−137	81
CaQrJt	4034,52	3184,83	3758,58	−27	15	−7
Jt	5010,75	3491,28	4854,94	−44	28	−3
QrJt	2207,97	5986,71	4802,95	63	−25	54
CaJt	2207,97	2405,7	2282,89	8	−5	3
V	0,99	1822,86	3677,6	100	50	100

3.2 Analysis of Transition Probabilities and Model Validation

Using all factors as covariates or retaining the five factors that are not correlated according to Pearson's r correlation coefficient (Temperature Seasonality: Bio4, precipitation of Driest Quarter: Bio17, Distance from human settlement, settlement density and distance from Forest edge) gave the same results (Table 3). Such fact could be explained by RF robustness toward correlated data. In general, we notice that the higher scores were recorded for numbers of trees equal to 50 and 100 trees. In addition, we conclude that the five non correlated parameters considerably affects the evolution of the forest species change between 1990 and 2004. As explained in the methods, the predictive model was used to predict 2018 forest cover state. The model returns simulated maps of forest species cover distribution for the year 2018, relative to each of RF model number of trees. The predicted maps were compared to existing 2018 forest cover maps and the validation was based on the use of Kappa coefficient. Validation results are given in Table 4.

We notice that there was no evident difference between the values achieved by Kappa Coefficient. Hence, we could conclude that the number of trees used in the random forest model does not influence the results of the simulation. Contrariwise, the thresholds experimented for each rule did influence the model.

Table 3. Transition probability of forest species cover areas for each number of trees.

Transition of species between 1990 and 2004	Area size	Training score					
		n = 10	n = 20	n = 30	n = 40	n = 50	n = 100
Ca–CaJt	223,56	0,79	0,79	0,8	0,8	0,8	0,81
Ca–QrJt	121,68	0,86	0,86	0,87	0,87	0,88	0,88
Ca–CaQrJt	509,13	0,82	0,82	0,83	0,84	0,84	0,84
Ca–CaQr	5945,58	0,42	0,42	0,45	0,46	0,47	0,48
CaQr–CaJt	1658,25	0,53	0,53	0,55	0,56	0,57	0,58
CaQr–QrJt	987,21	0,76	0,76	0,77	0,78	0,78	0,79
CaQr–CaQrJt	2459,7	0,32	0,32	0,35	0,36	0,37	0,38
CaQr–Qr	167,22	0,59	0,59	0,61	0,62	0,62	0,63
Jt–V	992,25	0,45	0,45	0,48	0,48	0,49	0,5
Jt–QrJt	166,77	0,45	0,45	0,48	0,49	0,49	0,5
QrJt–V	827,01	0,38	0,38	0,41	0,42	0,43	0,45
QrJt–Jt	589,23	0,45	0,45	0,47	0,49	0,41	0,51
QrJt–Qr	162,81	0,54	0,54	0,56	0,57	0,57	0,58
CaJt–QrJt	1338,66	0,24	0,24	0,27	0,28	0,29	0,31
CaJt–CaQrJt	188,28	0,24	0,24	0,27	0,29	0,3	0,32
CaJt–Qr	191,61	0,58	0,58	0,6	0,61	0,61	0,62

Table 4. Kappa's coefficient for each RF number of trees

Trees number	n = 10	n = 20	n = 30	n = 40	n = 50	n = 100
Kappa de Cohen	0.7213	0.72126	0.72127	0.72125	0.7212	0.7211

3.3   Model Comparison and Predictions for 2034

By comparing the areas occupied by each class in both simulated and observed maps, we observe some differences (Table 5):

Table 5. Comparison of the actual and simulated forest cover species in 2018

Classes	Observed 2018 (ha)	Simulated 2018 (ha)	Difference simulated observed (ha)	Simulated observed in 2018 (Δ%)
Ca	2007.85	1703.07	−304.78	−17,90
CaQr	7138.62	6513.03	−625.59	−9,61
Qr	233.1	1473.03	+1239.93	84,18
CaQrJt	3758.58	3707.1	−51.48	−1,39
Jt	4854.94	5069.34	+214.4	4,23
QrJt	4802.95	4768.92	−34.03	−0,71
CaJt	2282.89	2305.71	+22.82	0,99
V	3677.6	3200.67	−476.93	−14,90

The model did closely estimate the area of CaQr, CaQrJt, Jt, QrJt and CaJt. On the other hand, it did largely overestimated the area of Qr about −18% and miss estimated the area occupied by Ca, CaQr and V. In addition, the

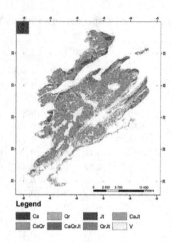

Fig. 3. Forest species cover predicted for 2032.

Table 6. Comparison of forest cover species change between 2018 and 2032

Class	Observed 2018 (ha)	Simulated 2032	Rate change (%)
Ca	2007.85	170.82	−91.49
CaQr	7138.62	9144.99	+28.11
Qr	233.1	3249.45	+1294.02
CaQrJt	3758.58	8444.61	+124.68
Jt	4854.94	1243.62	−74.39
QrJt	4802.95	2405.52	−49.92
CaJt	2282.89	1254.51	−45.05
V	3677.6	2881.44	−21.65

model has predicted a regression of Ca, CaQr, CaQrJt, QrJt and V and then the area affected by the change were localized. It predicted Qr, Jt and CaJt progression. The model overestimated some species classes and underestimated others because we did not take into account the interventions carried out by the administration in terms of silvicultural interventions on holm oak and cedar stands, the reforestation actions undertaken including silvicultural vacuums, the awareness campaigns conducted for the benefit of the populations bordering the forest and finally the forest police actions to preserve and conserve the forest heritage. Theses interventions must be deeply analyzed in order to assess their impacts on explaining the occurred forest cover changes.

With regard to the predictions, the model returned the forest species cover state for 2032 (Fig. 3 and Table 6).

The model predicted a strong regression for Ca and Jt cover and a medium regression for QrJt, CaJt, and V. It did also predicate a very strong progression for Qr and CaQrJt and medium progression for CaQr. The majority of forest species cover change are depending on their positioning to settlements and to forest's boundaries. Stratum that will show regression trends are located near human settlements and not far from forest edge. On the other hand, stratum with progressive trends are located far from human settlements and to forest's boundaries. Such finding seems to be coherent with literature and forest mangers declarations.

4   Conclusion

The current study was based on an integrated approach that combines remote sensing and GIS to simulate and predict plausible forest species cover changes for Sidi M'Guild Forest for the years 2032 using Cellular Automata (CA)-Random Forest regression (RF) model. The initial forest species cover map (1990), the transition potential maps (1990–2004) and the 1990–2004 transition probabilities were used to train RF model. Model was validated using actual and predicted 2018 forest species cover. The overall agreement between the two maps was 72% for each number of RF's trees used. The future 2032 projections indicate a strong regression of Cedar atlas with −91.49% and thuriferous juniper cover with −74.39% and a medium regression of holm oak and thuriferous juniper mixture with −49.92%, atlas cedar and thuriferous juniper mixture with −45.05%, and sylvatic and asylvatic vacuums with 21.65%, a very strong progression of holm oak with +1294.02%, and of atlas cedar, holm oak and thuriferous juniper mixture with +124.68% and medium progression of atlas cedar and holm oak mixture with +28.11% by 2032. The majority of forest species cover changes depends on their location to settlements and to forest's boundaries. Regression are located near human settlement and forest boundaries. As cedar is considered as national heritage, these findings could be useful for decision makers and for managers to review their strategies in order to ensure the sustainability of cedar as natural capital.

References

1. Costanza, R., et al.: Quality of life: an approach integrating opportunities, human needs, and subjective well-being. Ecol. Econ. **61**(2–3), 267–276 (2007)
2. Bettinger, P., Boston, K., Siry, J., Grebner, D.: Forest Management and Planning, p. 362. Academic Press, Cambridge (2017)
3. Bettinger, P., Chung, W.: The key literature of, and trends in, forest-level management planning in North America, 1950–2001. Int. For. Rev. **6**, 40–50 (2004)
4. Giri, C.P., Pengra, B.W., Long, J.B., Loveland, T.R.: Next generation of global land cover characterization, mapping and monitoring. Int. J. Appl. Earth Obs. Geoinf. **25**, 30–37 (2013)
5. Tena, T.M., Mwaanga, P., Nguvulu, A.: Impact of land use/land cover change on hydrological components in chongwe river catchment. Sustainability **11**, 6415 (2019)
6. Mertens, B., Lambin, E.: Modelling land cover dynamics: integration of fine-scale land cover data with landscape attributes. Int. J. Appl. Earth Obs. Geoinf. **1**, 48–52 (1999)

7. Yang, F., Liu, Y., Xu, L., Li, K., Hu, P., Chen, J.: Vegetation-ice-bare land cover conversion in the oceanic glacial region of tibet based on multiple machine learning classifications. Remote Sens. **12**(6), 999 (2020)
8. Abdi, A.M.: Land cover and land use classification performance of machine learning algorithms in a boreal landscape using Sentinel-2 data. GISci. Remote Sens. **57**, 1–20 (2019)
9. Akubia, J.E.K., Bruns, A.: Unravelling the frontiers of urban growth: spatio-temporal dynamics of land- use change and urban expansion in Greater Accra Metropolitan Area, Ghana. Land **8**(9), 131 (2019)
10. Fu, X., Wang, X.Y., Yang, J.: Deriving suitability factors for CA-Markov land use simulation model based on local historical data. J. Environ. Manag. **206**, 10–19 (2018)
11. Islama, K., Rahmanb, M.F., Jashimuddinc, M.: Modeling land use change using Cellular Automata and Artificial Neural Network: the case of Chunati Wildlife Sanctuary, Bangladesh. Ecol. Ind. **88**, 439–453 (2018)
12. Mustafa, A., Rienow, A., Saadi, I., Cools, M., Teller, J.: Comparing support vector machines with logistic regression for calibrating cellular automata land use change models. Eur. J. Remote Sens. **51**, 391–401 (2018)
13. Liu, X., et al.: A future land use simulation model (FLUS) for simulating multiple land use scenarios by coupling human and natural effects. Landsc. Urban Plan. **168**, 94–116 (2017)
14. Gounaridis, D., Chorianopoulos, I., Symeonakis, E., Koukoulas, S.: A Random Forest-Cellular Automata modelling approach to explore future land use/cover change in Attica (Greece), under different socio-economic realities and scales. Sci. Total Environ. **646**, 320–335 (2019)
15. Gounaridis, D., Chorianopoulos, I., Koukoulas, S.: Exploring prospective urban growth trends under different economic outlooks and land-use planning scenarios: the case of Athens. Appl. Geogr. **90**, 134–144 (2018)
16. Moukrim, S., et al.: Agrofor. Syst. **93**(4), 1209–1219 (2019)
17. Breiman, L., Friedman, J., Stone, C.J., Olshen, R.A.: Classification and Regression Trees. Chapman and Hall/CRC, Boca Raton (1984)
18. Breiman, L.: Bagging predictors. Mach. Learn. **24**(2), 123–140 (1996). https://doi.org/10.1007/BF00058655
19. Dietterich, T.: An experimental comparison of three methods for constructing ensembles of decision trees: bagging, boosting and randomization. Mach. Learn. **32**, 1–22 (1998)
20. Zhao, Q., Yu, S., Zhao, F., Tian, L., Zhao, Z.: Comparison of machine learning algorithms for forest parameter estimations and application for forest quality assessments. Forest Ecol. Manag. **434**, 224–234 (2019)
21. Srinet, R., Nandy, S., Patel, N.R.: Estimating leaf area index and light extinction coefficient using Random Forest regression algorithm in a tropical moist deciduous forest, India. Ecol. Inform. **52**, 94–102 (2019)
22. Kamusoko, C., Gamba, J.: Simulating urban growth using a Random Forest-Cellular Automata (RF-CA) model. Int. J. Geo-Inf. **4**(2), 447–470 (2015)
23. Zapf, A., Castell, S., Morawietz, L., Karch, A.: Measuring inter-rater reliability for nominal data - which coefficients and confidence intervals are appropriate? BMC Med. Res. Methodol. **16**, 93 (2016). https://doi.org/10.1186/s12874-016-0200-9

Machine Learning Application and Innovation

Incep-EEGNet: A ConvNet for Motor Imagery Decoding

Mouad Riyad[(⊠)], Mohammed Khalil, and Abdellah Adib

Networks, Telecoms and Multimedia Team, LIM@II-FSTM,
B.P. 146, 20650 Mohammedia, Morocco
riyadmouad1@gmail.com, mohammed.khalil@univh2c.ma, adib@fstm.ac.ma

Abstract. The brain-computer interface consists of connecting the brain with machines using the brainwaves as a mean of communication for several applications that help to improve human life. Unfortunately, Electroencephalography that is mainly used to measure brain activities produces noisy, non-linear and non-stationary signals that weaken the performances of Common Spatial Pattern (CSP) techniques. As a solution, deep learning waives the drawbacks of the traditional techniques, but it still not used properly. In this paper, we propose a new approach based on Convolutional Neural Networks (ConvNets) that decodes the raw signal to achieve state-of-the-art performances using an architecture based on Inception. The obtained results show that our method outperforms state-of-the-art filter bank common spatial patterns (FBCSP) and ShallowConvNet on based on the dataset IIa of the BCI Competition IV.

Keywords: Deep learning · Electroencephalography · Convolutional neural network · Brain-computer interfaces

1 Introduction

Brain-computer interfaces (BCI) link machines and human brains with the brainwaves as mean of communication for several purposes [1]. The necessity of such a link is crucial to automatize several tasks such as the prediction of epilepsy seizure, or the detection of neurological pathologies. Also, it commonly uses brain signals as a control signal for devices such as keyboards or joysticks, which can improve the quality of life of severely disabled patients, or many non-medical applications such as video games, controlling a robot or authentication [13]. The most used sensor is electroencephalography (EEG) that relies on electrodes placed in the scalp to detect the variation of electrical activity. It processes the collected data with signal processing techniques to keep important features. Then, machine learning take a decision depending on the use case.

The most well-known applications are related to Motor Imagery (MI) [15]. It is a neural response that is produced when a person performs a movement or just imagine it. Unfortunately, the signals are intrinsically non-stationary, non-linear, and noisy [13]. Overcoming those problems requires the use of sophisticated

© Springer Nature Switzerland AG 2020
A. El Moataz et al. (Eds.): ICISP 2020, LNCS 12119, pp. 103–111, 2020.
https://doi.org/10.1007/978-3-030-51935-3_11

algorithms that requires human intervention (e.g. the eye blink elimination) and computational power that can be constraining. Deep Learning permits to waive a solution to all the previously cited obstacles [9]. It extracts the features automatically without human-engineered features and classifies in the same process which enables end-to-end approaches. Several other advances in new activation function, regularization, training strategies, and data augmentation yielded to state-of-the-art performances in several fields [3,7,10]. Also, it is possible to explain the decision of deep classifiers by advance visualization methods such as weight visualization to discover the learned features.

In this paper, we propose a new convolutional neural network (Convnet) architecture based on Inception for motor imagery classification. It allows to process the data with parallel process In our approach, we use the multivariate raw signal as input with a bandpass filter as preprocessing. Therefore, we use the same first block of [12] but with higher complexity which increases the capacity of the network. Then, an Inception block will extract temporal features more efficiency which improves the performance and speeds up the learning despite the depth to reduce the degradation problem [18]. To test our approach, we use dataset IIa from the BCI Competition IV [19]. As a baseline, we compare with FBCSP and ShallowConvNet which are the state-of-the-art techniques [2]. We investigate some visualization techniques to examine the ability of our networks to extract relevant features.

The rest of the paper is organize as follows: We presents some related works in Sect. 2. We introduce our method in Sect. 3. In Sect. 4, we evaluate the performances and visualize the learned features. Section 5 discuss the result and conclude the paper.

2   Related Works

The first interesting approach was a ConvNet that uses raw EEG data for P300 speller application [6]. It uses convolutional layers that extract temporal and spatial features. It is inspired from Filter Banks Common Spatial Pattern (FBCSP) [2]. A convolution is performed with a kernel of size $(1, n_t)$, then an other convolution with a kernel with a size $(C, 1)$ where C is the number of the channels. Then, it use a softmax layer to classifies the features extracted. [17] introduced similar architectures for MI. ShallowConvNet is a shallow convnet that is composed with the two convolutional layers then the classification layers. DeepConvNet is a deep architecture that includes more aggregation layer after the convolutional layer. ShallowConvNet outperforms state-of-the-art FBCSP. [12] proposed EEGNet as a compact version of the existing methods. It relies on Depthwise convolutional and separable convolution which permitted to reduce the number of the parameter using 796 parameters only for the EEGNet 4, 2. EEGNet performs lower than ShallowConvNet since it was not trained with the same data augmentation (cropped training) suggested by [17]. Also, cropped training requieres a huge time to train which can be problematic in that cas of a takes a huge time to train, for one subjects compared with EEGNet.

3   Method

3.1   EEG Proprieties and Data Representation

MI yields on the apparition of fluctuation of the amplitude of the neuro-signals generated in the primary sensorimotor cortex [14]. It appears as an increase and a decrease of amplitude that target specific frequency bands that are related to motor activities. They are called Event-Related Synchronization (ERS) and Event-Related Desynchronization (ERD). The μ and β bands are present respectively in [8, 13] Hz and the beta band [13, 30] Hz are the targeted pattern. As input, each trial is turned into a matrix of $\mathbb{R}^{C \times T}$ where C represents the number of electrodes and T represents the number of time samples. We sample our data at 128 Hz and we use the segment [0.5–2.5] s after cue.

3.2   Incep-EEGNet

We propose Incep-EEGNet as it is illustrated in Fig. 1. It is a multistage ConvNet that is based on Inception [18]. It is composed as follows:

The first part is the same as EEGNet from [12]. They base it on two convolutional layers that act as temporal and spatial filter as act similarly to FBCSP, which is a widely used approach. We use a temporal convolutional layer with F kernel of size $(1, tx)$ with padding. This layer will learn to extract relevant temporal features as it act as a FIR filter. We choose a size of 32 which correspond to a duration of 0.25 s of a signal sampled at 128 Hz. A second convolution is used to extract the spatial feature. It relies on Depthwise convolution that produces the number of feature maps per input which reduces considerably the computational cost. It is a convolution with a size of $(C, 1)$ where C represents the number of channels. Also, we use batch normalization after each convolution and activation after the second one. This layer will allow only the important electrodes to contribute to the decision and learning frequency-specific spatial filter with Depthwise convolution where it controls the number of connections by the depth parameter D.

In the second part, we introduce the novelty of this architecture which is an inception based block. This block comes as a solution to the inconvenience of EEGNET that is too shallow and too compact, which restricts the capacity of the networks leading to overfitting in most cases. Even with a deeper network, the performance still low because of a degradation problem for DeepConvNet. Hence, we suggest to use an inception stage based That will learn features from several branches:

- A convolutional branch with a convolution with a kernel size of $(1, 7)$.
- A convolutional branch with a convolution with a kernel size of $(1, 9)$.
- A branch with a pointwise convolution with a kernel size of $1, 1$ with a stride of $(1, 2)$
- A branch with an average pooling with a kernel size of

We merge the output of the different branches by stacking them along with the feature map dimension. We apply batch normalization and an activation. The use of dropout restricted only after final the activation cause we observed no improvement. Each convolutional branch include a pointwise convolution that reduces the number of feature map to 64 and an average pooling layer with a size of $(1, 2)$.

In the final part, we use an additional convolutional layer with a $F * D$ kernel with a size of $(1, 5)$ along with batchnormalization, activation, and dropout. We use an Global AveragePooling layer to reduce the number of parameters to $2 * F$. Then, we use Softmax classification with 4 units that represent the 4 classes of the dataset.

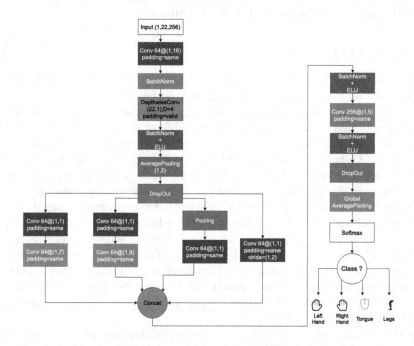

Fig. 1. Architecture of the proposed system with layers hyperparameters

3.3   Hyperparameters and Training

Our implementation uses publicly available codes of preprocessing based on *braindecode* [17]. We trained deep learning methods on a NVIDIA P100 1.12.0. We train our method by optimizing the categorical cross-entropy using ADAM Optimizer [11] with Nesterov. Dropout probability is 0.5 as advised by [3]. We use a batch size of 64 as for EEGNET [12]. We fix the network parameter to $F = 64$ and $D = 4$. Exponential Linear Unit (ELU) is chosen as the activation [7]. We train our ConvNets as follows: We train for 100 epochs with a learning rate (Lr) of 5×10^{-4}. At the end of the training, we retrain it for 50 epochs and Lr set to 1×10^{-4} with the merged training and validation set. Once again, we do the same operation for 30 epochs and a Lr set to 2×10^{-5}. Similar training was done for ShallowConvNet [17].

4 Experiment

4.1 Dataset

As a dataset, we use the dataset IIa from the BCI competition IV [19]. It contains EEG data of four MI tasks (right hand, left hand, foot, and tongue imagined movements) from nine subjects. It uses a set of 22 electrodes placed on the scalp. The recording was on two different sessions where the first was defined as a training set and the second one as a testing set. The subjects are asked to performs 288 MI tasks per session (72 trials for each class) after a cue that was. The original data is sampled at 240 Hz and filtered with a bandpass filter between 0.1 Hz and 100 Hz. We add additional preprocessing to the data as described in [17]. We resample the signals at 128 Hz and filter with a bandpass filter between 1 Hz and 32 Hz. We use 20% of the training set as a validation set. We use a cropping data augmentation by extracting the segments [0.3, 2.3] s, [0.4, 2.4] s, [0.5, 2.5] s, [0.6, 2.6] s, [0.7, 2.7] s post cue only on the training set (1152 trials). The validation and testing set contain only [0.5, 2.5] s segment to prevent leaking (for validation set) that can compromise the training. Therefor, the input will have a shape of 22×256.

4.2 Results

To assert the performances of our method, we compare with FBCSP, Riemannian geometry [4], Bayesian optimization [5], and ShallowNet [17]. Table 1 shows the results of the classification of our method and the baselines in terms of accuracy. It shows that the proposed method outperforms the baselines for several subjects (S2, S3, S5, S6, S7, S9). However, BO got better results for S1 and S8, when ShallowNet performs better for S4. On the other hand, FBCSP2 and RG did not achieve higher results. For an advanced evaluation, we conduct statistical testing with the Wilcoxon test. To evaluate the significance of the results on the mean value. It shows that our method has a statistically significant difference compared with BO with $p < 0.05$. Comparing with FBCSP2 and RG, the difference is highly significant with $p < 0.01$.

Table 2 shows the results of the classification of our method and the baselines in terms of kappa. The result shows that our method outperforms for most of the subjects. It only failed to outperform FBCSP1 for S2 and ShallowNet for S4. Once Again, FBCSP2 and RG got bad results. Statistical testing shows that the increase in mean kappa is statistically significant with $p < 0.05$ for FBCSP1, MDRM, and ShallowNet. For the other methods, the difference is highly significant at $p < 0.01$.

Table 1. Classification accuracy (%) comparaison of our methods and the baselines,

	BO	FBCSP2	RG	ShallowNet	Incep-EEGNet
S1	**82.120**	75.694	77.778	75.347	78.472
S2	44.860	44.792	43.750	43.056	**52.778**
S3	86.600	85.069	83.681	80.208	**89.931**
S4	66.280	63.542	56.597	**68.056**	66.667
S5	48.720	59.028	47.917	58.681	**61.111**
S6	53.300	36.458	47.569	49.306	**60.417**
S7	72.640	86.111	78.472	85.417	**90.625**
S8	**82.330**	79.167	79.861	77.778	82.292
S9	76.350	82.639	81.250	80.556	**84.375**
Average	68.133	68.056	66.319	68.711	**74.074**
p-values	0.038	0.008	0.008	0.011	1.000

Table 3 and Table 4 show the confusion matrix of Incep-EEGNet and FBCSP2 respectively. They show that both methods have difficulties to classify foot classes. Also, they confuse between right-hand and left-hand classes. Performances of our method are better than the reference.

Table 2. Kappa values comparison of our methods and the baselines

	FBCSP1	2nd	MDRM	FBCSP2	RG	ShallowNet	Incep-EEGNet
S1	0.680	0.690	0.750	0.676	0.704	0.671	**0.713**
S2	**0.420**	0.340	0.370	0.264	0.250	0.241	0.370
S3	0.750	0.710	0.660	0.801	0.782	0.736	**0.866**
S4	0.480	0.440	0.530	0.514	0.421	**0.574**	0.556
S5	0.400	0.160	0.290	0.454	0.306	0.449	**0.481**
S6	0.270	0.210	0.270	0.153	0.301	0.324	**0.472**
S7	0.770	0.660	0.560	0.815	0.713	0.806	**0.875**
S8	0.750	0.730	0.580	0.722	0.731	0.704	**0.764**
S9	0.610	0.690	0.680	0.769	0.750	0.741	**0.792**
Average	0.570	0.514	0.521	0.574	0.551	0.583	**0.654**
p-values	0.021	0.008	0.021	0.008	0.008	0.011	1.000

Figure 2a represents the Fourier transform of a temporal filter learned in the first convolution. It was designed to extract the temporal features of the EEG signals. As it was expected, Incep-EEGNet learned exactly the frequencies that are involved in the MI neural response. Also, we observe that there is a peak at 55 Hz, which can indicate that MI may be also characterized by this band as was

Table 3. Confusion matrix of Incep-EEGNet

		Predicted			
		L	R	F	T
Actual	L	**80.40**	8.80	5.09	5.71
	R	15.59	**74.38**	4.48	5.56
	F	9.88	7.56	**65.43**	17.13
	T	10.80	5.56	7.56	**76.08**

Table 4. Confusion matrix of FBCSP

		Predicted			
		L	R	F	T
Actual	L	73.30	15.28	4.78	6.64
	R	14.97	73.92	5.71	5.40
	F	8.64	13.43	56.02	21.91
	T	11.57	11.27	8.18	68.98

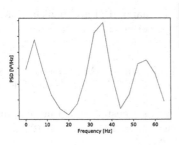   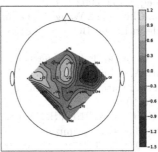

(a) Temporal filter visualization   (b) Spatial filter visualization

Fig. 2. Sample of relevant convolutional weights.

reported by [8]. Figure 2b shows a spatial filter reconstructed by interpolation of the weights. The scale in the right is from 1 to −1. It shows that Incep-EEGNet extracts the signals from the electrodes C3, CZ, and C4. It happens that those electrodes cover the part of the brain that is responsible for the movement of the hands and the feet.

5   Discussion and Conclusion

Designing ConvNets for BCI applications may be problematic. The existing approaches need an intensive data augmentation, and to be Shallow. Deep ConvNets are defective and lacks performances. Therefore, we built the Incep-EEGnet which is a modified EEGNET with a greater number of feature map that increases the complexity of the model where it outperforms state-of-the-art methods. To diminish any problem of degradation, we use an inception block that has several branches that offer an efficient feature extraction layer. The pointwise convolution works as a residual connection that prevents from vanishing gradient problems. Incep-EEGNet outperforms FBCSP, RG, and several ConvNets. Indeed, CSP techniques are considered state-of-the-art techniques for their efficiency, but as drawbacks, they are sensitive to noises, artifacts, and need larger datasets [16]. RG relies on and representation of the data that does not take into account the frequential features as its authors praise. But, it lowers its performances compared with FBCSP and ConvNets. ConvNet methods perform better and faster in the same conditions if we wisely use them. The overall performances are still low for several subjects highlighting a strong incompatibility between some subjects.

References

1. Abdulkader, S.N., Atia, A., Mostafa, M.S.M.: Brain computer interfacing: applications and challenges. Egypt. Inform. J. **16**(2), 213–230 (2015). https://doi.org/10.1016/j.eij.2015.06.002
2. Ang, K.K., Chin, Z.Y., Wang, C., Guan, C., Zhang, H.: Filter bank common spatial pattern algorithm on BCI competition IV datasets 2a and 2b. Front. Neurosci. **6**, 39 (2012)
3. Baldi, P., Sadowski, P.J.: Understanding dropout, p. 9
4. Barachant, A., Bonnet, S., Congedo, M., Jutten, C.: Multiclass brain-computer interface classification by Riemannian geometry. IEEE Trans. Biomed. Eng. **59**(4), 920–928 (2012)
5. Bashashati, H., Ward, R.K., Bashashati, A.: User-customized brain computer interfaces using Bayesian optimization. J. Neural Eng. **13**(2), 026001 (2016). https://doi.org/10.1088/1741-2560/13/2/026001. 00007
6. Cecotti, H., Graser, A.: Convolutional neural networks for P300 detection with application to brain-computer interfaces. IEEE Trans. Pattern Anal. Mach. Intell. **33**(3), 433–445 (2011). https://doi.org/10.1109/TPAMI.2010.125
7. Clevert, D.A., Unterthiner, T., Hochreiter, S.: Fast and accurate deep network learning by exponential linear units (ELUs). In: International Conference on Learning Representations (ICLR) (2016)
8. Dose, H., Møller, J.S., Iversen, H.K., Puthusserypady, S.: An end-to-end deep learning approach to MI-EEG signal classification for BCIs. Expert Syst. Appl. **114**, 532–542 (2018). https://doi.org/10.1016/j.eswa.2018.08.031. 00015
9. Goodfellow, I., Bengio, Y., Courville, A.: Deep Learning: Adaptive Computation and Machine Learning. The MIT Press, Cambridge (2016)
10. Ioffe, S., Szegedy, C.: Batch normalization: accelerating deep network training by reducing internal covariate shift. In: Bach, F., Blei, D. (eds.) Proceedings of the 32nd International Conference on Machine Learning. Proceedings of Machine Learning Research, vol. 37, pp. 448–456. PMLR, Lille, July 2015. 16886
11. Kingma, D.P., Ba, J.: Adam: a method for stochastic optimization. In: Proceedings of the 3rd International Conference on Learning Representations (ICLR), December 2014
12. Lawhern, V.J., Solon, A.J., Waytowich, N.R., Gordon, S.M., Hung, C.P., Lance, B.J.: EEGNet: a compact convolutional neural network for EEG-based brain-computer interfaces. J. Neural Eng. **15**(5), 056013 (2018). https://doi.org/10.1088/1741-2552/aace8c
13. Ortiz-Rosario, A., Adeli, H.: Brain-computer interface technologies: from signal to action. Rev. Neurosci. **24**(5) (2013). https://doi.org/10.1515/revneuro-2013-0032
14. Pfurtscheller, G., Neuper, C.: Motor imagery and direct brain-computer communication. Proc. IEEE **89**(7), 1123–1134 (2001). https://doi.org/10.1109/5.939829
15. Pfurtscheller, G., Neuper, C.: Movement and ERD/ERS. In: Jahanshahi, M., Hallett, M. (eds.) The Bereitschaftspotential: Movement-Related Cortical Potentials, pp. 191–206. Springer, Boston (2003). https://doi.org/10.1007/978-1-4615-0189-3_12. 00054

16. Reuderink, B., Poel, M.: Robustness of the common spatial patterns algorithm in the BCI-pipeline. Technical report, University of Twente (2008). 00042
17. Schirrmeister, R.T., et al.: Deep learning with convolutional neural networks for EEG decoding and visualization. Hum. Brain Mapp. **38**(11), 5391–5420 (2017). https://doi.org/10.1002/hbm.23730. Convolutional Neural Networks in EEG Analysis
18. Szegedy, C., Vanhoucke, V., Ioffe, S., Shlens, J., Wojna, Z.: Rethinking the inception architecture for computer vision. In: 2016 IEEE Conference on Computer Vision and Pattern Recognition (CVPR), pp. 2818–2826. IEEE (2016). 01916
19. Tangermann, M., et al.: Review of the BCI competition IV. Front. Neurosci. **6** (2012). https://doi.org/10.3389/fnins.2012.00055

Fuzzy-Based Approach for Assessing Traffic Congestion in Urban Areas

Sara Berrouk[1]([✉]), Abdelaziz El Fazziki[2], and Mohammed Sadgal[2]

[1] Faculty of Science and Technology, Cadi-Ayyad University, Marrakesh,
Morocco
berrouk.sara@gmail.com
[2] Faculty of Science, Cadi-Ayyad University, Marrakesh, Morocco
{elfazziki,sadgal}@uca.ma

Abstract. The very rapid evolution of urban areas leads to a reflection on the citizens' mobility inside the cities. This mobility problem is highlighted by the increase in terms of time, distance and social and economic costs, whereas the congestion management approach implemented rarely meets the road users' expectations. To overcome this problem, a novel approach for evaluating urban traffic congestion is proposed. Factors such as the imprecision of traffic records, the user's perception of the road's level of service provided and variation in sample data are mandatory to describe the real traffic condition. To respond to these requirements, a fuzzy inference-based method is suggested. It combines three independent congestion measures which are: speed ratio, volume to capacity ratio and decreased speed ratio into a single composite measure which is the congestion index. To run the proposed fuzzy model, the traffic dataset of Austin-Texas is used. Although it is still not possible to determine the best congestion measure, the proposed approach gives a composite aspect of traffic congestion by combining and incorporating the uncertainty of the three independent measures.

Keywords: Congestion measures · Fuzzy inference systems · Congestion estimation

1 Introduction

Traffic congestion, not restricted to but particularly predominant in big cities, is considered as the most prominently compounding issue related to traffic engineering and urban arranging, with clear ramifications on urban economy, environment, and way of life [1]. Traffic in urban areas keeps on developing especially in large cities of developing countries, which are portrayed by substantial monetary and population extension. This normally requires serious transportation of products and travelers, expanding interest for individual vehicular proprietorship that in the course of the most recent decade has seen exponential development around the world [2]. Catching congestion ends up significant in this regard, yet rather fundamental.

In the past few years, the traffic congestion phenomenon has been widely and differently described as a physical condition that occurs in traffic streams which leads to

A. El Moataz et al. (Eds.): ICISP 2020, LNCS 12119, pp. 112–121, 2020.
https://doi.org/10.1007/978-3-030-51935-3_12

restrained movement, prolonged delays and decreased speeds [3]. Its definition is conventionally classified based on four aspects: Speed, capacity, travel time/delay and the congestion cost [4]. As a result, the level of congestion on urban roads is measured and evaluated uniformly. In this sense, a novel approach for evaluating urban traffic congestion is proposed. Since the phenomenon of congestion does not have a uniform definition, its measuring on the different road segments should not be done uniformly. A fuzzy inference-based method is suggested that incorporate the uncertainty of the independent congestion measures. It takes into consideration the variation in the speed at peak and non-peak hours, the relationship between the vehicle counts and the capacity of roadway segments as well as the ratio between the average speed and the maximum permissible speed. As a result, this study makes it possible to tackle the vague concept of congestion from multiple aspects.

The remainders of this paper are organized as follows: Sect. 2 presents the literature review. The new fuzzy-based approach and its calculation process are proposed in Sect. 3. Section 4 illustrates the experimentation and results. the conclusion is described in Sect. 5.

2 Literature Review

2.1 Traffic Congestion Evaluation Methods

Traffic congestion is considered to be a relative measure compared to other traffic flow parameters. The traveler perception of the service provided by the transportation system at a specific time determines the presence of congestion, which requires the definition of congestion indicators [4]. As indicated in [5], the Level Of Service (LOS) is the best observational pointer of jams in the transportation system, where the proportions of supply and demand are analyzed and classified in one of six classes that range from A (free flow) to F (highly congested). There is no limit determined for LOS to portray the congestion state, however, obviously the LOS F is characterized as the most noticeably terrible condition of stream and speaks to blocked stream. Even though a few reports are utilizing other levels of services (D and E) as a congested stream, LOS F is commonly acknowledged as a condition of traffic stream and hence LOS is the most fitting congestion indicator. [6] provides a review of the congestion measures, the three regularly utilized indicators of traffic congestion are examined in the following paragraphs.

The roadway congestion index (RCI), first introduced by David Schrank in [7], is an indicator of the congestion severity in a wide area. The RCI compares between the daily vehicle miles per lane-mile (DVMT) of an area and the total predicted vehicle miles in congested conditions in that specific area, both measured depending on the road type (freeway or arterial roadway), Eq. 1 shows how the RCI is computed:

$$RCI = \frac{\begin{array}{l}(\text{Freeway DVMT}) \times (\text{Freeway Daily\_vehicule\_mile}) \\ + (\text{Arterial DVMT}) \times (\text{Arterial Daily\_vehicule\_mile})\end{array}}{\left(\begin{array}{l}13,000 \times \text{Freeway Daily\_vehicule\_mile} \\ + 5,000 \times \text{Arterial Daily\_vehicule\_mile}\end{array}\right)} \quad (1)$$

The RCI accepts a capacity limit of 13,000 DVMT on freeways and 5,000 DVMT on head arterials. An RCI estimation of 1.0 or more reflects a bothersome congestion level. But the fact that it's hard for road users to relate their travel experience to this index can be considered as a drawback because it infects the ability to forecast future conditions. This indicator is incapable to infer the vital traffic improvement plans to avoid traffic congestion since it is an area-wide measure.

The Relative Delay Rate (RDR) and the Congestion Index (CI) are two well-used congestion indicators, introduced respectively by Lomax in [4] and Taylor in [8]. They both translate the road users' perception of the traffic flow quality relatively to a perfect or tolerable condition. The RDR is calculated using Eq. 2:

$$\frac{\text{Relative}}{\text{Delay Rate}} = \frac{\left[\begin{array}{c}\text{Actual travel rate (min per mile)} \\ -\text{Tolerable travel rate (min per mile)}\end{array}\right]}{\text{Tolerable travel rate (min per mile)}} \tag{2}$$

where the travel rate is the travel time per segment length. The congestion index is calculated using the Eq. 3:

$$\text{Congestion index} = \frac{\left(\begin{array}{c}\text{Actual travel} \\ \text{time of section}\end{array} - \begin{array}{c}\text{FreeFlow} \\ \text{travel time}\end{array}\right)}{\text{FreeFlow travel time}} \tag{3}$$

Where the free-flow travel time represents the time spent traveling at the mentioned speed limit. An estimation of zero of the congestion index shows a low degree of congestion and considered to be near free-flow condition. An estimation higher than 2 of Eq. (3) reflects a severely congested condition. Most of the time, the use of the congestion index can be beneficial when traffic conditions are compared on several roadways since the records can be collected independently for various road segments.

2.2   Limitations in Traffic Congestion Assessment Methods

In this section, the limitations in assessing urban traffic congestion are mentioned. First, the imprecision of traffic data records and observations gathered in both the supply and demand sides. In the supply-side parameters, a single value describing the traffic state can be misleading since the traffic signal timing and geometric design parameters changes according to the road section and the study period. For the demand-side, parameters such as travel time, vehicle counts and delay can also be imprecise. We can conclude that uncertainty is associated with all congestion measures no matter how accurate their representation of the actual traffic state is. Second, the notion of acceptable or tolerable traffic conditions is fuzzy because it differs according to the roadway section and the road user's experience. Therefore, it is impossible to use a unique value to describe it. Third, to measure the traffic congestion, stepwise measures can be used such as the Level Of Service (LOS). These measures can be misleading in case the traffic situation is close to a threshold because often the situation will exceed the latter. This problem can be touched when the studied road section knows constantly changing geometric features and traffic parameters. Fourth, measures of congestion in the literature are mainly related to single traffic parameters; often it is travel speed or delay. However, road congestion is a vague concept, and the traveler impression to describe it is influenced by many factors. In this sense, a combination of more than one measure to evaluate the traffic condition is required. Fifth, the road user's opinion of what is tolerable or intolerable traffic condition should be incorporated when measuring road congestion. This boundary is not clear because of the opinion, the travel condition and between localities differ among travelers. Thus, an ideal measure should take into consideration the flexibility that represents the conditions of the locality.

To sum up, traffic congestion is a complex phenomenon that can be described in different manners and caused by multiple factors. Therefore, it is necessary to take into account the imprecise quantities and the traveler's perception in the process of assessing the level of congestion. Considering the problems aforementioned, in this paper, a fuzzy-based method is proposed to determine the level of traffic congestion. It is believed that the proposed method can take into consideration the following requirements of an ideal congestion measuring approach:

1. Integration of the imprecise nature of traffic data,
2. Integration of the subjective notion of travelers perceptions of traffic congestion (acceptable or inacceptable service quality),
3. And finally, the combination of multiple measures to form a composite measure.

3 A Congestion Measure Based on the Fuzzy Inference System

The proposed fuzzy process, illustrated in Fig. 1, is used to measure the congestion levels by using traffic data (such as vehicle counts, speed, etc.). This process provides a solution for ambiguous and uncertain problems. The procedure involves the calculation of the values of the input parameters, the categorization of the input parameters values into distinguishing groups, the definition of the various congestion states, and the determination of the congestion index values.

3.1 Data Acquisition and Preparation

To validate our proposal, the choice fell on the Radar traffic Counts [9] and Travel Sensor [10] datasets provided by the transportation department of Austin City. The choice of these datasets was not arbitrary: the dataset offers traffic parameters needed to conduct our study (such as volume, speed, etc.) and they put the light on the inter-sections in Austin having a heavy traffic flow which will reflect the benefits of our proposed system in detecting the most congested intersections and road segments. The output of the system can be so beneficial in the making of short and long terms decisions from the authorities. The sensors cover a total of 17 intersections, each having a specific number of lanes which made it possible to analyze the congestion in about 76 road segments. The traffic count and speed data are gathered from a specific

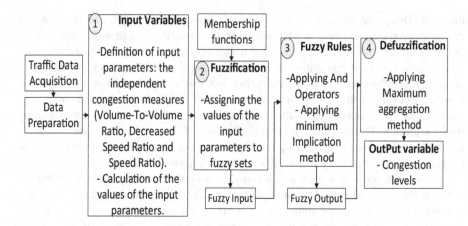

Fig. 1. The proposed fuzzy logic process for congestion evaluation

type of sensors deployed by the city of Austin which are the Wavetronix radar sensors. These data have been collected since the creation of the dataset in November 2017 with an interval of 15 min; which led to a huge amount of data (about 4.65 M records). In this study, one month of data (from 01-March-2019 to 31-March-2019) is used to test the proposed system and minimize its running time. The peak and non-peak speed data are extracted from the dataset considering that the morning peak-hours extend from 9:30 am to 11:30 am, the evening peak hours extend from 5:30 pm to 7:30 pm and the non-peak hours extend from 7:00 am to 9:00 am and from 12:00 am to 5:00 pm. Other data was added to the dataset to fulfill the need of this study such as the capacity and the maximum permissible speed of each segment.

3.2   Input Parameters

The three measures are used to describe congestion, which are: volume-to-capacity ratio, the decreased speed ratio, and the speed ratio. The Volume to capacity ratio measure (Eq. 4) reflects the ratio between the counts of vehicles passing through a road segment and the maximum number of vehicles that that road segment can support. This value ranges from 0 to values sometimes greater than 1. A value near to "0" represents the best traffic state (free flow), and a value near or greater than 1 represents the worst traffic state (severely congested).

$$V/C \, Ratio = \frac{volume \, of \, vehicles}{capacity \, of \, road \, segment} \qquad (4)$$

The second measure is the decreased speed ratio (Eq. 5) that denotes the rate of reduced speed of the vehicle due to a congested situation. This measure represents the traffic condition of the road network for non-peak and peak periods. This value ranges from 0 to 1, 0 being the best condition when the Peak average speed is bigger than or equal to the Non-Peak average speed, and 1 being the worst condition when Peak's average speed is near 0.

$$\text{Decreased speed ratio} = \frac{NonPeakAvgSpeed - PeakAvgSpeed}{NonPeakAvgSpeed} \qquad (5)$$

The third measure is the speed ratio (Eq. 6); it reflects the ratio between the average speed and the maximum allowed speed on a road segment. It is used as an evaluation indicator of the traffic state in urban areas. The speed ratio value ranges from 0 to 1, three threshold values (0.25, 0.5, and 0.75) are adopted to be the classification criterion of the traffic condition. The extreme values 0 and 1 reflect the worst and the best conditions of traffic respectively.

$$\text{Speed Ratio} = \frac{AvgSpeed}{SpeedMax} \qquad (6)$$

3.3 Fuzzification Phase

After computing the congestion measures, a domain transformation called fuzzification is launched. This process presents the first stage of the fuzzy logic processing. It consists of assigning the values to fuzzy sets that have vague boundaries. The fuzzy sets are classes of road conditions, simplified by a natural language such as "moderate congestion". The values of the volume-to-capacity ratio are grouped into three classes (Low, Moderate and High). Figure. 2 shows the given values of the v/c ratio and their mapping to the three classes with the help of the membership functions.

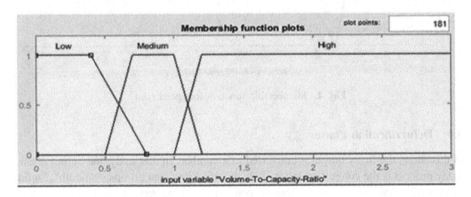

Fig. 2. Membership function for volume to capacity ratio

For the decreased speed ratio measure, the values are categorized into six comprehensive categories by the traveler. The six categories range from A to F and the defined membership functions for each one of the six classes in which the value of the decreased speed ratio is given in Fig. 3.

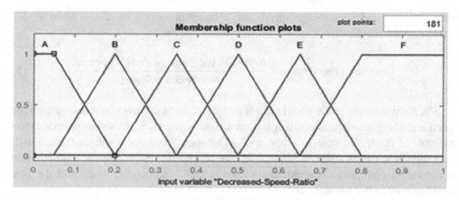

Fig. 3. Membership function for the decreased speed ratio

In like manner, four classes are identified to classify the values of the speed Ratio. The classes are Free Flow, Moderate, High and Very High. Figure 4 shows the membership functions of the four classes in which the value of the speed ratio is given and the compatibility of this value with a specific class.

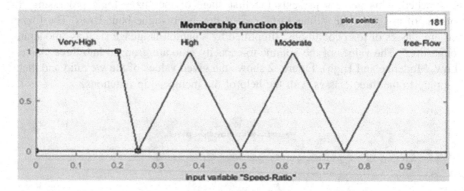

Fig. 4. Membership function for speed ratio

3.4 Defuzzification Phase

In this stage, the fuzzy output is converted to a number. In this study, the output of the fuzzy process is the congestion index. It is classified into four groups: "smooth", "mild congested", "congested" and "heavily congested". It ranges from 0 to 1, with 0 reflecting the best traffic condition and 1 reflects the worst. Figure 5 shows the membership function of the congestion index. The four congestion index classes are defined on the scale from 0 to 1 and each one is assigned to a natural comprehensible language term (see Fig. 5).

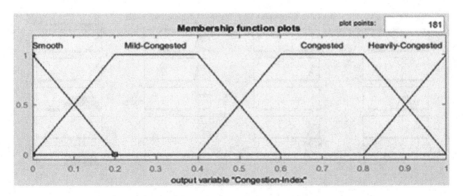

Fig. 5. Membership function for the proposed congestion index

3.5 Fuzzy Rule Base

Numerous fuzzy rules are performed to combine the three input parameters and to obtain a natural-language and comprehensive congestion measure. The fuzzy rules contain two parts. The first part is the antecedent; it is the one following "IF". And the second part is the consequent; it is the one following "Then". The fuzzy rules for each transport mode are described in the following way:

Rule: **IF** Volume-To-Capacity Ratio is A And Decreased Speed Ratio is B And Speed Ratio is C **Then** the congestion level is D.

Where A, B, C, and D each denote the level of congestion as mentioned in Figs. 2, 3, 4 and 5.

4 Experimentation and Results

4.1 Congestion Index Computing: Application of FIS

In this paper, MATLAB Toolbox [11] for fuzzy modeling is used to estimate the level of traffic congestion. The proposed fuzzy architecture is designed with three congestion measures as inputs, congestion index as the only output and seventy-two fuzzy rules. Having specific estimations of the volume-to-capacity ratio, the decreased speed ratio and the speed ratio as inputs for the rules, a match is done by a particular rule between the input values and the antecedent of the rule using the aforementioned membership functions (Figs. 2, 3 and 4). Doing so, the minimum estimation of the membership function levels among the three is considered as the truth of the rule's antecedent. Furthermore, this estimation affects the truth of the rule's consequent. In the case where the values of the inputs belong to more than a class, all possible rules can be either used or fired and the result is formed by combining the consequents of the possible. How the rules are executed and the result is computed is shown in Fig. 6.

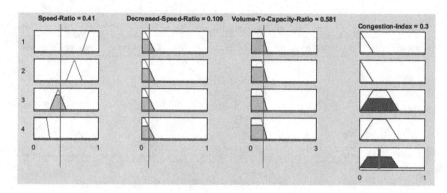

Fig. 6. The calculation of the congestion index using the proposed fuzzy model

4.2 Results and Discussion

Table 1 shows an example of the congestion index generation for some segments of the study area. It may be observed that using either one of the input parameters cannot give a real picture of the traffic congestion. The combination of all three parameters captures the real status of the condition. It is observed that if one of the input parameters is much higher and the remaining two parameters are on the lower side as is the case of segment 10, the congestion effect is moderate. However, on the traffic link 23, all the three parameters are of the noticeable amount resulting higher congestion effect. Similarly, the links subjected to lower values of speed reduction rate, very low-speed rate and v/c ratio resulting in lower congestion index such as segment 40. The proposed model gives a more realistic and detailed congestion picture compared to that obtained in traditional methods by considering a single parameter.

Table 1. The caption of the generated congestion index for some road segments

Segment number	Vehicle counts	Average speed	Non-peak average speed	Peak average speed	Volume to capacity ratio	Decreased speed ratio	Speed ratio	Resulting congestion index
3	69.35	34.48	36.5	24.25	0.73	0.33	0.86	0.299
5	72.73	26.69	37.2	16.7	1.72	0.55	0.67	0.7
10	70.89	32.20	31.04	13.5	1.13	0.56	0.81	0.54
17	111.71	32.08	27.2	15.4	1.04	0.44	0.802	0.41
23	75.78	33.4	37.5	5.02	2.05	0.86	0.74	0.906
40	103.16	42.07	40.25	22.37	0.95	0.44	0.84	0.3
69	89.08	22.14	33.5	17.42	2.07	0.92	0.4	0.934
71	68.15	37.93	42.17	29.37	1.74	0.43	0.68	0.7
75	52.03	26.72	32.54	20.37	1.83	0.59	0.48	0.91

5   Conclusion

The contribution in this paper consists of modeling a fuzzy-based method for urban traffic congestion measuring. It combines three inputs which are traditionally used as measures of traffic congestion into one output which is a composite congestion measure. Each one of the independent measures, the speed ratio, the volume to capacity ratio and the decreased speed ratio, describes a particular aspect of the traffic conditions using precise thresholds. Doing so, the error possibility in measurement is high which misleads the representation of the real traffic condition. The suggested fuzzy inference approach takes into consideration every little variation in the input congestion measures and their combination provides a more accurate representation of the traffic state. The accuracy of such a model is higher because it is based on natural-language rules which reflect perfectly the traveler perception of the traffic condition. The suggested method can be used to evaluate the congestion over road sections, arterials, or highway road network. In terms of perspectives, we are attempting to expand the proposed model to incorporate other factors playing a considerable role in changing the traffic condition such as parking in streets, roads structure, etc. Further work will make possible to redefine the membership functions and the rules more precisely.

References

1. Wang, H., et al.: Joint link-based credit charging and road capacity improvement in continuous network design problem. Transp. Res. Part A: Policy Pract. **67**, 1–14 (2014)
2. Li, Z., Huang, J.: How to mitigate traffic congestion based on improved ant colony algorithm : a case study of a congested old area of a metropolis. Sustainability **11**(4), 1140 (2019)
3. Weisbrod, G., Vary, D., Treyz, G.: Economic Implications of Congestion, Washington, DC (2001)
4. Lomax, T., et. al.: Quantifying congestion. NCHRP Report 398, Transportation Research Board 1 and 2, Washington DC (1997)
5. Cottrell, B.W.D.: Empirical freeway queuing duration model. J. Transp. Eng. **127**(1), 13–20 (2001)
6. Rao, K.R., Rao, A.M.: Measuring urban traffic congestion - a review. Int. J. Traffic Transp. Eng. **2**(4), 286–305 (2012)
7. Schrank, D.L., Turner, S.M., Lomax, T.J.: Estimates of Urban Roadway Congestion – 1999, vol. 1, no. 2. Texas A&M University, College Station (1993)
8. Taylor, M.A.P.: Exploring the nature of urban traffic congestion: concepts, parameters, theories and models. In: 16th Meeting of the Australian Road Research Board, p. 16 (1992)
9. Radar Traffic Counts Dataset of Austin-Texas city (2016). https://data.austintexas.gov/Transportation-and-Mobility/Radar-Traffic-Counts/i626-g7ub Accessed 01 Oct 2019
10. Travel Sensor Dataset of Austin-Texas City (2016). https://data.austintexas.gov/Transportation-and-Mobility/Travel-Sensors/6yd9-yz29 Accessed 01 Oct 2019
11. Matlab Toolbox. https://www.mathworks.com/help/thingspeak/matlab-toolbox-access.html Accessed 20 March 2020

Big Data and Reality Mining in Healthcare: Promise and Potential

Hiba Asri[1(✉)], Hajar Mousannif[2], Hassan Al Moatassime[1], and Jihad Zahir[2]

[1] OSER Laboratory, Cadi Ayyad University, Marrakesh, Morocco
asri.hiba@gmail.com, hassan.al.moatassime@gmail.com
[2] LISI Laboratory, Cadi Ayyad University, Marrakesh, Morocco
mousannif@uca.ma, j.zahir@uca.ac.ma

Abstract. Nowadays individuals are creating a huge amount of data; with a cell phone in every pocket, a laptop in every bag and wearable sensors everywhere, the fruits of the information are easy to see but less noticeable is the information itself. This data could be particularly useful in making people's lives healthier and easier, by contributing not only to understand new diseases and therapies but also to predict outcomes at earlier stages and make real-time decisions. In this paper, we explain the potential benefits of big data to healthcare and explore how it improves treatment and empowers patients, providers and researchers. We also describe the capabilities of reality mining in terms of individual health, social network mapping, behavior patterns and treatment monitoring. We illustrate the benefits of reality mining analytics that lead to promote patients' health, enhance medicine, reduce cost and improve healthcare value and quality. Furthermore, we highlight some challenges that big data analytics faces in healthcare.

Keywords: Big data · Reality mining · Healthcare · Predictive analytics · Machine learning

1 Introduction

Individuals are creating torrents of data, far exceeding the market's current ability to create value from it all. A significant catalyst of all this data creation is hyper-specific sensors and smart connected objects (IoT) showing up in everything from clothing (wearable devices) to interactive billboards. Sensors are capturing data at an incredible pace.

In the specific context of healthcare, The volume of worldwide healthcare data is expected to grow up to 25,000 petabytes by 2020 [1]. Some flows have generated 1,000 petabytes of data now and 12 ZBs are expected by 2020 from different sources, such as Electronic Healthcare Record (HER), Electronic Medical Records (EMR), Personal Health Records (PHR), Mobilized Health Records (MHR), and mobile monitors. Moreover, health industries are investing billions of dollars in cloud computing [2]. Thanks to the use of HER and EMR, an estimated number of 50% of hospitals will

© Springer Nature Switzerland AG 2020
A. El Moataz et al. (Eds.): ICISP 2020, LNCS 12119, pp. 122–129, 2020.
https://doi.org/10.1007/978-3-030-51935-3_13

integrate analytics solutions, such as Hadoop, hadapt, cloudera, karmasphere, MapR, Neo, Datastax, for health big data [3].

This paper gives insight on the challenges and opportunities related to analyzing larger amounts of health data and creating value from it, the capability of reality mining in predicting outcomes and saving lives, and the Big Data tools needed for analysis and processing. Throughout this paper, we will show how health big data can be leveraged to detect diseases more quickly, make the right decisions and make people's life safer and easier.

The remainder of this paper will be organized as follows:

- Section 2 present the context of reality mining in healthcare system.
- In Sect. 3, we discuss capabilities of reality mining in terms of individual health, social network mapping, behavior patterns and treatment monitoring.
- Sections 4 highlights benefits of reality mining to patients, researchers and providers.
- Section 5 numbers advantages and challenges of big data analytics in the healthcare industry.
- Conclusions and directions for future work are given in Sect. 6.

2   Reality Mining in Healthcare

Reality Mining is about using big data to study our behavior through mobile phone and wearable sensors [4]. In fact every day, we perform many tasks that are almost routines. Cataloguing and collecting information about individuals helps to better understand people's habits. Using machine learning methods and statistical analytics, reality mining can now give a general picture of our individual and collective lives [7].

New Mobile phones are equipped with different kinds of sensors (e.g. motion sensors, location sensors, and haptic sensors). Every time a person uses his/her cell-phone, some information is collected. The Chief Technology Officer of EMC corporation, a big storage company, estimates that the amount of personal sensor information storage will balance from 10% to 90% in the next decade [5]. The most powerful way to know about a person's behavior is to combine the use of some software, with data from phone and from other sources such as: sleep monitor, microphone, accelerometers, camera, Bluetooth, visited website, emails and location [4].

To get a picture of how reality mining can improve healthcare system, here are some examples:

- Using special sensors in mobiles, such as accelerometers or microphone, some diagnosis data can be extracted. In fact, from the way a person talks in conversations, it is possible to detect variations in mood and eventually detect depression.
- Supervising a mobile's motion sensors can contribute to recognize some changes in gait, and could be an indicator of an early stage of Parkinson's disease.
- Using both healthcare sensor data (sleep quality, pressure, heart rate...) and mobile phone data (age, number of previous miscarriage...) to make an early prediction of miscarriage [6].

Communication logs or surveys recorded by mobiles or computers give just a part of the picture of a person's life and behavior. Biometric sensors can go further to track blood pressure, pulse, skin conductivity, heartbeats, brain, or sleep activity. A sign of depression, for instance, can be revealed just by using motion sensors that monitor changes in the nervous system (brain activity) [1].

3 Reality Mining Capabilities

3.1 Individual Health

Currently, the most important source of reality mining is mobile phones. Every time we use our smart phone, we create information. Mobile phone and new technologies are now recording everything about the physical activity of the person. While these data threat the individual privacy, they also offer a great potential to both communities and individuals.

Authors in [7, 9] assert that the autonomic nervous brain system is responsible of the change of our activity levels that can be measured using motion sensors or by audio; it has been successfully used to screen depression from speech analysis software in mobile phone.

Authors in [10] assert that mobile phones can be used to scale time-coupling movement and speech of the person, which is an indication of a high probability of problems in language development.

Unaware mimicry between people (e.g., posture changes) can be captured using sensors. It is considered as trustworthy indicators of empathy and trusts; and manipulated to strongly enhance compliance. This unconscious mimicry is highly mediated by mirror neurons [11].

Authors in [12] show that several sensors can also detect and measure fluidity and consistency of person's speech or movement. These brain function measurements remain good predictors of human behaviors. Hence, this strong relationship helps for diagnosis of both neurology and psychiatry.

3.2 Social Network Mapping

Besides data from individual health, mobile phones have the capability to capture information about social relationship and social networks. One of the most relevant applications of reality mining is the automatic mapping social network [4]. Through mobile phone we can detect user's location, user's calls, user's SMS, who else is nearby and how is moving thanks to accelerometers integrated in new cell phone.

Authors in [5] describe three type of self-reported: self-reported reciprocal friends when both persons report to other as friend, self-reported non-reciprocal friends when one of both reports to other as a friend and self-reported reciprocal non-friend when no one reports to other as a friend. This information has been shown to be very useful for identifying several important patterns.

Another study use pregnant woman's mobile phone health data like user's activity, user's sleep quality, user's location, user's age, user's Body Mass Index (BMI)among others, considered as risk factors of miscarriage, in order to make an early prediction of miscarriage and react as earlier as possible to prevent it. Pregnant woman can track her state of pregnancy through a mobile phone application that authors developed [22].

Another good example of network mapping is the computer game named Dia-BetNet. It is a game intended for young diabetics to help them keep track of their food quality, activity level and blood sugar rates [13].

3.3 Behaviour Patterns and Public Health Service

Behavior pattern involves how a person lives, works, eats etc., not a place, age or residence. Reality mining has the potential to identify behaviors with the use of classification methods that are very useful to predict health patterns [15]. Understanding and combining different behavior patterns of different populations is critical since every subpopulation has its own attitudes and profiles about their health choices.

Google Flu Trends represents a good example to model health of a large population. Just by tracking terms typed in World Wide Web (WWW) searches and identifying the frequency for words related to influenza as illnesses, an influenza outbreak is detected indirectly. In the U.S., Google searches prove an intense correlation between those frequencies and the incidence of estimated influenza based on cases detected by Centers of Disease Control and Prevention (CDC).

Also, with GPS and other technologies, it is easily to track the activities and movements of the person. Location's logs present a valuable source to public health in case of infectious diseases such as tuberculosis, anthrax, and Legionnaires disease. In fact, location logs can help in identifying the source of infections and government may react for preventing further transmission.

3.4 Treatment Monitoring

Once a patient takes his treatment, which is pharmaceutical, behavioral, or other, doctors and clinicians have to monitor their patient's response to treatment. Reality mining information used for diagnosis can be also relevant data for monitoring patient response to treatment and for comparing. Characteristics such as behavior, activity and mobility could be collected in real-time and they could be useful for clinicians to change or adjust treatment depending on patient's response, and in some cases it could be a more effective care with a lower cost.

A concrete example of this is Parkinson's patients. Real-time data are gathered through wearable accelerometers that integrate machine learning algorithms to classify movements' states of Parkinson's patients and to get the development of those movements.

Two significant studies exist in literature to classify dyskinesia, hypokinesia and bradykinesia (slow movements) for seven patients [16]. Data are collected using different sources: wearable accelerometers, videos and clinical observations. Results of studies show that bradykinesia and hypokinesia are the two main classes identified with a high accuracy. Another classification is made to classify patients who feel off or about having dyskinesia.

4 Reality Mining Benefits to Health Care

By combining big data and reality mining, we rang from single to large hospital network. The main benefits can be summarized into detecting diseases at earlier stages, detecting healthcare abuse and fraud faster, and reducing costs. In fact, Big data market also contributes up to 7% of the global GDP and reduces 8% of healthcare costs. Big data analytics improve health care insights in many aspects:

4.1 Benefits to Patients

Big data can help patients make the right decision in a timely manner. From patient data, analytics can be applied to identify individuals that need "proactive care" or need change in their lifestyle to avoid health condition degradation. A concrete example of this is the Virginia health system Carillion Clinic project [17], which uses predictive models for early interventions.

Patients are also more open to giving away part of their privacy if this could save their lives or other people's lives. "If having more information about me enables me to live longer and be healthier", said Marc Rotenberg of the electronic Privacy Information Center," then I don't think I'm going to object. If having more information about me means my insurance rates go up, or that I'm excluded from coverage, or other things that might adversely affect me, then I may have a good reason to be concerned" [18].

4.2 Benefits to Researchers and Developers (R & D)

Collecting different data from different sources can help improving research about new diseases and therapies [21]. R & D contribute to new algorithms and tools, such as the algorithms by Google, Facebook, and Twitter that define what we find about our health system. Google, for instance, has applied algorithms of data mining and machine learning to detect influenza epidemics through search queries [19, 20]. R & D can also enhance predictive models to produce more devices and treatment for the market.

4.3 Benefits to Healthcare Providers

Providers may recognize high risk population and act appropriately (i.e. propose preventive acts). Therefore, they can enhance patient experience. Moreover, approximately 54% of US hospitals are members in local or regional Health-Information Exchanges (HIEs) or try to be in the future. These developments give the power to access a large array of information. For example, the HIE in Indiana connects currently 80 hospitals and possess information of more than ten million patients [1] (Fig. 1).

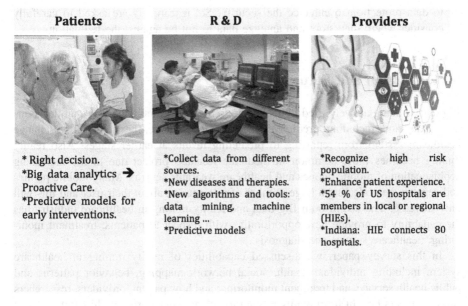

Patients	R & D	Providers
* Right decision. *Big data analytics ➔ Proactive Care. *Predictive models for early interventions.	*Collect data from different sources. *New diseases and therapies. *New algorithms and tools: data mining, machine learning ... *Predictive models	*Recognize high risk population. *Enhance patient experience. *54 % of US hospitals are members in local or regional (HIEs). *Indiana: HIE connects 80 hospitals.

Fig. 1. Summary of reality mining benefits to healthcare.

5 Limitations and Challenges of Data Analytics in Healthcare Industry

Although big data analytics enhance the healthcare industry, there are some limitations to the use of big data in this area:

1. The source of data from organizations (hospital, pharmacies, companies, medical centers...) is in different formats. These organizations have data in different systems and settings. To use this huge amount of data, those organizations must have a common data warehouse in order to get homogeneous information and be able to manage it. However, having such systems requires huge costs.
2. Quality of data is a serious limitation. Data collected are, in some cases, unstructured, dirty, and non-standardized. So, the industry has to apply additional effort to transform information into usable and meaningful data.
3. A big investment is required for companies to acquire staff (data scientists), resources and also to buy data analytics technologies. In addition, companies must convince medical organizations about using big data analytics.
4. Using data mining and big data analytics requires a high level of expertise and knowledge. It is a costly affair for companies to hire such persons.
5. Due to serious constraints regarding the quality of collected data, variations and errors in the results are not excluded.
6. Data security is in big concern and researchers paid more attention on how we can secure all data generated and transmitted. Security problems include personal privacy protection, financial transactions, intellectual property protection and data protection. In some developing and developing countries, they propose laws related

to data protection to enhance the security. So, researchers are asked to carefully consider where they store and analyze data to not be against the regulations.

6 Conclusion and Future Work

Big data is being actively used in healthcare industry to change the way that decisions are made; and including predictive analytics tools, have the potential to change healthcare system from reporting to predicting results at earlier stages. Also reality mining becomes more common in medicine research project due to the increasing sophistication of mobile phones and healthcare wearable sensors. Many mobile phones and sensors are collecting a huge number of information about their user and this will only increase. The use of both Big data and reality mining in healthcare industry has the capability to provide new opportunities with respect to patients, treatment monitoring, healthcare service and diagnosis.

In this survey paper, we discussed capabilities of reality mining in healthcare system including individual health, social network mapping, behavior patterns and public health service, and treatment monitoring; and how patient, providers, researchers and developers benefit from reality mining to enhance medicine. We highlight as well several challenges in Sect. 5 that must be addressed in future works. Adoption of big data and reality mining in healthcare raises many security and patient privacy concerns that need to be addressed.

References

1. Kim, S.G.: Big data in healthcare hype and hope TT - 의료 분야에 있어서 대용량 데이터의 활용. 한국 Cad/Cam 학회 학술발표회 논문집, vol. 2013, pp. 122–125 (2013)
2. Koumpouros, Y.: Healthcare administration. IGI Global (2015)
3. Raghupathi, W., Raghupathi, V.: Big data analytics in healthcare: promise and potential. Health Inf. Sci. Syst. 2(1) (2014). Article number. 3. https://doi.org/10.1186/2047-2501-2-3
4. Eagle, N., Pentland, A.S.: Reality mining: sensing complex social systems. Pers. Ubiquit. Comput. 10(4), 255–268 (2006). https://doi.org/10.1007/s00779-005-0046-3
5. Eagle, N., Pentland, A.S., Lazer, D.: Inferring friendship network structure by using mobile phone data. Proc. Natl. Acad. Sci. 106(36), 15274–15278 (2009)
6. Asri, H., Mousannif, H., Al Moatassime, H.: Real-time miscarriage prediction with SPARK. Procedia Comput. Sci. 113, 423–428 (2017)
7. Asri, H., Mousannif, H., Al Moatassime, H., Noel, T.: Big data in healthcare: challenges and opportunities. In: 2015 International Conference on Cloud Technologies and Applications (CloudTech), pp. 1–7. IEEE, June 2015
8. Stoltzman, W.T.: Toward a social signaling framework: activity and emphasis in speech. Doctoral dissertation, Massachusetts Institute of Technology (2006)
9. France, D.J., Shiavi, R.G., Silverman, S., Silverman, M., Wilkes, M.: Acoustical properties of speech as indicators of depression and suicidal risk. IEEE Trans. Biomed. Eng. 47(7), 829–837 (2000)
10. Cowley, S.J.: Naturalizing language: human appraisal and (quasi) technology. AI Soc. 28(4), 443–453 (2013). https://doi.org/10.1007/s00146-013-0445-3

11. Stel, M., Hess, U., Fischer, A.: The role of mimicry in understanding the emotions of others. In: Emotional Mimicry in Social Context, pp. 27–43. Cambridge University Press (2016)
12. Crispiani, P., Palmieri, E.: Improving the fluidity of whole word reading with a dynamic co-ordinated movement approach. Asia Pac. J. Dev. Differ. 2(2), 158–183 (2015)
13. Asri, H., Mousannif, H., Al Moatassime, H.: Reality mining and predictive analytics for building smart applications. J. Big Data 6(1) (2019). Article number. 66. https://doi.org/10.1186/s40537-019-0227-y
14. Pouw, I.H.: You are what you eat: serious gaming for type 1 diabetic persons, Master's thesis, University of Twente (2015)
15. Asri, H., Mousannif, H., Al Moatassime, H., Noel, T.: Using machine learning algorithms for breast cancer risk prediction and diagnosis. Procedia Comput. Sci. 83, 1064–1069 (2016)
16. Klapper, D.A.: Use of wearable ambulatory monitor in the classification of movement states in Parkinson's disease. Doctoral dissertation, Massachusetts Institute of Technology (2003)
17. IBM News room: IBM predictive analytics to detect patients at risk for heart failure - United States, 19 February 2014
18. Bollier, D.: The Promise and Peril of Big Data, pp. 1–66 (2010)
19. Ghani, K.R., Zheng, K., Wei, J.T., Friedman, C.P.: Harnessing big data for health care and research: are urologists ready? Eur. Urol. 66(6), 975–977 (2014)
20. Lazer, D., Kennedy, R., King, G., Vespignani, A.: The parable of Google Flu: traps in big data analysis. Science 343, 1203–1205 (2014)
21. Asri, H., Mousannif, H., Al Moatassime, H.: Big data analytics in healthcare: case study-miscarriage prediction. Int. J. Distrib. Syst. Technol. (IJDST) 10(4), 45–58 (2019)
22. Asri, H., Mousannif, H., Al Moatassime, H.: Comprehensive miscarriage dataset for an early miscarriage prediction. Data Brief 19, 240–243 (2018)

A Dataset to Support Sexist Content Detection in Arabic Text

Oumayma El Ansari[⊠], Zahir Jihad, and Mousannif Hajar

LISI Laboratory, Faculty of Sciences Semlalia, Cadi Ayyad University,
Marrakech, Morocco
ansari.oumaima@gmail.com,
{j.zahir,mousannif}@uca.ac.ma

Abstract. Social media have become a viral source of information. This huge amount of data offers an opportunity to study the feelings and opinions of the crowds toward any subject using Sentiment Analysis, which is a struggling area for Arabic Language. In this research, we present our approach to build a thematic training set by combining manual and automatic annotation of Arabic texts addressing Discrimination and Violence Against Women.

Keywords: Sentiment analysis · Arabic language

1 Introduction

Violence Against Women (VAW) is one of the most commonly occurring human rights violations in the world, and Arab region is no exception. In fact, UNWOMEN [3] reports that 37% of Arab women have experienced some form of violence in their lifetime. It's been demonstrated that discriminatory attitudes are still posing a challenge for women's status in the Arab States.

There is a variety of methods to measure attitudes and opinions towards a subject. Data generated by Internet activity, especially Social Media activity where users have the freedom of speech, is an interesting data source that can be used to evaluate the public opinion regarding both Discrimination (DAW) and Violence Against Women (VAW).

Sentiment Analysis uses data, typically from Social Media, to analyze the crowds' feelings toward a certain subject. It's the task of determining from a text if the author is in favor off, against or neutral toward a proposition or target [4]. In other words, it classifies a subjective text into different polarities: Positive (e.g. it's the best phone ever!), Negative (e.g. it sucks!) and Neutral (e.g. the new version is out) [5].

There are two main approaches to Sentiment Analysis: Machine learning (ML) based and lexicon-based. ML methods consist of using pre-annotated datasets to train classifiers. In lexicon-based methods, the polarity of a text is derived from the sentiment value of each single word in a given dictionary. ML based methods usually give higher accuracy [6], however these methods require a good quality training set to provide accurate and precise classifiers.

© Springer Nature Switzerland AG 2020
A. El Moataz et al. (Eds.): ICISP 2020, LNCS 12119, pp. 130–137, 2020.
https://doi.org/10.1007/978-3-030-51935-3_14

We are interested in applying Machine learning based Sentiment Analysis to the topic of 'Violence and Discrimination against Women'. However, there is, as far as we know, no existing annotated dataset featuring Arabic texts addressing this topic.

In fact, building an annotated training set is a fastidious and time consuming task requiring the involvement of human annotators. In this research, we present our approach to build a thematic training set by combining manual and automatic annotation of Arabic texts addressing Discrimination and Violence against Women.

In this work, we make three main contributions:

A. We develop an initial training set [7] that contains Arabic texts related to Discrimination and Violence Against Women and annotated by humans,
B. We propose a method that automatically extends the initial training set. In fact, we use the initial training set to generate a list of key expressions, and use them to produce a new expanded training with roughly the same characteristics.
C. We analyze the inheritance level of polarities from the Initial Dataset to the new collected data.

The remainder of this paper is organized as follows: Sect. 2 is a description of the general process. Section 3 describes the steps we followed to generate key expression. Section 4 represents the phase of building the new extended training set and the results of our work.

2 The Approach

In this approach, we develop an initial training set that contains Arabic texts related to Discrimination and Violence Against Women and annotated by human volunteers. Then, we use this initial training set to retrieve two lists (positive and negative) of key-expressions that represent the most significant terms used by Arab-speaking internet users to express themselves either negatively or positively towards Discrimination and Violence Against Women. Based on the generated lists we build a new training set on which we studied the inheritance of polarities (Fig. 1).

The general process is composed of two main phases: A. the automatic generation of key expressions. B. Building and analyzing the new training set.

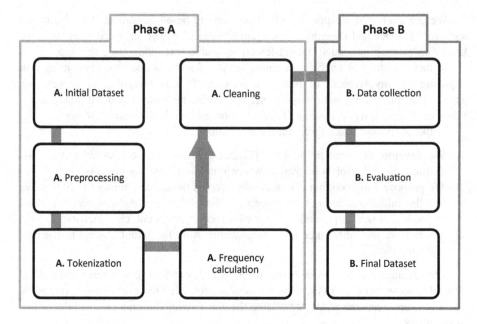

Fig. 1. The general process

3 Generation of Key Expressions

3.1 Initial Dataset

The starting point of this research is an initial pre-annotated training set which consists of a sample of raw text data containing tweets and some YouTube comments. A number of tweets in the sample were collected during the International Women's Day in 2018. Human annotators had to choose between six labels to annotate the different pieces of text:

- Off-Topic: If the text doesn't address a topic related to women's and girls' rights or realities,
- Neutral: If the text presents neutral information,
- Positive: If the text provides a positive position or opinion,
- Negative: If the text provides a negative position or opinion,
- Mixed: If the text provides a mix of positive and negative positions or opinions.

The labels 'Neutral', 'Positive', 'Negative' and 'Mixed' apply only for text data that actually addresses a topic related to women's and girls' rights.

3.2 Preprocessing

Data Cleaning. First, we started by extracting separately the positive and negative comments from the dataset. We retrieved a smaller set of 518 comments: 292 positive and 226 negative.

The preprocessing phase is slightly delicate for morphological languages such as Arabic. We started with cleaning data from any noises in order to have a clean Arabic piece of texts ready to be processed instead of noisy comments. Below all the components that we eliminate from data:

a - Diacritics: These diacritics express the phonology of the language and, in contrary of English, Arabic words could be read with or without their diacritics. Example:

المرأة سياقة سِيَاقَةُ المَرْأَةُ

b - Emojis: Users on social media tend to frequently use Emojis to express their opinions and feelings. In our case we removed all the emoticons and symbols from our set.

c - Punctuation: in the field of Natural Language Processing, the presence of punctuation affect directly the treatment of texts. To avoid bad results, we eliminated any presence of punctuation in our comments.

d - Numbers: We removed all the numerical digits.

Normalization and Stop Words Removal. Normalization is the process of transforming words to a standard format. In Arabic language, the phonetic sound of [i] [ɑ] [u] could be written in multiple syntactic forms, it depends on its location in the word. We implemented a normalized format to this sounds to reduce conflicts during the treatment of the texts (Table 1).

Table 1. Normalized letters.

Letters	normalization
ا / آ / أ / إ / ٱ	ا
ى	ي
ئ / ؤ	ء
ئ	ء

Stop words removal helps to eliminate unnecessary text information. We used a predefined set of a 3616 Arabic stop words to clean data. However, we faced a huge difficulty in this step because most of the time Arabic internet users tend to skip putting a space between a short stop word and the word that follows, the system consider it as one different word. Example: و is a stop word but it will not be removed as its attached to المرأة:

المَرْأَةُ   و   والمرأة

3.3   Frequency Calculation

Now as we cleaned our data, we must prepare it for the next step that consists of finding key expressions based on frequencies calculation.

Tokenization is the task of chopping a text into pieces, called tokens. Tokens do not always refer to words or terms but it could represent any sequence of semantic expressions. Depending on the typology of our work, we've choose to fragment the text into words, then we calculated the frequency of all the expressions of one to five words (n-grams) in order to retrieve the most significant expressions toward our topic.

At this stage, and whether for negative or positive tokenized text, we retrieved the top 20 most frequent expressions for each n-gram (n = 1 → 5). We have end up with 100 expressions for each polarity.

3.4   Final Key Expressions

The preprocessing phase that we carried out was not sufficient to have good result. In fact, the 200 expressions that we retrieved were not very convincing due to many parameters, in what follows the actions that we did to filter the expressions:

Remove "Religious" Expressions. Arab speakers tend usually to use general purpose religious expressions in their discourse, the same case for Arabic internet users, which explains the strong presence of such terms in the retrieved expressions. During pre-processing, we replace these religious expressions by the keyword "RE".

Example:

حسبي الله ,الرسول صلى الله عليه و سلم

Remove Named Entities Expressions. The collected data is usually influenced by major events that invade social networks during the time of scrapping which leads to a redundancy of a person's name or an organization, for example, in collected data:

الشيخ فوزان ,الملك سلمان

Remove Insults. Insults are frequently used by net surfers especially in social media platforms. Example:

غبية

Remove Expressions Present in Both Negative and Positive Lists. In order to leave only significant expressions toward either positive and negative polarity, it's obvious to remove expressions present in both lists at once.

Normalizing Identic Expressions. This step is about conserving only one of various expressions that give the same meaning. For example both "سياقة" and "قيادة" are synonyms to the word "drive":

سياقة المرأة   قيادة المرأة

Final Lists. After filtering the 200 expressions, we kept 6 expressions for each polarity: RE = Religious Expression

Positive	Negative
ضرب المرأة'Hit the woman'	قرن بيوتكن'Stay home'
عقلية ذكورية 'Male mentality'	تطليق المرأة'get divorced'
امساك بمعروف RE	تكشف وجهها'Show her face'
تسريح بإحسان RE	تلتزم بالحجاب'keep the Hijab'
ختان المرأة'Female circumcision'	مطيعة'Obedient women'
معاملة الرجل للمرأة 'Man treats woman'	تبرجن تبرج RE

4   Building the New Expanded Dataset

In phase A, we generated key-expressions, based on an initial dataset annotated by humans. Each polarity (positive and negative) is represented by 6 key expressions.

When we use these expressions as seeds to collect new data, we expect the collected data to have the same polarity as the seeds. In this section, we describe the data collection process and try to answer this question: Did the new data inherit the polarity of the key-expressions used to collect it?

4.1   Data Collection

Using Twitter Developer API, we collected tweets using the pre-generated list of key expressions. We finally retrieved 1172 tweets: 573 negative, 599 positive.

4.2   Evaluation

We assess the quality of the expanded dataset by direct human judgment. This consists of human volunteers reviewing and annotating the collected texts using one of four labels:

- Off-Topic: If the text doesn't relate to our topic,
- Positive: If the text represents a positive opinion toward women,
- Negative: If the text represents a negative opinion toward the topic,
- Neutral: For texts that describes a neutral information or opinion.

With this step, we were able to distinguish between true and false negative texts (i.e. positive texts).

4.3   Analysis

Data Collected with Positive Key Expressions. With these key expressions we retrieved good results, the majority of the collected tweets are true positive with a percentage as high as 86%. Negative tweets represent only 4% (Fig. 2).

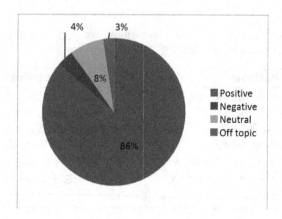

Fig. 2. Results for data collected with positive keywords

Data Collected with Negative Key Expressions. In this case, the results were not satisfying. True negative tweets represented only 42% of the whole collected data, the proportion of neutral tweets was remarkably high with a percentage of 34%. In fact, a substantial number of the retrieved tweets in the negative set consisted of verses of Quran, which were annotated as Neutral (Fig. 3).

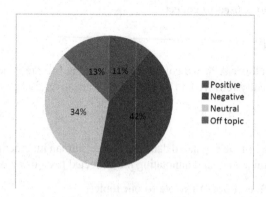

Fig. 3. Results for data collected with negative keywords

5 Conclusion

In this work, we presented an approach to build a training set by combining manual ant automatic annotation of Arabic text. The preprocessing was very challenging due to the complexity of the language and the typology of data in social media.

The obtained results were of good quality: we built a final training set of 1690 entries. However, even if the approach gives excellent results for positive key-expressions, the inheritance of polarities must be improved for negative key-expressions which we will further investigate in future works.

References

1. https://unstats.un.org/sdgs/indicators/indicators-list/. Accessed Feb 2019
2. Vaitla, B., et al.: Big data and the well-being of women and girls: applications on the social scientific frontier (2017)
3. http://arabstates.unwomen.org/en/what-we-do/ending-violence-against-women/facts-and-figures. Accessed Feb 2019
4. Mohammad, S., Kiritchenko, S., Sobhani, P., Zhu, X., Cherry, C.: A dataset for detecting stance in tweets. In: Proceedings of the Tenth International Conference on Language Resources and Evaluation (LREC 2016), pp. 3945–3952, May 2016
5. Abdul-Mageed, M., Diab, M.T.: AWATIF: a multi-genre corpus for modern standard arabic subjectivity and sentiment analysis. In: LREC, vol. 515, pp. 3907–3914, May 2012
6. Taboada, M.: Sentiment analysis: an overview from linguistics. Ann. Rev. Linguist. 2, 325–347 (2016)
7. Zahir, J.: Mining the web for insights on violence against women in the MENA region and Arab states (2019)

Multistage Deep Neural Network Framework for People Detection and Localization Using Fusion of Visible and Thermal Images

Bushra Khalid$^{(\boxtimes)}$, Muhammad Usman Akram , and Asad Mansoor Khan

National University of Science and Technology, Islamabad, Pakistan
bkhalid.ce16ceme@gmail.com, usmakram@gmail.com, asad.m.khan12@gmail.com

Abstract. In Computer vision object detection and classification are active fields of research. Applications of object detection and classification include a diverse range of fields such as surveillance, autonomous cars and robotic vision. Many intelligent systems are built by researchers to achieve the accuracy of human perception but could not quite achieve it yet. Convolutional Neural Networks (CNN) and Deep Learning architectures are used to achieve human like perception for object detection and scene identification. We are proposing a novel method by combining previously used techniques. We are proposing a model which takes multi-spectral images, fuses them together, drops the useless images and then provides semantic segmentation for each object (person) present in the image. In our proposed methodology we are using CNN for fusion of Visible and thermal images and Deep Learning architectures for classification and localization. Fusion of visible and thermal images is carried out to combine informative features of both images into one image. For fusion we are using Encoder-decoder architecture. Fused image is then fed into Resnet-152 architecture for classification of images. Images obtained from Resnet-152 are then fed into Mask-RCNN for localization of persons. Mask-RCNN uses Resnet-101 architecture for localization of objects. From the results it can be clearly seen that Fused model for object localization outperforms the Visible model and gives promising results for person detection for surveillance purposes. Our proposed model gives the Miss Rate of 5.25% which is much better than the previous state of the art method applied on KAIST dataset.

Keywords: Object detection · Object localization · Mask-RCNN

1 Introduction

In the present age of technology, security of individuals is one of the most important concerns. Offices, schools, hospitals and organizations are provided with complete security measures to a avoid any kind of security breach. These security measures include security personnels, Close Circuit Television cameras (CCTV)

© Springer Nature Switzerland AG 2020
A. El Moataz et al. (Eds.): ICISP 2020, LNCS 12119, pp. 138–147, 2020.
https://doi.org/10.1007/978-3-030-51935-3_15

and surveillance. CCTV cameras can be used in a number of other applications as well, such as offense detection, public safety, crime prevention, quick emergency response, management and for the reduction of fear of crime in public [1]. Surveillance cameras can also be used for monitoring traffic flow and to keep an eye on the staff in regards to the complaints received [2]. While monitoring a vicinity there are many types of surveillance cameras to cater different weather and lighting conditions. During the day light visible cameras come in handy for clear observation while utilizing the bright light of the sun. Thermal or Long Wave InfraRed (LWIR) cameras on the other hand becomes useful when there is a bad lighting condition, storm, fog or dark scene. Visible cameras compliment with a source of information where thermal cameras identify the presence of persons, animals and weapons. Heat signatures along with visible information becomes more informative during surveillance. Intruders and robbers can be identified during night time and even if they have hidden themselves behind any solid object, thermal cameras nullify the effect of occlusion.

Section 2 contains a brief overview of related work. Section 3 gives an overview of datasets used in experimentation. Section 4 states the details of the proposed methodology. Experiments and results are shown in Sect. 5 and Conclusion with future work is stated in Sect. 6.

2   Related Work

In feature level classification objects are classified, but not detected and localized. Convolution neural networks are capable of object detection and excelled in this field over the past few years. Object detection caught the eyes of researchers in 2014 after the introduction of the basic object detection technique introduced by Girshick [3]. After the comparison of results of the Regional Convolutional Neural Network (RCNN) and Histogram Of Gradient (HOG) based classifiers, it was clear that CNN outperformed other techniques with a clear margin. This literature review is divided into two parts: first we discuss about visible and thermal data sets and how they are utilized in various applications and then object detection and localization is discussed.

2.1   Visible and Thermal Image Data

Visible images consisting of three channels Red Green Blue (RGB) when paired with thermal or infrared images are known as multi-spectral images. The infrared image can be obtained by infrared camera which represents the thermal radiations within the range of 0 to 255. The difference between visible and infrared image is that visible images are highly sensitive to lighting conditions and they contain fixed pattern information [4]. While thermal images do not contain any kind of texture information and thermal images displays the heat map of any object whose temperature is above absolute zero. Thermal images are categorized into five types. Near Infrared Image (NIR), Short Wave Infrared (SWIR), Mid Wave Infrared (MWIR), Long Wave Infrared (LWIR) and Far Infrared (FIR).

NIR ranges from 0.75 to 1.4, SWIR ranges from 1.4 to 3, MWIR 3 to 8, LWIR 8 to 15, FIR 15 to 1000 μm wavelength.

2.2 Object Detection and Localization

Object detection is the center of attention in computer vision and it is one of the basic building blocks. A number of researchers have worked on object detection more specifically human detection to avoid collision events while driving autonomous cars or during the movement of robots. Till late 90 s detection of objects was totally based on visual images and visual scenarios. But now this problem is shifting towards multi-spectral identification, i.e. object detection using images of different spectrums or modalities. Regional CNN (RCNN) [5] takes image as an input, identifies the region of interest and provides bounding box output and also a classification score. The mechanism behind RCNN is that it creates a lot of bounding boxes or square boxes of different sizes of the image by the selective search method [SSM]. The Selective search method propagates square and rectangular windows of different sizes over the image and select the boxes in which adjacent pixels show a pattern of potential object. To address the issues of RCNN, Girshick proposed an improved model in 2015 which is known a Fast-RCNN [3]. The first problem which was solved in Fast-RCNN was that all the separate models were combined as one model. RCNN calculated separate feature vectors for each proposal, but Fast-RCNN combined all the feature vectors of an image in one vector. That feature vector is then used by CNN for classification and bounding box regression. These two solutions turned out to be effective in terms of speed. Region Of Interest (ROI) pooling is the process in which an image is converted from h × w matrix to H × W matrix by applying max pooling, but the image is dealt in the form of small windows and each window is max-pooled to get the output image window and the whole image likewise. The solution to the above two problems was proposed in Faster-RCNN [6]. Faster-RCNN is a combined model which provides a classification score and bounding boxes with the help of RPN and Fast-RCNN. CNN is pre-trained on imagenet for classification. For the generation of Region proposals, Region proposal Network RPN is a fine-tuned end-to-end. Proposals having IOU overlap greater than 0.7 are positive while the ones having IOU overlap less than 0.3 are negative. Fast-RCNN is trained using RPNs from Region Proposal Network. RPN is then trained using Fast-RCNN. They have some shared convolution layers. After this Fast-RCNN is fine-tuned, but only the unique layers of it are fine-tuned.

3 Dataset

The data set we are using for training CNNs in this paper is KAIST multi-spectral data set. KAIST multi-spectral data is acquired by mounting the visible and thermal cameras on a car and an additional beam splitter is used to align the LWIR and RGB image data [7]. Different images are taken in different lighting condition to observe the effect of light on the scene and on the object detection. Figure 1 shows randomly chosen images from KAIST multi-spectral data set and it ground truth bounding box representation. KAIST consists of 95000 Visual-Thermal image pairs. The size of every image is 640 × 512 and they are aligned geometrically. The data set is divided into 60 to 40 ratio for training and testing.

Fig. 1. Ground truth representation of KAIST dataset

Apart from KAIST multi-spectral data, we are using another local dataset for testing purpose which consisted of visible thermal pairs of 20 images. This dataset contained some images which showed a clear difference of a person's presence in the scene. In the visible image a person might be hidden behind the bush or might be wearing a camouflage which is invisible to the naked eye. In such cases Thermal images come in handy. Figure 2 shows some random images from the local dataset.

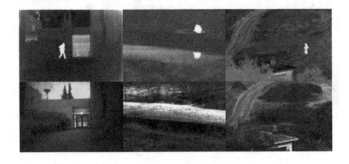

Fig. 2. Visible and thermal image pairs from local dataset

4   Proposed Methodology

Our proposed methodology consists of three main modules. The first module is a fusion module which consists of two encoders for visible and thermal image encoding. Both modules take the input images and after encoding the feature maps carry out the process of fusion. Once the features are fused the fused feature

Table 1. Dataset

Parameters	Training	Testing
Division	60%	40%
Images	50.2 k	45.1 k
Day	33399	29179
Night	16788	15962
Pedestrian	45.1 k	44.7 k

vector is transferred to the decoder block which decodes it back to image content. From this module a final fused image is obtained which is then transferred to the next module called ResNet block. This block takes the fused image and classifies it as a person class or a no person class. Images classified as no person class are discarded while the ones having person class are transferred to the third and final module of image localization. These three models are explained in detail below. Figure 3 shows the flow diagram of our proposed technique which clearly shows all the three blocks of the model (Table 1).

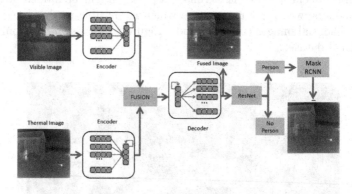

Fig. 3. Proposed image fusion and localization framework

4.1   Fusion and Classification of Visible and Thermal Images

The network architecture used in encoder part of our proposed model is a Siamese architecture [11]. Siamese architecture is a neural network, which is different from usual neural networks. It does not classify the input fed into it, rather it is designed to take decisions by comparing the similarities and dissimilarities of multiple inputs. Siamese architecture consists of two sub networks whose feature maps are compared and final encoded output is decided on the basis of that comparison. In the siamese network architecture two sister networks with same properties take same or different inputs and these inputs pass through convolution layers for feature extraction. Once the features are extracted from both inputs, then they are passed through a contrastive loss function which actually calculates the similarities and differences between the inputs.

For the classification of images we are using ResNet-152 [8] architecture. A fused image obtained from encoder-decoder architecture is passed through the convolution layers of ResNet architecture for its classification. If the network classifies the image as a person then there is a person or persons present in the frame. Images classified as persons are passed through the localization network for localization of all the objects such as persons present in the image.

4.2 Localization of Persons

In this section we will discuss the model we are using for object detection and localization. Mask-RCNN [9] is an extension of Faster-RCNN. Apart from object detection and classification Mask-RCNN also produces instance segmentation masks. Mask-RCNN introduces a new branch for the instance segmentation which outputs the masks of detected objects. The instance segmentation branch is a fully connected network, which provides pixel to pixel segmentation on the basis of the Region of Interests. A fully connected network is a network in which every node is connected to every other node. As detecting and overlaying a mask over the object is much more complex than just drawing the bounding boxes around them, Mask-RCNN introduces a new layer called ROI-Align layer in place if ROI-pooling layer. Figure 4 represents the architecture of Mask-RCNN.

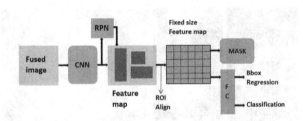

Fig. 4. Mask-RCNN architecture

RPN takes a sliding window and propagates it throughout the image. 2k anchors are formed in the result. RPN usually takes about 10 ms to skim an image, but in case of Mask-RCNN it might take longer than that as the input image to Mask-RCNN is relatively of bigger size hence it needs more anchors. Anchors are divided into two classes based on the IOU overlap. The foreground class is highly likely to contain an object while background class does not. Foreground anchors are then refined to get an exact bounding box of the object present inside the anchor. The problem of bounding box is solved, but the problem of classification still remains. To solve this problem ROI is used, the ROI takes the foreground object and classifies it into its actual class. Than ROI pooling is done to resize the image so that it can be sent into the classifier network.

The segmentation branch takes in the ROI results and gives the mask for objects in the output.

5   Experiments and Results

Our experimentation is divided into two parts. The first part explains the experimental setup used for fusion model and shows the results obtained from the fusion of thermal and visible images. The second part describes the experimental setup of detection and localization model and displays few results from localization part.

5.1   Detection and Localization Results

The image having persons present in it is then transferred into the Mask-RCNN model for localization. For training KAIST dataset provides bounding box annotations, but in case of Mask-RCNN we need binary masks for training. For this purpose, we used the VIA VGG annotator tool. Using this interface we created image masks for our data set. VIA tool saves json model for training and validation data sets separately. There is no need for extensive training by a huge bulk of data because Mask-RCNN model can utilize pre-trained weights of imagenet and MS-COCO. ResNet architectures need high computation power in the training process. NVIDIA GPU GTX-1080 Ti with 64 GB RAM is used in the

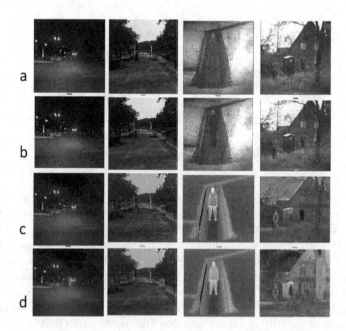

Fig. 5. KAIST: (a) Visible images (b) Localization of a (c) Fused images (d) Localization of c

training process. Training time taken by Mask-RCNN over 50.2 k images is 4 days, but once it is trained the testing timing Mask-RCNN provides promising results for localization of objects. Figure 5 shows the results of Mask-RCNN over KAIST multi-spectral dataset and local dataset. Images in the first two columns are from KAIST dataset while in images in the third and fourth column are from local dataset. Row a consists of Visible images (first column contains an image of night time while the second column contains images of day time and their respective detections are shown in row b. It can be seen that there are a lot of False detections in the visible image present in the first column of the row. While its respective fused image and its detection scan be seen in c and d. First column of row d shows that there is only one false detection in the fused image. From fourth column it is noted that the person present inside the box is hidden while in the fused image it can be seen partially. Detection results show that in the visible image the person inside the box remains undetected when passed through Mask-RCNN model while both the persons are detected in the fused image. There are some false detections in the fused image as the model is trained over KAIST and these test images are randomly chosen. For the calculation of miss rate, state of the art methodology and formula is used which can found in detail in paper proposed by Konig et al. [13]. Table 2 shows the graph of comparison of our proposed model with previous techniques. From this graph it can be seen that our proposed model outperformed previous methodologies in terms of Miss Rate. Miss Rate gives us the measure of accuracy by which an object was detected pixel by pixel.

Table 2. Comparison of MR% with previous studies

S. no	Author	Year	Technique	Data set	Miss rate
1	Wagner et al. [10]	2016	CNN	KAIST	43.80%
2	Liu et al. [12]	2016	Faster-RCNN	KAIST	37%
3	Konig et al. [13]	2017	Fusion RPN + BDT	KAIST	29.89%
4	Xu et al. [14]	2017	CMT-CNN	KAIST	10.69%
5	Ours	2019	Encoder-Decoder+ Mask-RCNN	KAIST	5.25%

6   Conclusion and Future Work

The purpose of this paper is to utilize modern technology and computer vision models for efficient surveillance and the provision of foolproof security in organizations, schools, hospitals and military zones. For this purpose, we are utilizing visible and thermal cameras to obtain images of the premises as well as heat maps of suspicious intruders. These images are then fused together to get a combined more informative output for detection of a doubtful presence. We are using Encoder-decoder CNN architecture for fusion of visible and thermal images and Resnet architecture for object detection and localization. Localization of object or persons is done using Mask-RCNN model which not only localizes the object, but also provides a mask for localized object. KAIST multi-spectral data are used for the training of CNNs and local dataset is also used in the testing process. When the results of visible detections are compared with results of Fused detections, it is clearly observed Fused model outperforms the detection and localization process by giving accurate masks for KAIST and comparatively better masks for local dataset than the visible model. The learning time of model can be improved by minimizing the layers of ResNet architecture.

References

1. Ratcliffe, J.H.: Video Surveillance of Public Places: Problem Oriented Guides for Police, Response Guides Series, No. 4, pp. 195–197. Center for Problem Oriented Policing, Washington DC (2006)
2. National Rail CCTV Steering Group and Others: National rail and underground closed circuit television (CCTV) guidance document: final version, vol. 30 (2010)
3. Girshick, R., Donahue, J., Darrell, T., Malik, J.: Region-based convolutional networks for accurate object detection and segmentation. IEEE Trans. Pattern Anal. Mach. Intell. **38**, 142–158 (2016)
4. Gade, R., Moeslund, T.B.: Thermal cameras and applications: a survey. Mach. Vis. Appli. **25**, 245–262 (2014). https://doi.org/10.1007/s00138-013-0570-5
5. Girshick, R., Donahue, J., Darrell, T., Malik, J.: Rich feature hierarchies for accurate object detection and semantic segmentation. In: Proceedings of the IEEE Conference on Computer Vision and Pattern Recognition, pp. 580–587 (2014)
6. Shaoqing, R., Kaiming, H., Girshick, R., Sun, J.: Faster R-CNN: towards real-time object detection with region proposal networks. IEEE Trans. Pattern Anal. Mach. Intell. **39**(6), 1137–1149 (2017)
7. Hwang, S., Park, J., Kim, N., Choi, Y., So, K.: Multispectral pedestrian detection: benchmark dataset and baseline. In: Proceedings of the IEEE Conference on Computer Vision and Pattern Recognition, pp. 1037–1045 (2015)
8. He, K., Zhang, X., Ren, S., Sun, J.: Deep residual learning for image recognition. In: Proceedings of the IEEE Conference on Computer Vision and Pattern Recognition, pp. 770–778 (2016)
9. He, K., Gkioxari, G., Dollar, P., Girshick, R.: Mask R-CNN. In: 2017 IEEE International Conference on Computer Vision (ICCV), pp. 2980–2988 (2017)
10. Wagner, J., Fisher, V., Herman, M., Behenke, S.: Multispectral pedestrian detection using deep fusion convolutional neural networks. In: 24th European Symposium on Artificial Neural Networks, Computational Intelligence and Machine Learning (ESANN) (2016)
11. Ram, P.K., Sai, S., Venkatesh, B.: DeepFuse: a deep unsupervised approach for exposure fusion with extreme exposure image pairs. In: Proceedings of the IEEE International Conference on Computer Vision (2017)
12. Liu, J., Zhang, S., Wang, S., Metaxas, D.: Multispectral deep neural networks for pedestrian detection. arXiv preprint arXiv:1611.02644 (2016)

13. Konig, D., Michael, A., Christian, J., Georg, L., Heiko, N., Michael, T.: Fully convolutional region proposal networks for multispectral person detection. In: Proceedings of the IEEE Conference on Computer Vision and Pattern Recognition Workshops (2017)
14. Xu, D., Ouyang, W., Ricci, E., Wand, X., Sebe, N.: Learning cross-modal deep representations for robust pedestrian detection. In: Proceedings of the IEEE Conference on Computer Vision and Pattern Recognition (2017)

Biomedical Imaging

Diagnosing Tuberculosis Using Deep Convolutional Neural Network

Mustapha Oloko-Oba and Serestina Viriri[✉]

School of Mathematics, Statistics and Computer Science,
University of KwaZulu-Natal, Durban, South Africa
mustaphasmo@yahoo.com, viriris@ukzn.ac.za

Abstract. One of the global topmost causes of death is Tuberculosis (TB) which is caused by mycobacterium bacillus. The increase rate of infected people and the recorded deaths from TB disease is as a result of its transmissibility, lack of early diagnosis, and inadequate professional radiologist in developing regions where TB is more prevalent. Tuberculosis is unquestionably curable but needs to be detected early for necessary treatment to be effective. Many screening techniques are available, but chest radiograph has proven to be valuable for screening pulmonary diseases but hugely dependent on the interpretational skill of an expert radiologist. We propose a Computer-Aided Detection model using Deep Convolutional Neural Networks to automatically detect TB from Montgomery County (MC) Tuberculosis radiographs. Our proposed model performed at 87.1% validation accuracy and evaluated using confusion matrix and accuracy as metrics.

Keywords: Chest radiograph · Convolutional neural network · Preprocessing · Feature extraction · Tuberculosis · Classification

1 Introduction

Tuberculosis (TB) is a contagious disease that is considered worldwide as a significant source of death from a single transmittable agent as well as among the top 10 sources of death [1,2]. The TB disease is caused by the mycobacterium tuberculosis bacillus which is easily contractible by having close contact with infected individuals. This disease mostly affects the lungs but can as well affects other parts of the body [3,4].

World Health Organization (WHO) estimated about 10 million individuals fell sick as a result of Tuberculosis disease in 2018 which resulted in about 1.2 million deaths from the previous 1.3 million deaths recorded in 2017 [2].

Tuberculosis disease is more prevalent in developing regions and can affect both males and females but more prominent in males. Among the total number of individuals infected with tuberculosis in 2017, 1 million cases were reported in children aged less than 14, 5.8 million cases in males, and 3.2 million cases in females [1].

© Springer Nature Switzerland AG 2020
A. El Moataz et al. (Eds.): ICISP 2020, LNCS 12119, pp. 151–161, 2020.
https://doi.org/10.1007/978-3-030-51935-3_16

Tuberculosis disease is certainly curable but needs to be detected early for appropriate treatment. Several lung examination techniques are available but the Chest radiographs conversationally known as chest X-ray or CXR for short is a prominent screening tool for detecting abnormalities in the lungs [5,6,14]. Basically TB manifestation can be detected on CXR, however, quality CXR imaging equipment along with skilled radiologists to accurately interpret the CXR is either limited or not available in TB prevailing regions [7,8].

A geographical report by the World Health Organization of most Tuberculosis cases for 2018 is shown in Table 1.

Table 1. Geographical TB cases (WHO, 2018)

Region	Cases (%)
South-East Asia	44
Africa	24
Western Pacific	18
Eastern Mediterranean	8
The Americas	3
Europe	3

Due to the deadly nature of TB disease and the rate at which it can easily be spread, WHO has laid emphasis on more proactive measures for continuous reduction of TB cases and deaths [9]. Also, the decision to embark on a mission to put an end the universal TB epidemic by the year 2030 is underway as contained in the 2019 Global Tuberculosis report [2].

The lack of skilled radiologist, high number of patients waiting in line to be screen and mostly outdated equipment which results in a high rate of errors in properly screening the CXR remain a major problem that requires prompt attention.

As a result, to profer solution to the issue of limited or lack of expert radiologist and misdiagnosis of CXR, we propose a Deep Convolutional Neural Networks (CNN) model that will automatically diagnose large numbers of CXR at a time for TB manifestation in developing regions where TB is most prevalent. The proposed model will eliminate the hassle of patients waiting in line for days to get screened, guarantee better diagnosis, performance accuracy and ultimately minimize cost of screening as opposed to the process of manual examination of the CXR which is costly, time-consuming, and prone to errors due to lack of professional radiologist and huge number of the CXR pilled up to be diagnosed.

2   Related Work

The evolution of computer-aided detection and investigative systems has offered a new boost attributed to the emerging digital CXR and the ability of computer

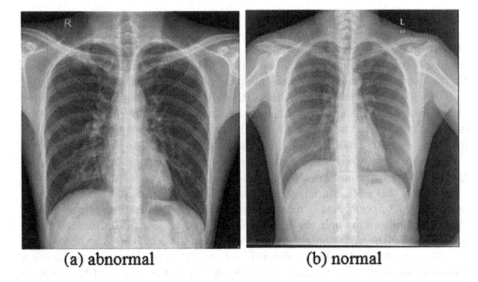

(a) abnormal   (b) normal

Fig. 1. Sample of the normal and abnormal chest X-ray.

vision for screening varieties of health diseases and conditions. Although much impressive research has been carried out in the last few years regarding computer-aided detection, nevertheless lots more finding is required in the field medical imaging to improve the existing methods and find convenient lasting solutions to deadly medical conditions as the case of Tuberculosis and many more.

A processing method that combines the Local Binary Pattern with laplacian of gaussian was employed in [10] for the manual detection of Tuberculosis nodules in CXR. This research centers on accentuating nodules by the Laplacian of Gaussian filter, lung segmentation, ribs suppression and the use of local binary pattern operators for texture classification. Computer-aided diagnosis system presented by [11] for screening Tuberculosis using two different Convolutional Neural Networks architectures (Alexnet and VGGNet) to classify CXR into positive and negative classes. Their experiment which is based on Montgomery and Shenzhen CXR datasets found VGGNet outperformed Alexnet as a result of a deeper network of VGGNet. The performance accuracy of 80.4% was obtained for Alexnet while VGGNet reached 81.6% accuracy. The authors conclude that improved performance accuracy can be achieved by increasing the dataset size used for the experiment.

One of the first research papers that utilized deep learning techniques on medical images is shown in [12]. The work was based on popular Alexnet architecture and transfer learning for screening the system performance on different datasets. The cross dataset performance analysis carried out shows the system accuracy of 67.4% on the Montgomery dataset and 83.7% accuracy on the Shenzhen dataset. A ConVnet model involving classifications of different manifestations of tuberculosis was presented in [13]. This work looked at unbalanced, less categorized X-ray scans and incorporate cross-validation with sample shuffling in training their model. The Peruvian Tuberculosis datasets comprising of a total of 4701 image samples with about 4248 samples marked as abnormal containing six manifestation of tuberculosis and 453 samples marked as normal were used for the experiment to obtain 85.6% performance accuracy.

CNN has also been applied by the authors of [15] for extracting discriminative and representative features from X-ray radiographs for the purpose of classifying different parts of the body. This research has exhibited the capabilities of CNN models surpassing traditional hand-crafted method of feature extraction.

An approach based on deep learning for the classification of chest radiographs into positive and negative classes is depicted in [16]. The CNN structure employed in this work consists of seven convolutional layers and three fully connected layers to perform classification experiments. The authors compared three variety optimizers in their experiments and found the Adam optimizer to perform better with validation accuracy of 0.82% and loss of 0.4013.

Other methods that have been utilized for TB detection and classification includes: Support Vector Machine [17,18], K-Nearest Neighbor [19], Adaptive Thresholding, Active Contour Model, and Bayesian Classifier [20], Linear Discriminant Analysis [21].

It is evident from the related work that more effort is required in dealing with the Tuberculosis epidemic that has continued as one of the topmost causes of death. In view of this, we have presented an improved performance validation accuracy concerning to detecting and classifying CXR for TB manifestation.

3   Materials

3.1   Datasets

The Montgomery County (MC) CXR dataset was employed in this research. The MC dataset is a TB specific dataset made available by the National Library of Medicine in conjunction with the Department of Health Services Maryland, U.S.A for research intent. This dataset composed of 58 abnormal samples labeled as "1" and 80 normal samples labeled as "0". All samples are of the size 4020 by 4892 pixels saved as portable network graphic (png) file format as shown in Fig. 1. This dataset is accompanied by clinical readings that give details about each of the samples with respect to sex, age, and manifestations. The dataset can be accessed at https://lhncbc.nlm.nih.gov/publication/pub9931 [23].

3.2   Preprocessing

Since deep neural networks are hugely dependent on large data size to avoid overfitting and achieve high accuracy [22], we performed data augmentation on the MC dataset as a way of increasing the size from 138 samples to 5000 samples. The following types of augmentation were applied: horizontal left and right flip with a probability = 0.3, random zoom = 0.3 with an area = 0.8, top and bottom flip $= 0.3$, left and right rotation $= 0.5$. Other preprocessing task employed here includes image resizing, noise removal and histogram equalization. The data augmentation procedure used in this work is not such that gives room for data redundancy.

4   Proposed Model

A model based on Deep Convolutional Neural Network (CNN) structure has been proposed in this work for the detection and classification of Tuberculosis. CNN models are based upon feed-forward neural network structures for automatic features selection and extraction as a result of taking advantage of the inherent properties of images. The depth of a CNN model has an impact on the performance of the features extracted from an image. CNN models have many layers but the Convolutional layer, MaxPooling, and the Fully connected layer are regarded as the main layers [15]. At the time of model training, diverse parameters are optimized in the convolution layers for extracting meaning features before is been pass on to the fully connected layer where the extracted features are then classified into the target classes which in this case is "normal and abnormal" classes.

Our proposed CNN structure is composed of feature extraction and features classification stages. The feature extraction stage consists of convolution layers, batch normalization, relu activation function, dropout, and max pooling while the classification stage contains the fully connected layer, flatten, dense and a softmax activation function. There are 4 convolution layers in the network for extracting distinct features from the input image with shape $224 \times 224 \times 3$ that is passed to the first convolutional layer learning 64, 3×3 filters, the same as the second convolutional layer. Both the third and fourth convolutional layers learn 128, 3×3 filters. Relu activation function and batch normalization were employed in all the convolutional layers but only the second and fourth layer uses max pooling with a 2×2 pooling size and 25% dropout. The fifth layer which is the fully connected layers output 512 feature that is mapped densely to 2 neurons required by the softmax classifier for classifying our images into normal and abnormal classes. The detail representation of our proposed TB detection model is presented in Table 2.

4.1   Convolution Layer

At each convolution layer, the feature maps from the preceding layer convolute with kernels which are then fed through the ReLu activation function to configure the feature output maps. Also, each output map can be formulated with respect to several input maps. This can be mathematically written as:

$$y_j^i = f\left(\sum\nolimits_{i \in N_j} y_i^{l-1} * M_{ij}^l + a_j^l\right)$$

where y_j^i depicts the jth output feature of the lth layer, $f(.)$ is a nonlinear function, N_j is the input map selection, y_i^{l-1} refer to the ith input map of $l-1^{th}$ layer, M_{ij}^l is the kernel for the input i and output map M in the jth layer, and a_j^l is the addictive bias associated with the jth map output.

Table 2. Details representation of our proposed TB detection model

Layers	Output size	Filter
INPUT IMAGE	$224 \times 224 \times 3$	
CONVO1	$112 \times 112 \times 64$	3×3
ACTN	$112 \times 112 \times 64$	
BATCHNORM	$112 \times 112 \times 64$	
CONVO2	$112 \times 112 \times 64$	3×3
ACTN	$112 \times 112 \times 64$	
BATCHNORM	$112 \times 112 \times 64$	
MAXPOOL	$56 \times 56 \times 64$	2×2
DROPOUT	$56 \times 56 \times 64$	
CONVO3	$56 \times 56 \times 128$	3×3
ACTN	$56 \times 56 \times 128$	
BATCHNORM	$56 \times 56 \times 128$	
CONV4	$56 \times 56 \times 128$	3×3
ACTN	$56 \times 56 \times 128$	
BATCHNORM	$56 \times 56 \times 128$	
MAXPOOL	$28 \times 28 \times 128$	2×2
DROPOUT	$28 \times 28 \times 128$	
FULLY CONN	512	
ACTN	512	
BATCHNORM	512	
DROPOUT	512	
SOFTMAX	2	

4.2   MaxPooling Layer

MaxPooling layer carryout a downsampling operation of the input map by calculating the maximum activation value in each feature map. The downsampling of MaxPooling is done to partly control overfitting and is formally written as:

$$y_j^l = f\left(\alpha_j^l down(y_i^{l-1} * M_{ij}^l) + a_j^l\right)$$

where α_j^l represent the multiplicative bias of every feature output map j that scale the output back to its initial range, $down(.)$ can be substituted for either $avg(.)$ or $max(.)$ over an $n \times n$ window effectively scaling the input map by n times in every dimension.

Our model is trained using the Stochastic Gradients Descents (SGD) optimizer with an initial learning rate set to 0.001, batch size of 22 samples, momentum equals 0.9, regularization L2 equals 0.0005 to control overfitting and training loss. The SGD optimizer is given below as:

$$\alpha = \alpha - n. \bigtriangledown_\alpha J(\alpha; x^{(i)}; y^{(i)})$$

where $\bigtriangledown_\alpha J$ is the gradient of the loss w.r.t α, n is the defined learning rate, α is the weight vector while x and y are the respective training sample and label.

4.3 Softmax Classifier

The Softmax classifier is used to process and classify the features that have been extracted from the convolutional stage. Softmax determine the probability of the extracted features and classify them into normal and abnormal classes defined as:

$$\sigma(q)_i = \frac{e^{q_i}}{\sum_{k=1}^{k} e_k^q}$$

where q mean the input vector to the output layer i that is depicted from the exponential element.

The structure of our model is presented in Fig. 2.

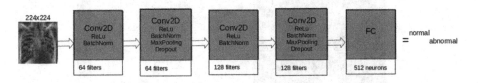

Fig. 2. Architecture of our ConvNet model.

5 Result

Training and validation of our model were performed on the Montgomery County (MC) TB specific dataset. The dataset which originally consists of 138 normal and abnormal samples were augmented to 5000 samples and split into 3,750 (75%) for training and 1,250 (25%) validation. The samples in the training set are indiscriminately selected i.e (shuffled) during the training to ensure features are extracted across all samples. Of course, the samples in the training set are

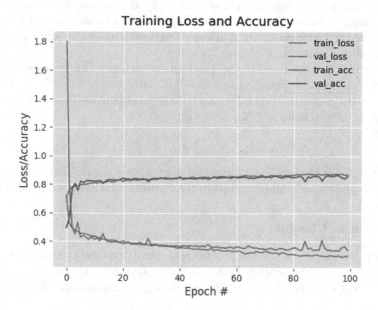

Fig. 3. Model accuracy and loss

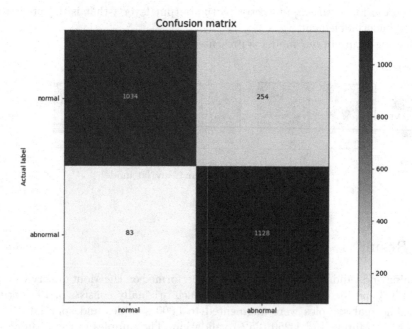

Fig. 4. Model confusion matrix

Table 3. Proposed model as compared

Ref	Accuracy (%)
23	82.6, 84.7
12	83.7
11	80.4, 81.6
16	82
13	85
24	83.7
Proposed	**87.1**

entirely different from the samples contained in the validation set which are samples that the model has not seen before during model training. We use confusion matrix and accuracy as the metrics for evaluating the model. The 224 × 224 sample is fed into the model which was trained for 100 iterations with a batch size of 22 for SGD optimizer at an initial 0.001 learning rate. The model employs cross entropy loss function for weight update at every iteration and use Softmax function for processing and classifying the samples into normal and abnormal classes labeled 0 and 1. The proposed model achieved 87.1% validation accuracy. The Performance accuracy of the model and confusion matrix are presented in Figs. 3 and 4.

6 Conclusion

Presented in this paper is a model that aids early detection of Tuberculosis using CNN structure to automatically extract distinctive features from chest radiographs and classify them into normal and abnormal categories. We did not test other architectures in this research; instead, their performance accuracy is reported in Table 3. The histogram equalizer we applied to enhance the visibility of the data samples, which makes the extracted features more evident, is one of the contributing factors responsible for the improved performance. We will consider the Shenzhen, JSRT, KIT, and Indiana datasets in our future work while we continue to aim for optimal performance accuracy.

References

1. World Health Organization: Global tuberculosis report. WHO (2018)
2. World Health Organization: Global tuberculosis report. WHO (2019)
3. Zaman, K.: Tuberculosis: a global health problem. J. Health Popul. Nutr. **28**(2), 111 (2010)
4. Baral, S.C., Karki, D.K., Newell, J.N.: Causes of stigma and discrimination associated with tuberculosis in Nepal: a qualitative study. BMC Public Health **7**(1), 211 (2007)

5. World Health Organization: Tuberculosis prevalence surveys: a handbook. Report No.: WHO/HTM/TB/2010.17. World Health Organization, Geneva (2011)
6. World Health Organization: Chest radiography in tuberculosis detection: summary of current WHO recommendations and guidance on programmatic approaches (No. WHO/HTM/TB/2016.20). World Health Organization (2016)
7. Lakhani, P., Sundaram, B.: Deep learning at chest radiography: automated classification of pulmonary tuberculosis by using convolutional neural networks. Radiology **284**(2), 574–582 (2017)
8. Pedrazzoli, D., Lalli, M., Boccia, D., Houben, R., Kranzer, K.: Can tuberculosis patients in resource-constrained settings afford chest radiography? Eur. Respir. J. **49**(3), 1601877 (2017)
9. Lönnroth, K., Migliori, G.B., Abubakar, I., et al.: Towards tuberculosis elimination: an action framework for low-incidence countries. Eur. Respir. J. **45**, 928–952 (2015)
10. Leibstein, J.M., Nel, A.L.: Detecting tuberculosis in chest radiographs using image processing techniques. University of Johannesburg (2006)
11. Rohilla, A., Hooda, R., Mittal, A.: TB detection in chest radiograph using deep learning architecture. In: Proceeding of 5th International Conference on Emerging Trends in Engineering, Technology, Science and Management (ICETETSM-2017), pp. 136–147 (2017)
12. Hwang, S., Kim, H.E., Jeong, J., Kim, H.J.: A novel approach for tuberculosis screening based on deep convolutional neural networks. In: Computer-Aided Diagnosis, vol. 9785, p. 97852W. International Society for Optics and Photonics (2016)
13. Liu, C., et al.: TX-CNN: detecting tuberculosis in chest X-ray images using convolutional neural network. In: IEEE International Conference on Image Processing (ICIP), pp. 2314–2318 (2017)
14. Konstantinos, A.: Testing for tuberculosis. Aust. Prescr. **33**(1), 12–18 (2010)
15. Srinivas, M., Roy, D., Mohan, C.K.: Discriminative feature extraction from X-ray images using deep convolutional neural networks. In: IEEE International Conference on Acoustics, Speech and Signal Processing (ICASSP), pp. 917–921 (2016)
16. Hooda, R., Sofat, S., Kaur, S., Mittal, A., Meriaudeau, F.: Deep-learning: a potential method for tuberculosis detection using chest radiography. In: IEEE International Conference on Signal and Image Processing Applications (ICSIPA), pp. 497–502 (2017)
17. Jaeger, S., Karargyris, A., Antani, S., Thoma, G.: Detecting tuberculosis in radiographs using combined lung masks. In: Annual International Conference of the IEEE Engineering in Medicine and Biology Society, pp. 4978–4981 (2012)
18. Karargyris, A., et al.: Combination of texture and shape features to detect pulmonary abnormalities in digital chest X-rays. Int. J. Comput. Assist. Radiol. Surg. **11**(1), 99–106 (2015). https://doi.org/10.1007/s11548-015-1242-x
19. Van Ginneken, B., Katsuragawa, S., ter Haar Romeny, B.M., Doi, K., Viergever, M.A.: Automatic detection of abnormalities in chest radiographs using local texture analysis. IEEE Trans. Med. Imaging **21**(2), 139–149 (2002)
20. Shen, R., Cheng, I., Basu, A.: A hybrid knowledgeguided detection technique for screening of infectious pulmonary tuberculosis from chest radiographs. IEEE Trans. Biomed. Eng. **57**(11), 2646–2656 (2010)
21. Hogeweg, L., Mol, C., de Jong, P.A., Dawson, R., Ayles, H., van Ginneken, B.: Fusion of local and global detection systems to detect tuberculosis in chest radiographs. In: Jiang, T., Navab, N., Pluim, J.P.W., Viergever, M.A. (eds.) MICCAI 2010. LNCS, vol. 6363, pp. 650–657. Springer, Heidelberg (2010). https://doi.org/10.1007/978-3-642-15711-0_81

22. Shorten, C., Khoshgoftaar, T.M.: A survey on image data augmentation for deep learning. J. Big Data **6**(1) (2019). Article number. 60. https://doi.org/10.1186/s40537-019-0197-0
23. National Center for Biomedical Communications. http://lhncbc.nlm.nih.gov

Semantic Segmentation of Diabetic Foot Ulcer Images: Dealing with Small Dataset in DL Approaches

Niri Rania[1($\boxtimes$)], Hassan Douzi[1], Lucas Yves[2], and Treuillet Sylvie[2]

[1] IRF-SIC Laboratory, Ibn Zohr University, Agadir, Morocco
rania.niri@edu.uiz.ac.ma
[2] PRISME Laboratory, Orléans University, Orléans, France

Abstract. Foot ulceration is the most common complication of diabetes and represents a major health problem all over the world. If these ulcers are not adequately treated in an early stage, they may lead to lower limb amputation. Considering the low-cost and prevalence of smartphones with a high-resolution camera, Diabetic Foot Ulcer (DFU) healing assessment by image analysis became an attractive option to help clinicians for a more accurate and objective management of the ulcer. In this work, we performed DFU segmentation using Deep Learning methods for semantic segmentation. Our aim was to find an accurate fully convolutional neural network suitable to our small database. Three different fully convolutional networks have been tested to perform the ulcer area segmentation. The U-Net network obtained a Dice Similarity Coefficient of 97.25% and an intersection over union index of 94.86%. These preliminary results demonstrate the power of fully convolutional neural networks in diabetic foot ulcer segmentation using a limited number of training samples.

Keywords: Diabetic Foot Ulcer (DFU) · Medical images segmentation · Deep learning · Fully convolutional networks · U-Net · Data augmentation

1 Introduction

Type 2 diabetes mellitus can seriously damage several body's organs over time. Diabetic foot ulceration is the most serious diabetes-related complication [1], and represents a major health problem. In a diabetic patient, this pathology is due to neuropathy, infection or peripheral arterial disease of the lower limb resulting in the formation of skin ulcers and may lead to subsequent lower limb amputation. It is estimated that a lower limb or part of a lower limb is lost somewhere in the world every 30 s as a consequence of diabetes [2]. Diabetic foot ulcers (DFU) have a negative impact on patient's quality of life and result in an important social and economic burden on the public health system. The early prevention and detection of diabetic foot ulcer and the use of an adequate treatment are the key to the management of diabetic foot and to prevent foot amputations.

The management of diabetic foot ulcers requires a prolonged assessment process and patients need frequent clinical evaluation to check regularly their ulcers. Therefore, ulcers monitoring is primordial to improve the healing rate and speed and to select an

© Springer Nature Switzerland AG 2020
A. El Moataz et al. (Eds.): ICISP 2020, LNCS 12119, pp. 162–169, 2020.
https://doi.org/10.1007/978-3-030-51935-3_17

efficient treatment. In current practice, DFU assessment is typically based on visual examination. Clinicians evaluate the wound healing status by manual measurements of the ulcer, including length, width, surface area and volume. There are limited manual methods for the assessment of diabetic foot lesions. Ulcer perimeters (length and width) are measured by using a simple ruler, surface area is approximated by tracing the wound outline on a transparent sheet and volume is obtained by filling the ulcer with a physiological liquid [3]. Moreover, these methods are uncomfortable, sometimes painful and are in direct contact with the ulcer bed which can carry high risk of infection and contamination [4]. In another hand the current traditional methods of wound healing assessment are often rely on the subjective diagnosis of the expert and are time consuming. Therefore, an automatic assessment tool is needed to assist clinicians for a more accurate management and optimal diabetic foot ulcer care.

Nowadays, the use of imaging technology for automatic DFU/wound assessment and measurements has increased considerably to become a common practice. Following this trend, many research works have started to perform wound assessment in medical environment using imaging devices [5]. The major advantages of using digital cameras or smartphones is that photography does not require contact with the skin and it also can help not only to measure the ulcer area but also to analysis the different types of tissue inside the wound bed. DFU assessment will be more objective, accurate and less time consuming for health professionals and patients. Therefore, these methods require the use of image processing and computer vision techniques for image analysis.

Several approaches based on Machine Learning (ML) techniques have been experimented on wound and DFU segmentation [6–9]. Mainly, these works performed the segmentation task using traditional ML algorithms such as SVM classifiers after manual feature extraction. The majority of these methods require the extraction of texture and color descriptors from images such as HOG, SIFT, LBP etc. These descriptors may be affected by light conditions, image resolutions and also skin shape and shades. Thus, the traditional ML methods are not robust due to their reliance on the handcrafted extracted features.

In contrast, deep learning (DL) methods do not require manual feature extraction, as this process is done automatically during the training phase. After the successful impact of these methods on various fields in science during the past few years, nowadays, DL approaches are widely used for many different medical image analysis tasks [10]. Recent approaches applied to wound/DFU use deep neural networks to segment or extract feature descriptors from wound images. Wang et al. [11] proposed a wound segmentation technique in an end-to-end style including wound segmentation and healing progress measurement. The research work is based on an encoder-decoder architecture to perform wound segmentation and was trained on a dataset containing 650 images. Goyal et al. [12] implemented a novel fast DL architecture for DFU classification called DFUNet which classified foot lesions into two classes (normal skin and abnormal one). In this work, they used a database of 397 images. In another work [13], Goyal et al. proposed a two-tier transfer learning method using fully convolutional networks (FCNs) [14] on a larger database of 705 images.

The methods based on Deep learning require massive amount of labelled training data. In biomedical field, images collection is not easily accessible and image annotation is a critical task. Since that the training set should be annotated only by medical experts, which is very time consuming, the training set is small. Our approach provides the use of DL methods for diabetic foot ulcer segmentation using a small database.

This research work is part of the STANDUP project http://www.standupproject.eu/ which aim to prevent diabetic foot ulceration at an early stage and also to help clinicians to monitor the ulcer healing over time using a portable and cost-effective system. This system is composed of a smartphone and a small thermal camera called FlirOne in order to combine color and thermal information for an accurate assessment. Temperature indicators from the thermal image can help to detect tissue infection and inflammation. The final system can be used as a low-cost device to help and assist health care professionals during the examination of DFU by automatically calculating the ulcer area after segmentation and to carry out wound tissue analysis based on color and temperature. Also, it could be used as self-management tool to engage patients with foot ulceration in their own wound care routine. For an optimal DFU diagnosis, this application can be used as telemedicine tool to transmit information about the healing status between clinicians and patients located far from health centers to avoid supplementary costs due to transportation, ambulances, etc.

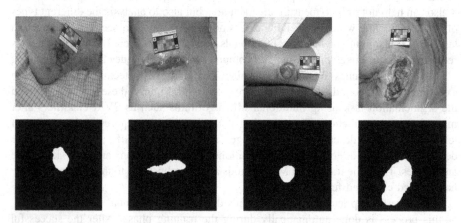

Fig. 1. Sample images from the CW training set illustration of high-resolution CW and the corresponding ground truth masks.

2   Methodology

2.1   DFU Database

In the present work, chronic wounds (CW) images from ESCALE database [15] were used as training set. This database contains 92 high resolution chronic wounds images acquired in several medical centers in France using different commercial digital cameras. A patch was included in the field of view as a scale factor for dimensional measurements and also to estimate the white balance for color calibration [16]. Considering the purpose of this work which is the segmentation of diabetic foot skin lesion, the method was trained on different chronic wounds images and tested only on DFU images. Test data set was obtained from diabetic patients during clinical exams of DFU at CHRO "Centre Hospitalier Regional d'Orleans", located in Orléans, France. The images were acquired free-handedly using a smartphone camera and without any strict protocol. The whole database has been annotated by experts into two classes i.e. ulcer area and healthy skin. The training set images are in RGB while their ground truth masks are binary (see Fig. 1).

2.2 Training Convolutional Neural Networks

Our objective is to classify DFU images at a pixel level classification into ulcer and non-ulcer classes. We performed the segmentation by using fully convolutional networks. These models are the state of the art of semantic segmentation which allows a pixelwise prediction [14]. The advantage of using FCNs is that the entire image can be directly used as an input to the network. Therefore, a class label will be assigned for each pixel of the image which leads to an efficient segmentation of the ulcer area. In our case, the ulcer area will be represented by white pixels, healthy skin and background by black pixels. Searching for a suitable model to our small database, we focused on U-Net proposed by Ronneberger et al. [17]. This neural network model was specially created to perform semantic segmentation of small datasets. It was designed for biomedical images, and its specific symmetric architecture produces precise segmentation using few images for training. Figure 2 shows the architecture of the U-Net.

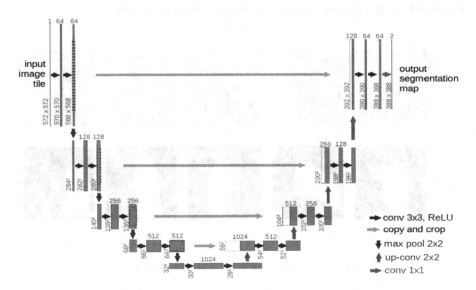

Fig. 2. U-Net architecture based on the paper by Olaf Ronneberger et al.

Since the images were taken in different hospital sites and with different cameras, they had different resolutions. Therefore, all the images were resized to a resolution of 512×512 to perform the U-net training. After that, the database was divided into two sets, resulting in a training set consisting of 92 CW images and validation set consisting of 22 DFU images. To perform our training, we used the Keras framework with TensorFlow backend [18]. The parameters of the kernels were optimized by using the adaptive moment estimation (Adam) with a learning rate set at 0.0001.

Two methods have been chosen to evaluate diabetic foot ulcer segmentation using the proposed approach based on U-Net, which are V-Net [19] and SegNet [20]. As an improvement of U-Net architecture, Milletari et al. proposed an architecture called V-Net to segment prostate MRI volumes. The tested model is the same architecture of the original V-Net but with modified convolution kernel to perform 2D image segmentation. SegNet network is a parallel encoder-decoder based architecture, for every encoder layer there is a corresponding decoder. Unlike U-Net, this network does not contain any skip connections. SegNet uses a pre-trained VGG16 as encoder. We used different image sizes to fit the original input size of each architecture.

2.3   Data Augmentation

The number of samples in our database is limited and not sufficient to feed a deep learning neural network. Therefore, due to the small size of our database, data augmentation is required to increase the size of training set and to avoid overfitting [21]. Different techniques have been used to randomly deform the input image and correspondent ground truth segmentation map. Each image was horizontally flipped, rotated, translated and zoomed (see Fig. 3). After data augmentation, the database size has been increased by four and contains now 368 images instead of 92 images.

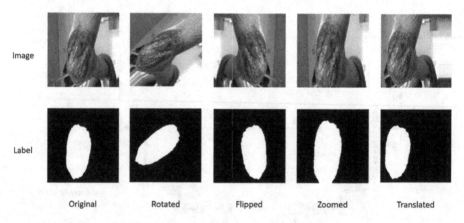

Fig. 3. An example of data augmentation results.

3   Results

3.1   Evaluation Metrics

In order to evaluate the segmentation results quantitatively, we choose *Intersection Over Union index (IoU)*, and *Dice Similarity Coefficient (DSC)* [22]. They are the most used metrics in the field of medical image segmentation. The DSC coefficient assesses the similarity between the predicted segmented ulcer region bounded by the three deep learning networks and manually labeled ground truth given by experts. These two-evaluation metrics are defined respectively by (Eq. 1) and (Eq. 2):

$$IoU = \frac{TP}{TP + FP + FN} \tag{1}$$

$$DSC = \frac{2TP}{2TP + FP + FN} \tag{2}$$

where TP stands for the true positive samples, FP for the false positive samples and FN for the false negative ones.

3.2 Experimental Results

As mentioned before, the purpose of our research work was to segment DFU images into ulcer and non-ulcer classes. Due to computational resources limitations and memory constraint, the U-net model was trained for just 5 epochs. The network did surprisingly well, after 5 epochs the calculated accuracy was about 95%. Overall, the qualitative results shown in Fig. 4 reveal the power of the network to generate a mask similar to the ground truth. As we can clearly notice that after overlaying the predicted mask on the original image of the ulcer, the network segment correctly the ulcer area with a high precision.

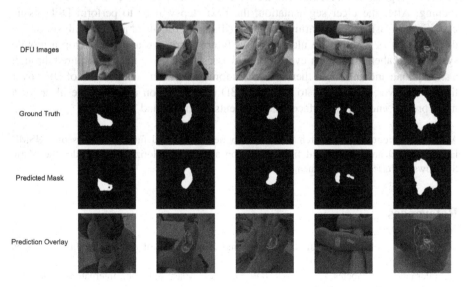

Fig. 4. An illustration of the predicted segmentation map of the proposed network on some images from our DFU validation set.

The quantitative evaluation of the segmentation results was performed on the 22 images of our DFU validation database. Table 1 compares the results of these three methods using test accuracy, mean IoU score and mean DSC coefficient. These results show that U-Net and V-Net methods give both good results compared to the SegNet network. We clearly notice that the best segmentation results are given by the U-Net architecture.

Table 1. DFU Segmentation results. Test Accuracy, Intersection over Union (IoU) and Dice coefficient (DSC) in %.

Method	Accuracy	IoU	DSC
U-Net	**94.96**	**94.86**	**97.25**
V-Net	92.52	92.17	95.74
SegNet	88.94	87.87	93.45

4 Conclusion

To summarize, in this preliminary work our objective was to find an accurate deep learning architecture for diabetic foot ulcer segmentation using a limited number of samples, as data collection and expert labeling are very expensive in the field of medical imaging. The current training database contains only 92 images of different chronic wounds images including foot pathologies. Different methods were tested and evaluated using 22 DFU images. The results show that the U-Net network produces the best DFU segmentation results with a DSC coefficient of 97% after only five epochs of training. After the ulcer segmentation, the next step will be to perform DFU tissue classification using deep learning to help clinicians for an objective analysis of the different tissues withing the ulcer bed. As future work, we aim to develop a mobile system for diabetic foot ulcer evaluation and assessment using images acquired through a smartphone and an add-on thermal camera for a more accurate diagnosis of DFU over time. Moreover, we attempt to perform a 3D reconstruction model of the ulcer for a more precise and robust surface measurements of the wound area.

Acknowledgment. This research work is supported by internal funding sources of PRISME laboratory, Orleans, France and from the European Union's Horizon 2020 under the Marie Sklodowska-Curie grant agreement No 777661.

References

1. American diabetes association: diagnosis and classification of diabetes mellitus. Diabetes Care Suppl 1, 81–90 (2014)
2. International Diabetes Federation: IDF Diabetes Atlas, 8th edn. International Diabetes Federation, Brussels (2017)
3. Humbert, P., Meaume, S., Gharbi, T.: Wound healing assessment. Phlebolymphology **47**, 312–319 (2014)

4. Keast, D.H., et al.: Measure: a proposed assessment framework for developing best practice recommendations for wound assessment. Wound Repair Regenerat. **12**(1), 1–17 (2004)
5. Mohafez, H., Ahmad, S.A., Roohi, S.A., Hadizadeh, M.: Wound healing assessment using digital photography: a review. JBEIM, 3(5) (2016). ISSN: 2055-1266
6. Kolesnik, M., Fexa, A.: Segmentation of wounds in the combined color-texture feature space. In: Medical Imaging 2004: Image Processing, vol. 5370, pp. 549–556. International Society for Optics and Photonics (2004). https://doi.org/10.1117/12.535041
7. Kolesnik, M., Fexa, A.: How robust is the SVM wound segmentation. In: Proceedings of the 7th Nordic Signal Processing Symposium, pp. 50–53 (2006)
8. Wannous, H., Lucas, Y., Treuillet, S.: Efficient SVM classifier based on color and texture region features for wound tissue images. In: Medical Imaging 2008: Computer-Aided Diagnosis, vol. 6915, p. 69152T. International Society for Optics and Photonics (2008). https://doi.org/10.1117/12.770339
9. Wang, L., et al.: Area determination of diabetic foot ulcer images using a cascaded two-stage SVM-based classification. IEEE Trans. Biomed. Eng. **64**(9), 2098–2109 (2017)
10. Litjens, G., et al.: A survey on deep learning in medical image analysis. Med. Image Anal. **42**, 60–88 (2017)
11. Wang, C., et al.: A unified framework for automatic wound segmentation and analysis with deep convolutional neural networks. In: 2015 37th Annual International Conference of the IEEE Engineering in Medicine and Biology Society (EMBC) (2015)
12. Goyal, M., Reeves, N.D., Davison, A.K., Rajbhandari, S., Spragg, J., Yap, M.H.: DFUNet: convolutional neural networks for diabetic foot ulcer classification. In: IEEE Transactions on Emerging Topics in Computational Intelligence, pp. 1–12. IEEE (2018)
13. Goyal, M., Yap, M.H., Reeves, N.D., Rajbhandari, S., Spragg, J.: Fully convolutional networks for diabetic foot ulcer segmentation. In: 2017 IEEE International Conference on Systems, Man, and Cybernetics (SMC) (2017)
14. Shelhamer, E., Long, J., Darrell, T.: Fully convolutional networks for semantic segmentation. In: IEEE Transactions on Pattern Analysis and Machine Intelligence, vol. 39, no. 4 (2017)
15. Lucas, Y., Treuillet, S., Albouy, B., Wannous, H., Pichaud, J.C.: 3D and color wound assessment using a simple digital camera. In: 9th Meeting of the European Pressure Ulcer Advisory Panel, Berlin (2006)
16. Wannous, H., et al.: Improving color correction across camera and illumination changes by contextual sample selection. J. Electron. Imag. **21**(2), 023015 (2012)
17. Ronneberger, O., Fischer, P., Brox, T.: U-net: convolutional networks for biomedical image segmentation. In: Navab, N., Hornegger, J., Wells, W.M., Frangi, A.F. (eds.) MICCAI 2015. LNCS, vol. 9351, pp. 234–241. Springer, Cham (2015). https://doi.org/10.1007/978-3-319-24574-4_28
18. Kingma, D.P., Ba, J.: Adam: a method for stochastic optimization. CoRR, vol. abs/1412.6980 (2014)
19. Milletari, F., Navab, N., Ahmadi, S.A.: V-net: fully convolutional neural networks for volumetric medical image segmentation. In: International Conference on 3D Imaging, Modeling, Processing, Visualization and Transmission (3DIMPVT) (2016)
20. Badrinarayanan, V., Kendall, A., Cipolla, R.: Segnet: a deep convolutional encoder-decoder architecture for image segmentation. IEEE Trans. Pattern Anal. **39**(12), 2481–2495 (2017)
21. Perez, L., Wang, J.: The Effectiveness of Data Augmentation in Image Classification using Deep Learning (2017). arXiv preprint arXiv:1712.04621
22. Lee, R.: Dice: measures of the amount of ecologic association between species. Ecology **26**(3), 297–302 (1945)

DermoNet: A Computer-Aided Diagnosis System for Dermoscopic Disease Recognition

Ibtissam Bakkouri$^{(\boxtimes)}$ and Karim Afdel

LabSIV, Department of Computer Science, Faculty of Science,
Ibn Zohr University, Agadir, Morocco
ibtissam.bakkouri@gmail.com, k.afdel@uiz.ac.ma

Abstract. The research of skin lesion diseases is currently one of the hottest topics in the medical research fields, and has gained a lot of attention on the last few years. However, the existing skin lesion methods are mainly relying on conventional Convolutional Neural Network (CNN) and the performance of skin lesion recognition is far from satisfactory. Therefore, to overcome the aforementioned drawbacks of traditional methods, we propose a novel Computer-Aided Diagnosis (CAD) system, named DermoNet, based on Multi-Scale Feature Level (MSFL) blocks and Multi-Level Feature Fusion (MLFF). Further, the DermoNet approach yields a significant enhancement in terms of dealing with the challenge of small training data sizes in the dermoscopic domain and avoiding high similarity between classes and overfitting issue. Extensive experiments are conducted on the public dermoscopic dataset, and the results demonstrate that DermoNet outperforms the state-of-the-art approaches. Hence, DermoNet can achieve an excellent diagnostic efficiency in the auxiliary diagnosis of skin lesions.

Keywords: Dermoscopic diseases · Dermoscopic pattern recognition · Multi-layer feature fusion · Multi-scale features · Multi-class classification

1 Introduction

Nowadays, skin disease is one of the most difficult diseases to cure [1]. It is well-known that more early diagnosis of skin lesions is a crucial issue for the dermatologists [2]. Over the last few years, skin lesion recognition in dermoscopy images based on deep learning approach has attracted the attention of many researchers. Authors of [4] presented a computer aided diagnosis system to classify abnormal skin lesions in dermoscopic image into melanocytic nevus, seborrheic keratosis, basal cell carcinoma or psoriasis. The proposed method was evaluated on clinical database originated from the dermatology department of Peking Union Medical College Hospital, and the feature extraction and classification were carried out by GoogleNet Inception-V3 based on domain expert knowledge.

© Springer Nature Switzerland AG 2020
A. El Moataz et al. (Eds.): ICISP 2020, LNCS 12119, pp. 170–177, 2020.
https://doi.org/10.1007/978-3-030-51935-3_18

Inspired by the layers in the deep networks [5], authors of [6] applied ResNet-50 to extract discriminative features and Support Vector Machine (SVM) with a chi-squared kernel to classify melanoma images. This method was evaluated on the International Skin Imaging Collaboration (ISIC) 2016 Skin lesion challenge dataset, and it reached promising performance. In [7], an error-correcting output coding classifier combined with SVM for the multi-class classification scenario was introduced. The authors investigated the pre-trained AlexNet convolutional neural network model as feature extractor. VGG-16 architecture with fine-tuning technique was suggested to recognize melanoma abnormality in [8]. The proposed method was evaluated on ISIC 2017 challenge dataset, yielding satisfactory results. In this context, the similar approach based on convolutional multi-layer feature fusion network and continuous fine-tuning of CNNs using VGG-16, ResNet-18 and DenseNet-121 architectures were suggested by authors of [9] in order to classify seven skin diseases. The proposed method was evaluated on the testing subset of the HAM10000, and it reached promising results. A convolutional neural-adaptive network for melanoma recognition was proposed by authors [10]. The authors tried to adapt the pre-trained low-level weights on dermoscopic dataset using AlexNet architecture and fine-tuning technique. The images were obtained from ISIC dataset, and the proposed system provided efficient results.

Despite the high achievement of deep CNN for skin lesion diagnosis, it needs some improvement concerning the multi-class classification task, and overcome many problems that confronted most of previous algorithms such as semantic feature ambiguity and overfitting issue. Therefore, to overcome the aforementioned drawbacks of traditional methods, we propose a new mechanism based on CNN for skin lesion recognition, called DermoNet, in order to significantly improve the effectiveness and efficiency of computer-aided diagnosis system. The main contribution, novelty and characteristic of DermoNet are summarized as follows.

- An effective multi-scale feature architecture, called MSFL, is introduced to formulate a robust dermoscopic lesion representation for coping with the complex morphological structures of abnormal skin tissues.
- A new mechanism of feature fusion in CNNs, named MLFF, is proposed and applied for skin lesion classification to overcome the problem of semantic feature ambiguity and high inter-class visual similarities between the classes, and deal with the problem of having small set of dermoscopic data samples.
- Assesment of performance of the proposed methodology on test subset against the backbone as MSFL network without MLFF architecture. five statistical measurements were used for comparison of performances including accuracy, sensitivity, specificity, precision and F1-Score metrics.

The proposed framework has been validated on the public Human Against Machine with 10000 training images (HAM10000) dataset [3], which is designed for skin lesion diagnosis, and has been compared to state-of-the-art skin pattern classification approaches. The performance evaluation analysis shows the potential clinical value of the proposed framework. To the best of our knowledge, no prior work exists on skin lesion characterization based on convolutional multi-scale feature fusion network that is investigated here.

The remainder of this paper is organized as follows: In Sect. 2, we detail the proposed approach. Then, we present the results of experiments realized on HAM10000 dataset in Sect. 3. Finally, we conclude this paper in Sect. 4.

2   Proposed DermoNet Framework

The main contribution of this study is to evaluate a novel CAD approach, called DermoNet, based on multi-scale CNN features combined with multi-level feature fusion for classifying the skin structures as Melanoma (MEL), Melanocytic Nevus (NV), Basal Cell Carcinoma (BCC), Actinic Keratosis (AKIEC), Benign Keratosis (BKL), Dermatofibroma (DF), or Vascular Lesion (VASC). Our proposed methodology based CAD system was developed in several stages. First, the representative square regions with size of 64×64 pixels were selected from the digitized dermoscopic images and enhanced by the pre-processing techniques. Then, we built MSFL blocks to extract the most significant features. Inspired by feature fusion, to resolve the challenge of dealing with limited availability of the training dermoscopic samples, improve the performance of our model and avoid overfitting, the multi-level feature fusion was proposed and the extracted features from the last layer of each MSFL block were preserved by applying MLFF architecture to avoid missing useful information during convolution. Finally, the multi-class classification was performed using two Fully Connected layers (FC-1 and FC-2). An overview of the proposed DermoNet approach is given in Fig. 1.

2.1   Pre-processing of Dermoscopic Data

As indicated in our previous work [9], pre-processing of the digital dermoscopic data is performed in four stages, including data size normalization technique, hair artifact removal, intensity correction process and class balancing to facilitate application of DermoNet algorithm. The determinant of the Hessian combined with pyramidal REDUCE decomposition was carried out to normalize data size. To reduce the effects of hair artifact and enhance dermoscopic image brightness, we employed a computerized method as described in [11] and local adaptive bi-histogram equalization [12], respectively. As the dermoscopic image volumes of different categories vary widely, data augmentation is needed to balance the image volumes of different classes. In this paper, we balanced the dataset by augmenting the data for the minority classes, applying geometric transformation techniques as indicated in [9]. Finally, after data preparation process, the number of dermoscopic images is expanded at least by a factor of 20.

2.2   Multi-Scale Feature Level (MSFL) Extraction

The suggested multi-scale CNN is quite different from previous conventional CNN [13–15]. Rather than using a stack of consecutive convolutional layers in

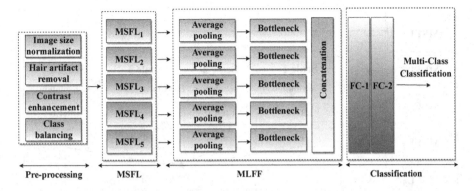

Fig. 1. Flow chart of the proposed DermoNet method. The multi-scale feature architecture consists of five MSFL blocks. The proposed MLFF architecture is composed by three types of layer: average pooling, bottleneck and concatenation layers.

each block, we used a multi-scale convolutional layer per block with a relatively strong depth throughout the whole net, introducing a merge layer to maintain the information flow and ensure that no feature is lost along the network. Multi-scale layers allow for preserving details of different sizes from the input images. Moreover, we used Leaky Rectified Linear Unit (LReLU) activation function to avoid the zero gradients in negative part and increase the bias shift. Batch Normalization (BN) was also used to fight the internal covariate shift problem. The proposed multi-scale CNN method was evaluated in five blocks denoted by $MSFL_n$, where n is the number of blocks. As shown in Fig. 2, in each MSFL block, four scales of convolutional layers with kernel shape of $[3 \times 3] \times 128$, $[5 \times 5] \times 128$, $[7 \times 7] \times 128$ and $[9 \times 9] \times 128$, LReLU activation function, BN and element-wise multiplication layers are involved. The multi-scale feature blocks were randomly initialized using the Xavier weight initialization algorithm and trained with the dermoscopic dataset.

2.3 Multi-Level Feature Fusion (MLFF)

Recent research in feature fusion in CNN has a great effect on pattern recognition which can cope with the problem of semantic feature ambiguity and high inter-class visual similarities between the classes and deal with the problem of having small set of dermoscopic data samples with very short calculation times and low computational costs. The proposed MLFF stores hierarchical multi-scale features where each level of MLFF preserves the output of the last layer of the MSFL block after applying average pooling operation with stride S = 2 and kernel size 2×2 and bottleneck layer with ratio $\alpha = 0.125$. The fusion layer is used to ensure maximum information flow between layers in the network. The MLFF method is first applied to skin lesion recognition problem and the model provides a solution to the most complex structures, like melanoma and basal cell carcinoma, perfectly. It can not only solve the overfitting issue like others

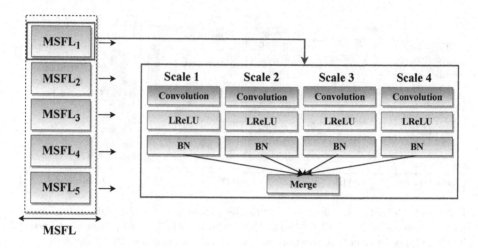

Fig. 2. Structure of a basic multi-scale feature block.

conventional methods, but also has a very simple structure which makes our model much easier to train.

3   Experimental Results

3.1   Data Acquisition

In our experiments, we evaluated the performance of our proposed system on HAM10000 dataset which is the official ISIC 2018 challenge dataset hosted by the annual MICCAI conference in Granada, Spain, but is also available to research groups who do not participate in the challenge [3]. This dataset was collected over a period of 20 years from two different sites, the department of dermatology at the medical university of Vienna, Austria and the skin cancer practice of Cliff Rosendahl in Queensland, Australia. It includes seven generic classes: MEL, NV, BCC, AKIEC, BKL, DF and VASC. The dataset consists of 10015 dermoscopic images with size of 600×450 showing skin lesions which have been diagnosed based on expert consensus, serial imaging and histopathology. The distribution is completely imbalanced, and then we applied the geometric transformation techniques as described in [9] to overcome class prevalence. After balancing process, we obtained 2000 images per class. The typical data set proportions are 60% for training, 20% for validation and 20% for testing [9].

3.2   Experiment Settings and Implementation

The proposed system was trained using Adam algorithm with learning rate $\alpha = 10^{-3}$, first moment-decay $\beta_1 = 0.9$ and second moment-decay $\beta_2 = 0.999$ with Xavier algorithm for weight initialization and LReLU activation function inserted after each convolutional layer. The softmax activation function was combined with the cross-entropy loss for training the network. The optimization ran for 30 epochs with mini batch size of 32 samples. To help our models generalize better in these circumstances, prevent over-fitting, we introduce dropout layer after the first fully connected layer with factor $\rho = 0.5$. The proposed method has been developed using the Keras library with python wrapper, and Compute Unified Device Architecture (CUDA) enabled parallel computing platform to access the computational resources of Graphics Processing Unit (GPU). The available hardware, used for training, is a PC with a Core i7 CPU, 8 GB RAM and a single NVIDIA GeForce GTX 1060 with 6 GB memory. Training dermoscopic images using the convolutional multi-scale feature fusion algorithm took about 6 h , 48 min and 43.19 s. For testing, on average, it takes 3.56 s in GPU and 5.37 s in CPU per image.

3.3   Classification Results

The evaluation of the performance of our proposed system was fulfilled with the statistical quality metrics such as accuracy (Acc), sensitivity (Sen), specificity (Spe), precision (Pre) and F1-Score (FSc). As shown in Table 1, we compare the results of the proposed method (MSFL+MLFF) with regard to MSFL without multi-level fusion based on the number of MSFL blocks ($MSFL_n$) using all five quality metrics. As indicated in Table 1, considering $MSFL_6$ without MLFF as a base model, $MSFL_6$ achieved the accuracy of 96.72%, sensitivity of 92.14%, specificity of 95.97%, precision of 91.85% and F1-Score of 91.99%. Compared to these results, $MSFL_5$+MLFF made improvements by 1.61%, 2.07%, 2.97%, 3.32% and 2.69% in terms of accuracy, sensitivity, specificity, precision and F1-Score, respectively. Noticeably, $MSFL_5$+MLFF shows the highest improvement for skin lesion classification, outperforming the two competing methods: MSFL at all levels and four instances of MSFL+MLFF including $MSFL_2$+MLFF, $MSFL_3$+MLFF, $MSFL_4$+MLFF and $MSFL_6$+MLFF.

3.4   Comparison with State of the Art

Since research on HAM10000 dermoscopic dataset is still relatively scarce, we tried to implement some existing algorithms used for skin lesion recognition and test them on our dermoscopic dataset. Then, the [7,9,10,16,17] were implemented and tested on HAM10000 dataset. The visual comparison given in Table 2 demonstrates that our best performing network outperforms the pre-trained AlexNet with an error-correcting output coding classifier combined with support vector machine [7], convolutional neural-adaptive networks [10], pre-trained GoogleNet inception-V3 [16], pre-trained very deep residual networks with 152 layers [17] and convolutional multi-layer feature fusion network with continuous fine-tuning of CNNs using VGG-16, ResNet-18 and DenseNet-121 architectures [9].

Table 1. Quantitative result of different MSFL blocks with and without MLFF on testing subset in terms of accuracy, sensitivity, specificity, precision and F1-Score metrics.

Methods	$MSFL_n$	Acc. (%)	Sen. (%)	Spe. (%)	Pre. (%)	FSc. (%)
MSFL	$MSFL_2$	86.37	81.49	85.19	80.66	81.07
	$MSFL_3$	89.11	83.70	82.56	80.01	81.81
	$MSFL_4$	92.89	89.36	92.84	90.25	89.80
	$MSFL_5$	96.34	91.68	95.76	92.12	91.89
	$MSFL_6$	**96.72**	**92.14**	**95.97**	**91.85**	**91.99**
MSFL + MLFF	$MSFL_2$	90.71	87.22	90.89	91.04	89.08
	$MSFL_3$	92.95	89.56	92.18	90.65	90.10
	$MSFL_4$	96.48	91.80	95.92	91.39	91.59
	$MSFL_5$	**98.33**	**94.21**	**98.94**	**95.17**	**94.68**
	$MSFL_6$	98.00	93.54	98.71	94.63	94.08

Table 2. Comparison between the proposed method and other algorithms using the HAM10000 dataset.

Methods	Acc. (%)	Sen. (%)	Spe. (%)	Pre. (%)	FSc. (%)
Method [16]	91.56	87.26	92.48	92.52	89.81
Method [7]	83.91	81.47	84.20	81.76	81.61
Method [10]	84.28	79.34	84.79	84.01	81.60
Method [17]	86.77	84.91	86.67	85.37	85.13
Method [9]	98.09	93.35	98.88	93.36	93.35
Ours	**98.33**	**94.21**	**98.94**	**95.17**	**94.68**

4   Conclusion

In this work, we have presented a novel CAD system, DermoNet, for practical skin lesion recognition. It is developed using the combination between MSFL and MLFF architectures. The aim of MSFL is to formulate a robust dermoscopic lesion representation for coping with the complex morphological structures of abnormal skin tissues, while the MLFF was applied to overcome the problem of semantic feature ambiguity, deal with the small set of dermoscopic data samples and avoid overfitting issue. We conducted extensive experiments on HAM10000 dataset, and the comparison of DermoNet model with existing methods shows that this adduced work outperforms the most representative state-of-the-art strategies. So, we can conclude that DermoNet may assist the dermatologists to make the final decision for the further treatment. In the future, we will continue our work in integrating useful features from an explainable CNN for other task recognition in medical imaging field.

References

1. Hollestein, L., Nijsten, T.: An insight into the global burden of skin diseases. J. Invest. Dermatol. **134**, 1499–1501 (2014)
2. Chen, M., Zhou, P., Wu, D., Hu, L., Hassan, M., Alamri, A.: AI-Skin: skin disease recognition based on self-learning and wide data collection through a closed-loop framework. Inf. Fusion **54**, 1–9 (2019)
3. Tschandl, P., Rosendahl, C., Kittler, H.: The HAM10000 dataset, a large collection of multi-source dermatoscopic images of common pigmented skin lesions. Sci. Data **5** (2018)
4. Zhang, X., Wang, S., Liu, J., Tao, C.: Towards improving diagnosis of skin diseases by combining deep neural network and human knowledge. BMC Med. Inform. Decis. Making **18** (2018)
5. He, K., Zhang, X., Ren, S., Sun, J.: Deep residual learning for image recognition. In: 2016 IEEE Conference on Computer Vision and Pattern Recognition (CVPR) (2016)
6. Yu, Z., et al.: Melanoma recognition in dermoscopy images via aggregated deep convolutional features. IEEE Trans. Biomed. Eng. **66**, 1006–1016 (2019)
7. Dorj, U.-O., Lee, K.-K., Choi, J.-Y., Lee, M.: The skin cancer classification using deep convolutional neural network. Multimed. Tools Appl. **77**(8), 9909–9924 (2018). https://doi.org/10.1007/s11042-018-5714-1
8. Thao, L., Quang, N.: Automatic skin lesion analysis towards melanoma detection. In: 2017 21st Asia Pacific Symposium on Intelligent and Evolutionary Systems (IES) (2017)
9. Bakkouri, I., Afdel, K.: Computer-aided diagnosis (CAD) system based on multilayer feature fusion network for skin lesion recognition in dermoscopy images. Multimed. Tools Appl. (2019). https://doi.org/10.1007/s11042-019-07988-1
10. Bakkouri, I., Afdel, K.: Convolutional neural-adaptive networks for melanoma recognition. In: Mansouri, A., El Moataz, A., Nouboud, F., Mammass, D. (eds.) ICISP 2018. LNCS, vol. 10884, pp. 453–460. Springer, Cham (2018). https://doi.org/10.1007/978-3-319-94211-7_49
11. Lee, T., Ng, V., Gallagher, R., Coldman, A., McLean, D.: Dullrazor®: a software approach to hair removal from images. Comput. Biol. Med. **27**, 533–543 (1997)
12. Tang, J., Mat Isa, N.: Adaptive image enhancement based on bi-histogram equalization with a clipping limit. Comput. Electr. Eng. **40**, 86–103 (2014)
13. Yong, L., Bo, Z.: An intrusion detection model based on multi-scale CNN. In: 2019 IEEE 3rd Information Technology, Networking, Electronic and Automation Control Conference (ITNEC) (2019)
14. Li, J., Zhang, R., Li, Y.: Multiscale convolutional neural network for the detection of built-up areas in high-resolution SAR images. In: 2016 IEEE International Geoscience and Remote Sensing Symposium (IGARSS) (2016)
15. Alhichri, H., Alajlan, N., Bazi, Y., Rabczuk, T.: Multi-scale convolutional neural network for remote sensing scene classification. In: 2018 IEEE International Conference on Electro/Information Technology (EIT) (2018)
16. Esteva, A., et al.: Dermatologist-level classification of skin cancer with deep neural networks. Nature **542**, 115–118 (2017)
17. Han, S., Kim, M., Lim, W., Park, G., Park, I., Chang, S.: Classification of the clinical images for benign and malignant cutaneous tumors using a deep learning algorithm. J. Invest. Dermatol. **138**, 1529–1538 (2018)

A New Method of Image Reconstruction for PET Using a Combined Regularization Algorithm

Abdelwahhab Boudjelal[1,2]([✉]), Abderrahim El Moataz[1], and Zoubeida Messali[2]

[1] Image Team, University of Caen Normandy and the ENSICAEN in the GREYC Laboratory, 6 Boulevard Maréchal Juin, 14050 Caen Cedex, France
{abdelwahhab.boudjelal,abderrahim.elmoataz}@unicaen.fr
[2] Electrical Engineering Laboratory LGE M'sila University, M'sila, Algeria
messalizoubeida@gmail.com

Abstract. Positron emission tomography (PET), is a medical imaging technique that provides functional information about physiological processes. The goal of PET is to reconstruct the distribution of the radioisotopes in the body by measuring the emitted photons. The computer methods are designed to solve the inverse problem known as "image reconstruction from projections." In this paper, an iterative image reconstruction algorithm ART was regularized by combining Tikhonov and total variation regularizations. In the first step, combined regularization algorithm of total variation and Tikhonov regularization was applied to the image obtained by ART algorithm in each iteration for background noise removal with preserving edges. The quality measurements and visual inspections show a significant improvement in image quality compared to other algorithms.

Keywords: Positron emission tomography · ART algorithm · SART algorithm · Total variation · Tikhonov

1 Introduction

Positron emission tomography (PET) [1,2] is a medical imaging modality that is used widely in clinical practice and scientific research as it provides quantitative and non-invasive information about biochemical and physiological processes in vivo. In PET [3], a small amount of a radioactive compound labeled with a radioisotope, called a radiotracer, is usually introduced into a patient's body through intravenous injection or inhalation, and then the spatial and sometimes also the temporal distribution of the radioisotope, the decay of which generates photons, is reconstructed from the photon measurements.

The emitted photons detected by the detectors are collected in set of projection data or sinogram. The aim of emission tomography is to reconstruct the spatial distribution of the radioisotope from the sinogram data by considering

A. El Moataz et al. (Eds.): ICISP 2020, LNCS 12119, pp. 178–185, 2020.
https://doi.org/10.1007/978-3-030-51935-3_19

geometrical factors, physical effects, and noise properties. The image reconstruction problem belongs to the class of inverse problems [4].

Reconstruction algorithms proposed in the literature can be divided into two classes: analytical and iterative.

The use of iterative image reconstruction algorithms [5–10] can circumvent all these shortcomings. Iterative reconstructions have the advantages of incorporating corrections for image-degrading factors to handle an incomplete, noisy, and dynamic data set more efficiently than analytic reconstruction techniques. The most widely used iterative algorithms in emission tomography are the ML-EM (maximum-likelihood expectation maximization) algorithm and its accelerated version OSEM (Ordered Subset EM). The ML-EM method was introduced by Dempster et al. in 1977 [11] and first applied to PET by Shepp and Vardi [12]. The algebraic reconstruction Technique (ART) [13], considered as an important class of iterative approaches [14], assume that the cross-sectional section consists of a set of unknown, and then establishes algebraic equations for the unknown in terms of Measured projection data.

The regularization techniques are generally divided into a projection method and a penalty method [15]. In this paper, we use the penalty method techniques which are the Tikhonov regularization [16,17], TV method [18,19] (Fig. 4). In this paper, we combine the ART algorithm TV + Tikhonov regularization techniques. The regularization of Tikhonov is the form of the L2-norm method of regularization on the data and the terms of regularization of the inverse problem.

2   Materials and Methods

2.1   The PET Imaging Model

The reconstruction from projections of PET images is a particularity of the general inverse problem of estimating the radioactive activity map related to a measurement p by

$$p_i = A_{i,j}x_j + n_i \tag{1}$$

In the process of ECT imaging reconstruction, x is the reconstructed image, p is the measurement of projection data, A is the system matrix whose component $A_{i,j}$ accounts for the probability of a photon emitted from pixel j being recorded into bin i and n_i is the random and scatter events that add a bias to each detector.

The aim of reconstructing a PET scan image is to provide a direct solution of x from the raw data collected during PET scanning p. However, PET data have an inherent stochastic character. There are uncertainties related to several aspects of PET physics, including the decay process, the effects of attenuation, the scattered coincidences and the additive random coincidences, n. To solve these problems, we propose a novel regularised ART algorithm for the image reconstruction combined with Tikhonov + TV regularization.

2.2   Algebraic Reconstruction Technique (ART)

The ART is sequential method, i.e., each equation is treated at a time, since each equation is dependent on the previous. The equation of ART [13] is given by

$$x_j^{(k+1)} = x_j^{(k)} + \alpha \frac{p_i - \sum_{n=1}^{N} A_{i,n} x_n^{(k+1)}}{\sum_{n=1}^{N} A_{i,n}^2} A_{i,j} \qquad (2)$$

where $x_j^{(k+1)}$ and $x_j^{(k)}$ are the current and the new estimates, respectively; $\sum_{n=1}^{N} A_{i,n} x_n^{(k+1)}$ is the sum weighted pixels along ray i; for the kth iteration; p_i is the measured projection for the ith ray, and α is the relaxation parameter.

2.3   The Proposed Regularization Method Combining Tikhonov with Total Variation

Given p and A, the image x^0 is recovered from the model

$$\min_x \sum_{i=1}^{n^2} \|D_i x\| + \frac{\mu}{2} \|Ax - p\|_2^2, \qquad (3)$$

where $D_i u \in R^2$ denotes the discrete gradient of x at pixel i, and the sum $\sum \|D_i x\|$ is the discrete total variation (TV) of x.

At each pixel an auxiliary variable $\mathbf{w}_i \in R^2$ is introduced to transfer $D_i u$ out of the non-differentiable term $\| \cdot \|_2$ as follows [20].

$$\min_{\mathbf{w}, x} \sum_i \|\mathbf{w}_i\|_2 + \frac{\mu}{2} \|Ax - p\|_2^2, \quad s.t. \quad \mathbf{w}_i = D_i x \qquad (4)$$

The simple quadratic penalty scheme was adopted, yielding the following approximation model of (4):

$$\min_{\mathbf{w}, u} Q_A(u, \mathbf{w}, \beta) \qquad (5)$$

The main challenge of the paper is that solving the problem (3) is through solving a series of the combined Tikhonov and total variation regularized image models

$$\min_{x=x_1+x_2} TV(x_1) + \frac{\beta}{2} \text{Tikhonov}(x_2) + \frac{\mu}{2} \|Ax - p\|_2^2, \qquad (6)$$

Therefore, the intermediate results are corresponding to different size of β and the final solution is corresponding to a huge β where the TV regularization dominates and the Tikhonov regularization is almost ignorable.

Let $x = x_1 + x_2$ and $D_i x_1 = \mathbf{w}$, then $D_i x_2 = D_i x - \mathbf{w}_i$, where $\mathbf{w}$ is the auxiliary variable [20]. Equation (5) based on operator splitting and the quadratic penalty, can be rewritten as

$$\min_{x_1,x_2} \sum_i \|D_i x_1\|_2 + \frac{\beta}{2} \sum_i \|D_i x_2\|_2^2 \tag{7}$$

$$+ \frac{\mu}{2} \|A(x_1 + x_2) - p\|_2^2,$$

Here we consider the recovered image x as a sum of two components: a piecewise constant component x_1 and a smooth component x_2. Correspondingly, (5) is considered as a combination of Tikhonov regularization and total variation regularization.

The proposed algorithm combines ART algorithm with Tikhonov + TV. The algorithm involves the following: The ART algorithm can be regularized by adding a penalty term in the denominator:

$$x_j^{(k+1)} = x_j^{(k)} + \alpha \frac{p_i - \sum_{n=1}^N A_{i,n} x_n^{(k+1)}}{\sum_{n=1}^N A_{i,n}^2 + \beta \frac{\partial \; \texttt{Tikhonov+TV}(x_j^k)}{\partial x_j^k}} A_{i,j} \tag{8}$$

The overall process is implemented in two steps by combining iterative reconstruction algorithm ART with the `Tikhonov+TV` regularization (8).

3   Performance Evaluation

To evaluate the reconstructed results two criterions are calculated for the four implemented algorithms in addition to the visual quality of the resulting reconstructed images, in addition to the relative norm errors and the visual quality of the reconstructed image. The relative norm error of the resulting images [21] is used and defined as:

Root Mean Squared Error (RMSE) [22] is very commonly used and makes for an excellent general purpose error metric for numerical predictions.

$$\text{RMSE} = \sqrt{\frac{1}{MN} \sum_{i=1}^M \sum_{j=1}^N (x(i,j) - \hat{x}(i,j))^2} \tag{9}$$

where $x(i,j)$ is the value of the pixel in the test image $\hat{x}(i,j)$ and the value of the pixel in the reconstructed image, N is the total number of pixels. Greater $RMSE$, means that the resulting reconstructed image is closer to the test image. We used also The filled contour plot displays isolines of the reconstructed images and the plot of profiles of reconstructed images comparisons.

We shall use the following phantoms in this paper. In Fig. 4, we have the digital Moby phantom [23] was used to simulate the few-view projection data. One typical frame of the phantom is shown in Fig. 4 with size of 256×256 pixels.

4   Results and Discussion

In this section, we compared the reconstruction results of the conventional ART algorithm, the proposed ART+ `Tikhonov+TV` (ART-TTV) algorithm and

Simultaneous algebraic reconstruction technique (SART) [24] for image reconstruction in emission tomography. Our first study involved comparing the noise magnitude in high and low count regions of results images. To do this we first examined variance as a function of position for images of phantoms reconstructed using three algorithms. To generate projection data, we simply add Poisson noise to each of the attenuated projections of the phantoms. The noisy projections are then used to reconstruct the 2D Phantoms (Fig. 1).

The resultant reconstructed images obtained from conventional ART, ART-TTV and SART algorithms with 30 iterations, are shown in Fig. 2. From this Figures, the visual quality of the reconstructed image of the phantom using the ART-TTV algorithm is comparable to the other methods. As compared to some other methods, the experimental algorithm preserves edges better. The effectiveness of noise removal for the test algorithm was comparable to that of SART method; however, the intensity in the ROI was appreciably higher in the latter. Compared to the other methods, the ART-TTV method generates a superior intensity profile while preserving the edges.

The resultant of quality measurements ($RMSE$) of reconstructed images obtained from these algorithms by varying the number of iterations, are shown in Fig. 3. The later demonstrates that ART-TTV is providing better quality measurements than that of conventional ART and SART. The number of iterations is much required in order to enhance the image quality.

To better compare these differences Fig. 4 plots the 1D line profile is the horizontal line that crosses the image in the two ROIs. Noisy images in uniform

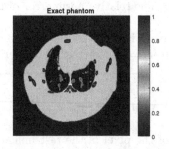

Fig. 1. Input image: Digital Moby phantom used in simulation study

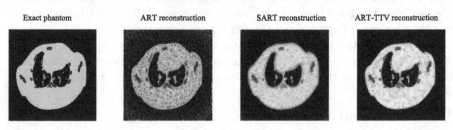

Fig. 2. Reconstructed images of the Moby phantom by different algorithms with 30 iterations.

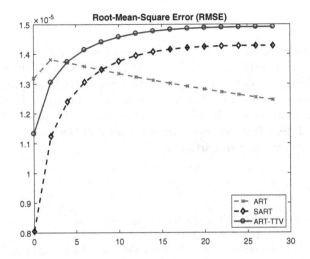

Fig. 3. RMSE vs iterations for ART, SART, and ART-TTV.

regions are shown as spikes as indicated by the SART; ART-TTV line profiles are closer to being noise free. The experimental method also nullifies aerial pixels by redistributing their values to pixels within the phantom. ART-TTV reconstructed the ideal profile more effectively than the other methods, and the image produced was a close approximation of the original phantom image.

All the visual-displays, the quality measurement and the line plots suggest that the proposed ART-TTV algorithm is preferable to the other algorithms. From all the above observations, it may be concluded that the proposed algorithm is performing better in comparison to conventional algorithms and provide a better reconstructed image.

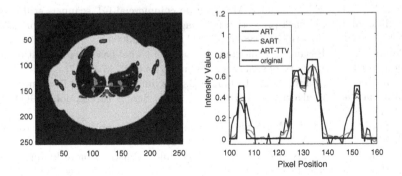

Fig. 4. Horizontal profiles through the images in Fig. 3 at the position of $y = 145$.

5 Conclusion

The new reconstruction algorithm has been presented for the reconstruction of projection data with insufficient iteration and data noisy. Under these conditions, the Digital Moby phantom simulation results verify that the proposed ART-TTV algorithm provides a clear improvement in reconstructed image quality and accuracy, compared with the other algorithms. The proposed algorithm produces an image, which is a close approximation of the original, improves its quality, and reduces noise and artifacts.

References

1. Phelps, M.E., Mazziotta, J.C., Schelbert, H.R.: Positron emission tomography. Los Alamos National Laboratory (1980)
2. Mazziotta, J.C., Schelbert, H.R.: Positron Emission Tomography and Autoradiography. Raven Press, New York (1985)
3. Turkington, T.G.: Introduction to PET instrumentation. J. Nucl. Med. Technol. **29**(1), 4–11 (2001)
4. Bertero, M., Boccacci, P.: Introduction to Inverse Problems in Imaging. CRC Press, Boca Raton (1998)
5. Fessler, J.A.: Penalized weighted least-squares image reconstruction for positron emission tomography. IEEE Trans. Med. Imaging **13**(2), 290–300 (1994)
6. Anastasio, M.A., Zhang, J., Xiaochuan Pan, Y., Zou, G.K., Wang, L.V.: Half-time image reconstruction in thermoacoustic tomography. IEEE Trans. Med. Imaging **24**(2), 199–210 (2005)
7. Boudjelal, A., Attallah, B., Messali, Z.: Filtered-based expectation - maximization algorithm for emission computed tomography (ECT) image reconstruction. In: 2015 3rd International Conference on Control, Engineering & Information Technology (CEIT), pp. 1–5. IEEE (2015)
8. Wernick, M.N., Aarsvold, J.N.: Emission Tomography: The Fundamentals of PET and SPECT. Academic Press, San Diego (2004)
9. Pan, X., Sidky, E.Y., Vannier, M.: Why do commercial CT scanners still employ traditional, filtered back-projection for image reconstruction? Inverse Prob. **25**(12) (2009). https://doi.org/10.1088/0266-5611/25/12/123009
10. Boudjelal, A., Messali, Z., Elmoataz, A.: A novel kernel-based regularization technique for PET image reconstruction. Technologies **5**(2), 37 (2017)
11. Dempster, A.P., Laird, N.M., Rubin, D.B.: Maximum likelihood from incomplete data via the EM algorithm. J. Roy. Stat. Soc.: Ser. B (Methodol.) **39**(1), 1–22 (1977)
12. Shepp, L.A., Vardi, Y.: Maximum likelihood reconstruction for emission tomography. IEEE Trans. Med. Imaging **1**(2), 113–122 (1982)
13. Gordon, R., Bender, R., Herman, G.T.: Algebraic reconstruction techniques (ART) for three-dimensional electron microscopy and X-ray photography. J. Theor. Biol. **29**(3), 471IN1477–476IN2481 (1970)
14. Boudjelal, A., Elmoataz, A., Lozes, F., Messali, Z.: PDEs on graphs for image reconstruction on positron emission tomography. In: Mansouri, A., El Moataz, A., Nouboud, F., Mammass, D. (eds.) ICISP 2018. LNCS, vol. 10884, pp. 351–359. Springer, Cham (2018). https://doi.org/10.1007/978-3-319-94211-7_38

15. Tehrani, J.N., Oh, T.I., Jin, C., Thiagalingam, A., McEwan, A.: Evaluation of different stimulation and measurement patterns based on internal electrode: application in cardiac impedance tomography. Comput. Biol. Med. **42**(11), 1122–1132 (2012)
16. Tikhonov, A.N., Arsenin, V., John, F.: Solutions of Ill-Posed Problems, vol. 14. Winston, Washington, DC (1977)
17. Tikhonov, A.N., Goncharsky, A.V., Stepanov, V.V., Yagola, A.G.: Numerical Methods for the Solution of Ill-Posed Problems, vol. 328. Springer, Dordrecht (2013). https://doi.org/10.1007/978-94-015-8480-7
18. Rudin, L.I., Osher, S., Fatemi, E.: Nonlinear total variation based noise removal algorithms. Phys. D **60**(1–4), 259–268 (1992)
19. Boudjelal, A., Messali, Z., Elmoataz, A., Attallah, B.: Improved simultaneous algebraic reconstruction technique algorithm for positron-emission tomography image reconstruction via minimizing the fast total variation. J. Med. Imaging Radiat. Sci. **48**(4), 385–393 (2017)
20. Wang, Y.: FTVd is beyond fast total variation regularized deconvolution. arXiv preprint arXiv:1402.3869 (2014)
21. Kak, A.C., Slaney, M.: Principles of Computerized Tomographic Imaging. Society for Industrial and Applied Mathematics, Philadelphia (2001)
22. Chai, T., Draxler, R.R.: Root mean square error (RMSE) or mean absolute error (MAE)? Arguments against avoiding RMSE in the literature. Geosci. Model Dev. **7**(3), 1247–1250 (2014)
23. Segars, W.P., Tsui, B.M.W., Frey, E.C., Johnson, G.A., Berr, S.S.: Development of a 4-D digital mouse phantom for molecular imaging research. Mol. Imaging Biol. **6**(3), 149–159 (2004)
24. Andersen, A.H., Kak, A.C.: Simultaneous algebraic reconstruction technique (SART): a superior implementation of the ART algorithm. Ultrason. Imaging **6**(1), 81–94 (1984)

Visualizing Blood Flow of Palm in Different Muscle Tense State Using High-Speed Video Camera

Ryo Takahashi[1](✉), Keiko Ogawa-Ochiai[2], and Norimichi Tsumura[3]

[1] Graduate School of Advanced Integration Science, Chiba University,
1-33 Yayoi, Inage, Chiba, Japan
takahashi0705@chiba-u.jp

[2] Department of Japanese Traditional (Kampo) Medicine,
Kanazawa University Hospital, 13-1 Takara, Kanazawa, Ishikawa, Japan

[3] Graduate School of Engineering, Chiba University, 1-33 Yayoi, Inage, Chiba, Japan

Abstract. In this paper, we propose a method to visualize blood flow of palm in different muscle tense state using RGB high-speed camera. Recently, new modalities are needed to develop a more accurate system to non-contact multi-modal affect analysis. Then, we focus on muscle tense. The muscle tense is caused by stress. Hence, the muscle tense is one of the effective modalities for non-contact multi-modal affect analysis. However, it is very difficult to measure muscle tense in the real environment because it requires a contact-type sensor. Therefore, we use iPPG to visualize the pulse wave during muscle tense from the skin video taken with the RGB video camera. As a result of this experiment, we found that it was possible to recognize the difference in pulse wave during muscle tense from the video that visualized the pulse wave. From this result, the realization of non-contact measurement of muscle tense can be expected.

Keywords: Pulse wave · Visualization · iPPG

1 Introduction

Non-contact methods for affect analysis have been developed in recent researches [1–3]. The methods use various modalities such as facial expression, voice, and heart rate. However, these modalities are less for stress assessment. We assume that muscle tense can be used as one of the modalities because the muscle tense is caused by stress. When a person to be state of nervous, it is known that it causes tense various parts of the body. By detecting this muscle tense state by any method, stress check and concentration measurement are possible. Therefore, measuring muscle tense is very important.

Currently, methods such as visual inspection and palpation by technicians are used to check the degree of muscle tone. However, differences occur for each technician in this method. Moreover, it is difficult to secure a technician and

© Springer Nature Switzerland AG 2020
A. El Moataz et al. (Eds.): ICISP 2020, LNCS 12119, pp. 186–193, 2020.
https://doi.org/10.1007/978-3-030-51935-3_20

diagnosis is difficult because the method is based on experience of technician. In addition, since contact-type measurement requires a dedicated device, it is difficult to realize it in general applications such as stress monitoring and measurement of concentration. From such a background, non-contact measurement of muscle tense is required.

In this research, therefore, we propose a non-contact, non-invasive, and low restrictive method of measurement for muscle tense state. And we verify that it is possible to measure muscle tense using this method. In order to realize this non-contact measurement of muscle tense, we focus on the blood flow. It is known that blood flow stagnates because muscles press blood vessels in muscle tense state. In previous studies, we proposed visualization of blood flow using an RGB camera to facilitate examination of diseases such as arteriosclerosis. Hence we thought that it was possible to distinguish from blood flow if part of the body was in a state of muscle tone using visualization of blood flow.

RGB Video   Melanin   Shading   Hemoglobin   Pulse Wave ᵀⁱᵐᵉ

Fig. 1. Procedure of acquiring pulse wave

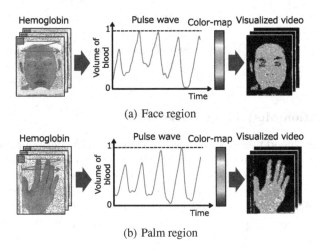

(a) Face region

(b) Palm region

Fig. 2. Procedure of visualization (Color figure online)

2   Proposed Method

In the proposed method, pulse wave is acquired from RGB video and visualized the pulse wave is created [4]. Then, we verify that whether it is possible to distinguish between muscle tense and not muscle tense from the video.

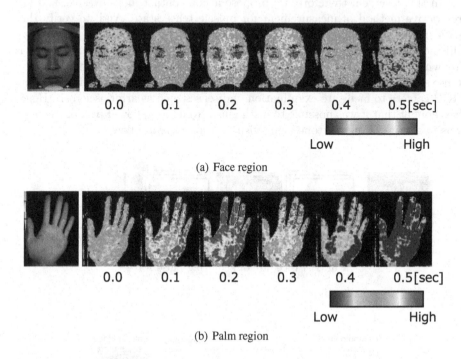

(a) Face region

(b) Palm region

Fig. 3. Result of visualization

2.1   Estimation Method for Pulse Wave

Pulse waves are blood pressure and volume changes in the peripheral vasculature associated with the beating of the heart. In general, a device called a contact type photo plethysmography is used to acquire a pulse wave. This device is attached to a fingertip and acquires reflected light by irradiating a fingertip with green light. Since hemoglobin in the blood absorbs this green light, it is possible to measure the pulse wave by measuring the intensity of the reflected light. However, in pulse wave acquisition by a photo plethysmography (PPG), it is necessary to contact the skin, and it is not possible to acquire spatial pulse wave distribution in the skin region.

A method called image-based photo plethysmography (iPPG) has been proposed as a method of acquiring this pulse wave without contact [5]. In this method, we focus on the fact that the G component of the video of the skin

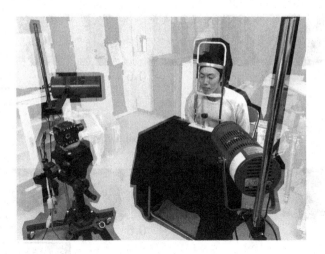

Fig. 4. Experimental setup

taken with the RGB camera is correlated with the blood volume. The method obtains pulse waves from time-series changes of G component. However, this method is not robust to fluctuations in illumination light and body movement, and the noise of acquired pulse waves is very large. Hence, many improved iPPG methods were proposed for robust measurement [6,7]. In this paper, we use a new method of iPPG using a method called pigment component separation [8]. In this method, first, hemoglobin components are extracted from RGB video using pigment component separation. Since the blood volume of the skin changes in association with the heart beats, it is possible to estimate the pulse wave by calculating the average pixel value of the hemoglobin component video in the skin area of the face or palm [9]. Since the pulse wave estimated by this method contains much noise, we need to extract only the band of 45 to 180 [bpm], which is the human heart rate using the band pass filter. The pulse wave can be acquired with high accuracy by this process. Figure 1 shows procedure of acquiring pulse wave from RGB video.

2.2 Visualization Method for Pulse Wave

In order to visualize muscle tone, it is necessary to visualize the blood flow state. At first, the skin video is divided into small regions such as 4×4 [px], and the hemoglobin component is extracted using pigment component separation to acquire the pulse wave. After that, pulse waves are acquired in the above-mentioned procedure from each divided area. The area in video is painted with pulse wave extracted from each area. The area is painted red in the high area of the pulse wave and blue in the low area. Figure 2 shows the correspondence between the pulse wave amplitude and the color map. By above processing, it is possible to create a pulse wave visualized video.

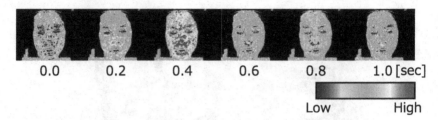

0.0   0.2   0.4   0.6   0.8   1.0 [sec]

Low   High

(a) Normal state (Right side: Normal, Left side: Normal)

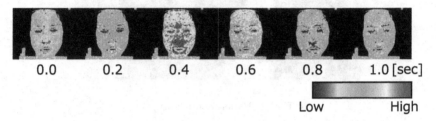

0.0   0.2   0.4   0.6   0.8   1.0 [sec]

Low   High

(b) Muscle tone state (Right side: Muscle tone, Left side: Normal)

Fig. 5. Visualization result of muscle tone state in the face area

2.3   Verification for the Proposed Method

Figure 3 shows the result of analysis of the video of the face and the palm. In addition, Fig. 3(a) shows the time change corresponding to one beat of the face area, and Fig. 3(b) shows the same time change of the palm area. It can be seen from Fig. 3(a) that in the area of the face, the area corresponding to the peak of the pulse wave transitions from the chin to the forehead. It is known that blood vessels in the face extend from the neck to forehead. This structure is consistent with the experimental results. It is considered that the experimental results are reasonable. It can be confirmed from Fig. 3(b) that the pulse wave is propagating around the blood vessel located at the shallow place also in the palm region. From the above results, it is thought that visualization of pulse wave propagation could be realized.

3   Experimental Methods

In the previous section, it is verified whether muscle tension can be identified by visualization of blood flow by using the pulse wave visualization method introduced. In this experiment, subjects do two patterns experiments to make muscle tone. The first is the action of biting teeth. The Subjects are asked to bite only one side tooth to make muscle tone. We take a video of the face at this time. The second is the action of applying power only in one hand. The muscle tone is made by applying power only in one hand. At this time, take a video of the hand. It is thought that the muscles of the jaw or palm that are putting in

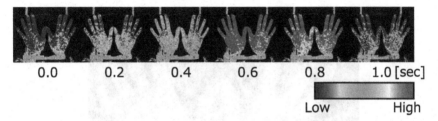

0.0   0.2   0.4   0.6   0.8   1.0 [sec]

Low   High

(a) Normal state (Right side: Normal, Left side: Normal)

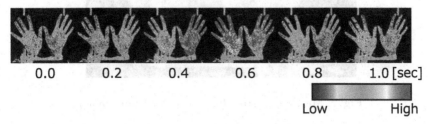

0.0   0.2   0.4   0.6   0.8   1.0 [sec]

Low   High

(b) Muscle tone state (Right side: Muscle tone, Left side: Normal)

Fig. 6. Visualization result of muscle tone state in the palm area (Color figure online)

force contract and blood flow stagnates. We visualize the blood volume of the face and palms at this time, and check whether it is possible to distinguish the person who is applying power in the teeth or hand.

In the experiment, we use a high-speed camera in the dark room, and we take a video of non-muscle tone and muscle tone state on the face and palm region. The time of taking video is 4 [s]. At this experiment, we set the frame rate of the high-speed camera 1000 [fps], and the resolution 1280×1024. Subject's head and palms are fixed on a base and we take a video in a stationary state. Figure 4 shows the experimental environment. We use a high speed camera MEMRECAM Q1m (nac Image Technology Inc.). It can take a video up to 2000 [fps] and 1280×1080 [px]. We also use two sets of artificial sun-lights SOLAX 100W (SERIC LTD.) as illumination.

4   Results

Figure 5 shows the time change of the visualized pulse wave of the face. Figure 5(a) shows a state in which the teeth are not bitten and muscle tension is not present. It can be shown in Fig. 5(a) that there is no difference in the propagation of the pulse wave in a state where muscle tone is not generated. And it was shown in Fig. 5(b) that no difference was observed in the pulse wave of the face even in the muscle tension state. This is thought to be due to the difference in the position of the masseter muscle and the facial artery.

Figure 6 shows the time change of the visualized pulse wave of the face. Figure 6(a) shows a state in which both hands are not muscle tone, and Fig. 6(b)

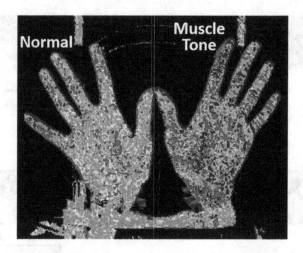

Fig. 7. Muscle tone state (0.4 [s]) (Color figure online)

is a state in which only the left hand is muscle tone. It can be shown in Fig. 6(a) that there is no difference in the propagation of the pulse wave in a state in which not muscle tone. And it was shown in Fig. 6(b) that the phases of pulse waves in part of the palm are largely different in the muscle tone state. Figure 7 shows enlarged view of 0.4 [s] of Fig. 6. It can be easily confirmed that the red and blue parts are mixed in the muscle tone state from Fig. 7. Focusing on a portion where the phase is different, it can be seen that only the blood vessel portion is different in phase. On the other hand, the area of blue and red decreases during muscle tone because the amplitude of pulse wave decreases. This is thought to be because blood vessels were compressed by muscle tone. From above result, we found that it was effective to measure the phase difference of the pulse wave in the palm for the discrimination of the muscle tone state by visualizing the pulse wave propagation.

5   Conclusion and Future Work

In this study, we estimate pulse wave by tracking the time change of hemoglobin component extracted using skin pigment separation. And the pulse wave propagation was visualized in a non-contact by painting the region of video according to the pulse wave value. In addition, it was confirmed that a phase difference is generated in the palm pulse wave in the muscle tone state by using visualization of pulse wave propagation. In the future, we will aim at the quantitative evaluation of muscle tone status using dispersion of phase difference of pulse wave. It will be expected that this makes it possible to determine chronic muscle tone and the like due to stress in the future.

References

1. Mansoorizadeh, M., Charkari, N.M.: Hybrid feature and decision level fusion of face and speech information for bimodal emotion recognition. In: CSICC 2009, pp. 652–657 (2009)
2. McDuff, M., Kaliouby, R.E., Cohn, J.F., Picard, R.W.: Predicting Ad liking and purchase intent: large-scale analysis of facial responses to Ads. IEEE Trans. Affect. Comput. 5(3), 223–235 (2015)
3. Okada, G., Masui, K., Tsumura, N.: Advertisement effectiveness estimation based on crowdsourced multimodal affective responses. In: Proceedings of the IEEE Conference on CVPR Workshops (2018)
4. Takahashi, R., Ochiai-Ogawa, K., Tsumura, N.: Spatial visualization of pulse wave propagation using RGB camera. In: Imaging and Applied Optics 2019 (COSI, IS, MATH, pcAOP), OSA Technical Digest. Optical Society of America (2019). Paper IW3B.6
5. Verkruysse, W., Svaasand, L.O., Nelson, J.S.: Remote plethysmographic imaging using ambient light. Opt. Express 16(26), 21434–21445 (2008)
6. Yang, Y., Liu, C., Yu, H., Shao, D., Tsow, F., Tao, N.: Motion robust remote photoplethysmography in CIELab color space. J. Biomed. Opt. 21(11), 117001 (2016)
7. Mcduff, D., Gontarek, S., Picard, R.: Improvements in remote cardiopulmonary measurement using five band digital camera. IEEE Trans. Biomed. Eng. 61(10), 2593–2601 (2014)
8. Fukunishi, M., Kurita, K., Yamamoto, S., Tsumura, N.: Non contact video based estimation of heart rate variability spectrogram from hemoglobin composition. Artif Life Robot. 22, 457–463 (2017). https://doi.org/10.1007/s10015-017-0382-1
9. Tsumura, N., et al.: Image-based skin color and texture analysis/synthesis by extracting hemoglobin and melanin information in the skin. ACM Trans. Graph. (TOG) 22, 770–779 (2003)

Deep Learning and Applications

Segmentation of Microscopic Image of Colorants Using U-Net Based Deep Convolutional Networks for Material Appearance Design

Mari Tsunomura[1(✉)], Masami Shishikura[2], Toru Ishii[2],
Ryo Takahashi[1], and Norimichi Tsumura[1]

[1] Graduate School of Advanced Integration Science,
Chiba University, 1-33 Yayoi, Inage, Chiba, Japan
`tsumura@faculty.chiba-u.jp`
[2] Central Research Laboratories, DIC Corporation,
631, Sakado, Sakura, Chiba, Japan

Abstract. In this study, U-Net based deep convolutional networks are used to achieve the segmentation of particle regions in a microscopic image of colorants. The material appearance of products is greatly affected by the distribution of the particle size. From that fact, it is important to obtain the distribution of the particle size to design the material appearance of products. To obtain the particle size distribution, it is necessary to segment particle regions in the microscopic image of colorants. Conventionally, this segmentation is performed manually using simple image processing. However, this manual processing leads to low reproducibility. Therefore, in this paper, to extract the particle region with high reproducibility, segmentation is performed using U-Net based deep convolutional networks. We improved deep convolutional U-Net type networks based on the feature maps trained for a microscopic image of colorants. As a result, we obtained more accurate segmentation results using the improved network than conventional U-Net.

Keywords: U-Net · Deep learning · Segmentation

1 Introduction

Colorants are used to design the material appearance of products. The material appearance is affected by the distribution of the particle size, which is one of the characteristics of colorants [1]. Therefore, it is important to measure the distribution of the particle size to obtain products that have the designer's intended material appearance.

It is necessary to segment the particle region in a microscopic image of colorants to obtain the distribution of the particle size. Conventionally, the segmentation of microscopic images of colorants is processed manually or using simple image processing, such as binarization with threshold values. However, if segmentation is performed manually, it takes a long time to segment the image, and the reproducibility of the segmentation results is low because the results differ depending on the operator. If segmentation is performed using simple image processing [2], correct segmentation results cannot be obtained in many cases because the color difference between the particles and the background in the microscopic image is not constant in most images.

Therefore, in this paper, we aim to achieve a segmentation method with high accuracy and high reproducibility. We focus on improving the accuracy of segmentation for large particles because such particles have a large effect on the material appearance of products. We use U-Net based deep convolutional networks to achieve our goal. Moreover, to perform segmentation with higher accuracy than conventional U-Net, we improve U-Net-type networks based on the feature maps trained on a microscopic image of colorants.

(a) Input data   (b) Output data

Fig. 1. Example of a dataset

2   Dataset

The input data are shown in Fig. 1(a). Art papers printed using ink mixed with carbon black pigment and resin by a proofer were captured using a scanning electron microscope.

The output data in Fig. 1(b) were created by manually segmenting the image in Fig. 1(a) using image editing software called CLIP STUDIO. In this study, the number of images for learning is 60.

3   Deep Convolutional Networks of U-Net Architecture

In this study, we use U-Net based deep convolutional networks for segmentation. We considered that U-Net is suitable for the segmentation of a microscopic image because it is generally used for the segmentation of various types of images [3–5].

The network architecture based on U-Net consists of an encoder path and decoder path, as shown in Fig. 2. U-Net has a skip connection that connects the learning result in each layer in the encoder path to the same depth layer in the decoder path. Using this feature, data can be up-sampled to retain detailed information. Each pass has a convolutional layer block, and each block has two convolutional layers with a filter size of 3×3. In the encoder, max pooling with a stride size of 2×2 is applied to the end of each block except the last block, so the size of the feature maps decreases from 512×512 to

32×32. In the decoder, up-sampling with a stride size of 2×2 is applied to the end of each block except the last block, so the size of the feature maps increases from 32×32 to 512×512. Zero padding is used to maintain the output dimension for all the convolutional layers of both the encoder and decoder paths. The number of channels increases from 1 to 1024 in the encoder and decreases from 1024 to 2 in the decoder. Adam is used for optimization; it is one of the most frequently used optimization algorithms in neural network learning. In the process of training, binary cross-entropy is used as the loss function of the network. It is achieved to perform segmentation of microscopic images of colorant by U-Net architecture shown in Fig. 2 [6].

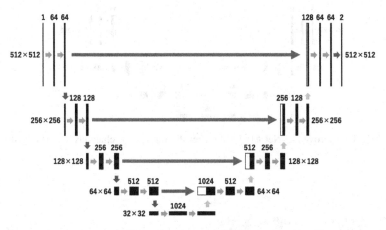

Fig. 2. Conventional U-Net

4 Improved U-Net Architectures

In this section, we improve the conventional U-Net architecture. We examined the problem of the conventional U-Net architecture from the feature maps, and improved the U-Net architectures. We can visualize the learning process by outputting the feature maps to show the output of the artificial neurons in the hidden layer [7].

4.1 Changing the Number of Channels

First, we changed the number of channels, as shown in Fig. 3. We segmented the particle region in the microscopic images for colorants using U-Net before and after changing the number of channels. We summarize the results of segmentation using each U-Net in Fig. 8 in Sect. 5.

Next, we output the feature maps for each U-Net. Figures 4(a) and (c) show the feature maps in the shallow and deep layers before the number of channels was changed. From these figures, we confirm that a large number of filters were not used for learning, particularly in deep layers. Figures 4(b) and (d) show the feature maps in the shallow and deep layers after the number of channels was changed. The ratio of filters that were not used for learning decreased significantly compared with the ratio before the number of channels was changed. From these results, we considered that the number of channels before the change was excessive. Therefore, after this section, we use the number of channels after the change shown in Fig. 3.

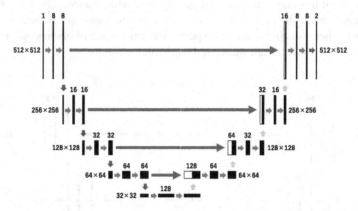

Fig. 3. Improved U-Net #1: U-Net after the number of channels was changed

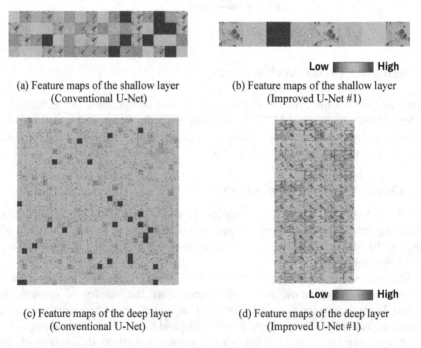

(a) Feature maps of the shallow layer
(Conventional U-Net)

(b) Feature maps of the shallow layer
(Improved U-Net #1)

(c) Feature maps of the deep layer
(Conventional U-Net)

(d) Feature maps of the deep layer
(Improved U-Net #1)

Fig. 4. Feature maps before and after the number of channels was changed

4.2 Improvement of the Deep Layer Architecture

When we segmented particles that were dark inside and did not have a sharp contour, we did not obtain appropriate results. Therefore, we had to improve the U-Net architectures to obtain highly accurate segmentation results for such particles.

Figure 5 shows the results of the feature map output when the particle image was learned. This feature maps shows that the contour of the particle was partially broken immediately after input and immediately before output, and was detected in the deepest layer. We consider that incorrect results were obtained because incorrect information as shown in Fig. 5(a) is passed to the decoder. Therefore, we propose two U-Net architectures that reduce the influence of information on the encoder side.

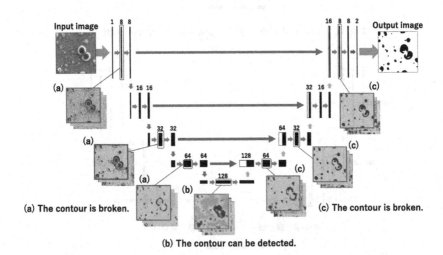

(a) The contour is broken.   (c) The contour is broken.

(b) The contour can be detected.

Fig. 5. Part of the feature maps of the hidden layer in improved U-Net #1

We propose two U-Net architectures:

(1) Improved U-Net #2: Delete skip connections in deep layers (Fig. 6).
(2) Improved U-Net #3: Halve the weights of the two skip connections that connect the deep layers (Fig. 7).

In improved U-Net #2, we deleted two skip connections that connected the deep layers. In improved U-Net #3, we halved the weights of the two skip connections that connected the deep layers. These proposed methods are written in python using Keras.

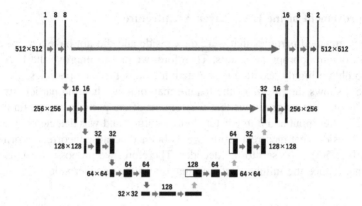

Fig. 6. Improved U-Net #2: Delete some skip connections

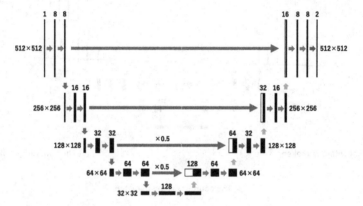

Fig. 7. Improved U-Net #3: halve the weights of the two skip connections that connect the deep layers

5   Results

Figure 8 shows the image processing results that removed particles smaller than 400 pixels from the segmentation results using each U-Net architectures. The purpose of this study is to improve the segmentation accuracy for large particles preferentially. Therefore, we did not consider small particles that have little effect on material appearance. We used root mean squares error (RMSE) as an evaluation index between the segmentation results and ground truth data. Table 1 shows the RMSE and standard deviation. Improved U-Net #1 had the highest accuracy. The accuracy of improved U-Net #3 increased compared with the conventional U-Net. By contrast, improved U-Net #2 had the lowest accuracy. However, the segmentation results in Fig. 8(b) were more accurate for improved U-Net #2 and conventional U-Net than improved U-Net #1 and #3. The segmentation results in Fig. 8(c) were more accurate for improved U-Net #2 than conventional U-Net. We consider that the accuracy of the improved U-Net #2 was

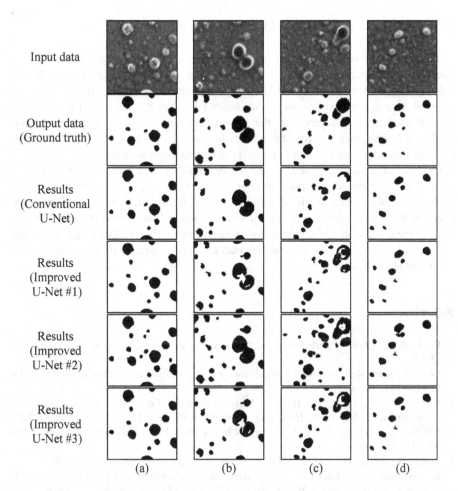

Fig. 8. Results of image processing after segmentation

Table 1. RMSE and standard deviation (average of 60 tested results)

	Conventional U-Net	Improved U-Net #1	Improved U-Net #2	Improved U-Net #3
RMSE	35.8975	35.2672	45.2187	35.5836
Standard deviation	6.8609	6.7671	7.8371	7.2075

low because many noise parts were classified as particles. From the standard deviation in Table 1, we consider that the improved U-Net #2 had the largest variation in results. From these results, we consider that the improved U-Net #2 can segment large particles with high accuracy, but also segment the noise parts as particles.

6 Conclusion and Future Work

In this study, we segmented microscopic images of colorants using improved deep convolutional networks of U-Net architectures. As a result of comparing the RMSE values of each method, improved U-Net #1, for which the number of channels was changed, had the highest accuracy, and improved U-Net #2, which had skip connections deleted in deep layers, had the lowest accuracy. However, the segmentation results for large particles were obtained with high accuracy for improved U-Net #2. We consider that improved U-Net #2 obtained the best segmentation results compared with other U-Nets because we focused on the segmentation of large particles in this study. In future work, we propose a more appropriate evaluation method. In addition, in this study, we used the data of 512×512 image size. In the future, we will examine the effect of the image size using data of different image size. Moreover, we need to achieve the separation and segmentation of contacting particles. As a result, we expect that the particle size distribution can be obtained more accurately so that it can be used for practical application.

References

1. Gueli, A.M.: Effect of particle size on pigments colour. Color Res. Appl. **42**(2), 236–243 (2017)
2. Otsu, N.: A threshold selection method from gray-level histograms. IEEE Trans. Syst. Man Cybern. **9**(1), 62–66 (1979)
3. Ronneberger, O., Fischer, P., Brox, T.: U-net: convolutional networks for biomedical image segmentation. In: Navab, N., Hornegger, J., Wells, W.M., Frangi, A.F. (eds.) MICCAI 2015. LNCS, vol. 9351, pp. 234–241. Springer, Cham (2015). https://doi.org/10.1007/978-3-319-24574-4_28
4. Dong, H., Yang, G., Liu, F., Mo, Y., Guo, Y.: Automatic brain tumor detection and segmentation using U-net based fully convolutional networks. In: Valdés Hernández, M., González-Castro, V. (eds.) MIUA 2017. CCIS, vol. 723, pp. 506–517. Springer, Cham (2017). https://doi.org/10.1007/978-3-319-60964-5_44
5. Zhang, Z.: Road extraction by deep residual U-Net. IEEE Geosci. Remote Sens. Lett. **15**(5), 749–753 (2018)
6. Implementation of deep learning framework - Unet, using Keras. https://github.com/zhixuhao/unet Accessed 17 Jan 2020
7. Horwath, J.P.: Understanding important features of deep learning models for transmission electron microscopy image segmentation (2019). arXiv:1912.06077

A Deep CNN-LSTM Framework for Fast Video Coding

Soulef Bouaafia[1(✉)] [ID], Randa Khemiri[1], Fatma Ezahra Sayadi[1] [ID],
Mohamed Atri[2] [ID], and Noureddine Liouane[3] [ID]

[1] Faculty of Sciences, University of Monastir, Monastir, Tunisia
{soulef.bouaafia,randa.khemiri}@fsm.rnu.tn, sayadi_fatma@yahoo.fr
[2] College of Computer Science, University of King Khalid, Abha, Saudi Arabia
matri@kku.edu.sa
[3] National Engineering School of Monastir, University of Monastir, Monastir, Tunisia
noureddine.liouane@enim.rnu.tn

Abstract. High Efficiency Video Coding (HEVC) doubles the compression rates over the previous H.264 standard for the same video quality. To improve the coding efficiency, HEVC adopts the hierarchical quadtree structured Coding Unit (CU). However, the computational complexity significantly increases due to the full search for Rate-Distortion Optimization (RDO) to find the optimal Coding Tree Unit (CTU) partition. Here, this paper proposes a deep learning model to predict the HEVC CU partition at inter-mode, instead of brute-force RDO search. To learn the learning model, a large-scale database for HEVC inter-mode is first built. Second, to predict the CU partition of HEVC, we propose as a model a combination of a Convolutional Neural Network (CNN) and a Long Short-Term Memory (LSTM) network. The simulation results prove that the proposed scheme can achieve a best compromise between complexity reduction and RD performance, compared to existing approaches.

Keywords: HEVC · CU partition · Deep learning · CNN · LSTM

1 Introduction

Over the last decades, we have noticed the success of deep learning in many application areas, where video and image processing has achieved a favorable outcome. The state-of-the-art video coding is High Efficiency Video Coding (HEVC), also known as H.265, which was standardized in 2013 [1]. Compared to its predecessor H.264/AVC standard, HEVC saves an average BitRate reduction of 50%, while maintaining the same video quality [2]. The hierarchical coding structure adopted in HEVC is the quadtree, including Coding Unit (CU), Prediction Unit (PU), and Transform Unit (TU) [3]. In this regard, the Coding Tree Unit (CTU) is the basic coding structure in which the size of the CTU is 64×64. A CTU can be divided into multiple CUs of different sizes from 64×64 with a depth of 0 to 8×8 with a depth of 3.

© Springer Nature Switzerland AG 2020
A. El Moataz et al. (Eds.): ICISP 2020, LNCS 12119, pp. 205–212, 2020.
https://doi.org/10.1007/978-3-030-51935-3_22

This exhaustive splitting continues until the minimum possible size of a CU is reached in order to find the optimal CU depth. This process is known as Rate-Distortion Optimization (RDO) which is required for each CTU. Due to the full RDO search, the HEVC computational complexity has considerably increased, making encoding speed a crucial problem in the implementation of HEVC. Accordingly, to enhance the coding efficiency of HEVC, fast algorithms have been proposed for reducing the HEVC complexity caused by the quadtree partition. These fast methods can be summarized into two categories: heuristic and learning-based schemes.

In heuristic methods, some fast CU decision algorithms have been developed to simplify the RDO process towards reducing HEVC complexity. For example, Cho et al. [4] developed a Bayesian decision rule with low complexity and full RD cost-based fast CU splitting and pruning algorithm. With regard to HEVC inter prediction, Shen et al. in [5] proposed a fast inter-mode decision scheme using inter-level and spatio-temporal correlations in which the prediction mode, motion vector and RD cost were found strongly correlated. To reduce the HEVC complexity, a look-ahead stage based fast CU partitioning and mode decision algorithm was proposed in [6]. Based on pyramid motion divergence, authors in [7] introduced a fast algorithm to split CUs at the HEVC inter coding.

On the other hand, the search of the optimal mode decision can be modeled as a classification problems. In this regard, researchers adopted learning-based methods in classifying CU mode decision in order to reduce the computational complexity. Shen et al. [8] proposed a CU early termination algorithm for each level of the quadtree CU partition based on weighted SVM. In addition, a fuzzy SVM-based fast CU decision method was proposed by Zhu et al. in [9] to improve the coding efficiency. To reduce the HEVC complexity with deep structure, in [10], authors developed a fast CU depth decision in HEVC using neural networks to predict split or non-split for inter and intra-mode. Reinforcement Learning (RL) and deep RL are also applied in video coding to learn a classification task, and to find the optimal CU mode decision. In this study, an end-to-end actor-critic RL based CU early termination algorithm for HEVC was developed to improve the coding complexity [11].

For video coding, the adjacent video frames exhibit similarity in video content, in which the similarity decreases along with temporal distance between two frames. Figure 1 shows the temporal correlation of CTU partition across HEVC video frames. In HEVC inter-mode, there are long and short-term dependencies of the CU partition across neighboring frames. It is in this context that we propose in this paper Long Short-Term Memory (LSTM) networks to study the temporal dependency of the CU partition across adjacent frames.

Based on our previous work presented in [12], our deep CNN structure has been proposed to predict the CU splitting in order to reduce the coding performance of inter-mode HEVC. A large-scale training database was established in [12]. Therefore, to predict the inter-mode CU partition, this paper proposes a CNN-LSTM-based learning approach where the aim is to reduce the HEVC complexity in terms of RD performance and encoding time.

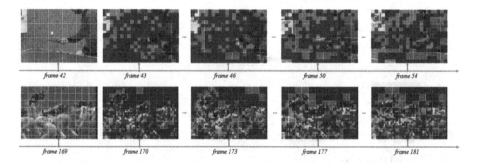

Fig. 1. Temporal similarity of CTU partition across HEVC video frames.

The paper is organized as follows: Sect. 2 introduces the proposed method, which reduces the HEVC complexity at inter prediction. The evaluation results are shown in Sect. 3. Section 4 concludes this paper.

2 Proposed Scheme

2.1 Database for Inter-mode

A large-scale database for CU partition of the inter-mode HEVC was established, to train the proposed model, as shown in [12]. However, to construct the database, we selected 114 video sequences with various resolutions (from 352×240 to 2560×1600) [15,16]. These sequences are divided into three subsets: 86 sequences for training, 10 sequences for validation, and 18 sequences for test. All sequences in our database were encoded by HEVC reference software using the Low Delay P (LDP) (using encoder_lowdelay_P_main.cfg) at four Quantization Parameters (QP) $\{22, 27, 32, 37\}$. Therefore, corresponding to different QPs and CU sizes (64×64, 32×32, and 16×16), 12 sub-databases were obtained under LDP configuration.

2.2 CNN-LSTM Network

According to the temporal correlation of the CU partition of adjacent frames, the proposed scheme is introduced in this section. The proposed LSTM network learns the long and short-term dependencies of the CU partition across frames. In our previous proposed method [12], all parameters of Deep CNN are trained over the residual CTU and the ground-truth splitting of the CTUs, then the extracted features $(FC_{1-l})_{l=1}^3$ of Deep CNN are the input of LSTM network at frame t. These features $(FC_{1-l})_{l=1}^3$ are extracted at the first fully connected layer of Deep CNN [12]. The proposed algorithm that combines CNN and LSTM is shown in Fig. 2.

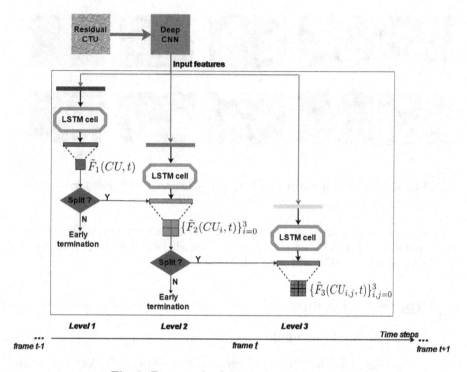

Fig. 2. Framework of the proposed algorithm.

The structure of the LSTM is composed of three LSTM cells corresponding to three levels splitting of each CU. Specifically, $\tilde{F}_1(CU, t)$ at level 1 indicates whether the CU of size 64×64 is split to sub-CUs of size 32×32. At level 2, $\{\tilde{F}_2(CU_i, t)\}_{i=0}^{3}$ determines the splitting of CUs from 32×32 to 16×16 and $\{\tilde{F}_3(CU_{i,j}, t)\}_{i,j=0}^{3}$ denotes the partitioning labels of CUs from 16×16 to 8×8. At each level, the LSTM cells are followed by a fully connected layer. In addition, the output features of the LSTM cells denote by $(FC'_l)_{l=1}^{3}$ at frame t. However, the next level LSTM cell is activated to make decisions on the next four CUs at the next level, when the CU of the current level is predicted to be split. Otherwise, the prediction on splitting the current CU is terminated early. Finally, the predicted CU partition of three levels is represented by the combination of $\tilde{F}_1(CU, t)$, $\{\tilde{F}_2(CU_i, t)\}_{i=0}^{3}$ and $\{\tilde{F}_3(CU_{i,j}, t)\}_{i,j=0}^{3}$.

The LSTM model learns the long short-term dependency of CTU partition across frames when the CTU partition is predicted. In fact, the LSTM cell consists of three gates: the input gate $\mathbf{i}_l(t)$, the forget gate $\mathbf{f}_l(t)$ and the output gate $\mathbf{o}_l(t)$. In particular, $FC_{1-l}(t)$ denotes the input features of the LSTM cell at frame t and $FC'_l(t-1)$ is the output features of the LSTM cell of frame t-1 at level l. In the following, these three gates are presented by:

$$\mathbf{i}_l(t) = \sigma(\mathbf{W}_i \cdot [\mathbf{FC}_{1-l}(t), \mathbf{FC}'_l(t-1)] + \mathbf{b}_i)$$

$$\mathbf{o}_l(t) = \sigma(\mathbf{W}_o \cdot [\mathbf{FC}_{1-l}(t), \mathbf{FC}'_l(t-1)] + \mathbf{b}_o) \qquad (1)$$

$$\mathbf{f}_l(t) = \sigma(\mathbf{W}_f \cdot [\mathbf{FC}_{1-l}(t), \mathbf{FC}'_l(t-1)] + \mathbf{b}_f)$$

where the sigmoid function denotes by $\sigma(\cdot)$. $\{\mathbf{W}_i, \mathbf{W}_o, \mathbf{W}_f\}$ are the weights and $\{\mathbf{b}_i, \mathbf{b}_o, \mathbf{b}_f\}$ are the biases for three gates. At frame t, the state $\mathbf{c}_l(t)$ of the LSTM cell can be updated by:

$$\mathbf{c}_l(t) = \mathbf{i}_l(t) \odot Tanh(\mathbf{W}_c \odot [\mathbf{FC}_{1-l}(t), \mathbf{FC}'_l(t-1)] + \mathbf{b}_c)$$
$$+ \mathbf{f}_l(t) \odot \mathbf{c}_l(t-1) \qquad (2)$$

where $\odot$ signifies the element-wise multiplication. The output of the LSTM cell $FC'_l(t)$ can be determined as follows:

$$\mathbf{FC}'_l(t) = \mathbf{o}_l(t) \odot \mathbf{c}_l(t) \qquad (3)$$

In the training phase, the LSTM model was trained from the training set of the inter database, which minimizes the loss function between the ground truth and the prediction of CTU partition. Here, the cross entropy is adopted as the loss function. Then, the Stochastic Gradient Descent algorithm with momentum (SGD) is used as a powerful optimization algorithm to update the network weights at each iteration and minimize the loss function.

3   Experimental Results

This section introduces the experimental results to evaluate the encoding performance of our proposed approach. Our experiments were integrated in HEVC reference test model HM, which were tested on JCT-VC video sequences from Class A to Class E at two QPs $\{22, 37\}$ using the LDP configuration. In order to validate the performance of the proposed scheme, all simulations were carried out on windows 10 OS platform with Intel ®core TM i7-3770 @ 3.4 GHz CPU and 16 GB RAM. We also use the NVIDIA GeForce GTX 480 GPU to dramatically improve speed during the training phase of the network model.

The Tensorflow-GPU deep learning framework was used in the training process. We first adopt a batch mode learning method with a batch size of 64 where the momentum of the stochastic gradient descent algorithm optimization is set to 0.9. Second, the learning rate was set to 0.01, changing every 2,000 iterations to train the LSTM model. Then, the LSTM length was set to $T = 20$. Finally, the trained model can be used to predict the inter-mode CU partition for HEVC.

For the test, to further enhance the RD performance and the complexity reduction at inter-mode, the bi-threshold decision scheme was adopted at three levels. Note that the upper and lower thresholds at level l represents by $\{\gamma_l\}_{l=1}^3$ and $\{\bar{\gamma}_l\}_{l=1}^3$. The predicted CU partition probability is output by the LSTM model at different levels ($P_l(CU)$). Therefore, the CU decides to be split when

Table 1. Performance comparison between the proposed scheme and the state-of-the-art approaches

Class	Sequence	Approaches	ΔBR (%)	ΔPSNR(dB)	ΔT (%)
A	Traffic	[12]	−0.458	−0.059	−56.66
		[13]	3.25	−0.081	41.8
		[14]	1.990	−0.052	−59.01
		[Proposed Scheme]	0.376	−0.101	−61.27
B	BasketballDrive	[12]	1.335	−0.031	−50.43
		[13]	2.01	−0.049	44.7
		[14]	2.268	−0.052	−53.92
		[Proposed Scheme]	1.528	−0.022	−52.05
C	PartyScene	[12]	−0.436	−0.076	−51.51
		[13]	3.12	−0.131	41.4
		[14]	1.011	−0.039	−48.27
		[Proposed Scheme]	0.494	−0.029	−58.48
D	BQSquare	[12]	−2.012	−0.128	−53.27
		[13]	4.15	−0.149	46.6
		[14]	0.770	−0.028	−48.85
		[Proposed Scheme]	0.219	−0.050	−59.52
E	Johnny	[12]	−0.181	−0.134	−64.43
		[13]	0.82	−0.052	48.7
		[14]	1.691	−0.038	−63.48
		[Proposed Scheme]	1.371	−0.028	−70.85
Average		[12]	−0.350	−0.085	−55.26
		[13]	2.67	−0.092	44.64
		[14]	1.546	−0.048	−54.70
		[Proposed Scheme]	**0.797**	**−0.046**	**−60.43**

$P_l(CU) > \gamma_l$; if $P_l(CU) < \overline{\gamma}_l$, the CU is not split. In this way, this considerably reduces the HEVC complexity by skipping the most redundant checking of RD cost.

The RD performance analysis is performed on the basis of the average PSNR ($\Delta PSNR$) gain and the average BitRate ($\Delta BitRate$) reduction. Additionally, the complexity reduction is the critical metric for the performance evaluation at HEVC inter-mode. Let ΔT the encoding time reduction. All performance metrics are written as:

$$\Delta PSNR = PSNR_P - PSNR_o \ (dB) \tag{4}$$

$$\Delta BitRate = \frac{BR_P - BR_o}{BR_o} \times 100 \ (\%) \tag{5}$$

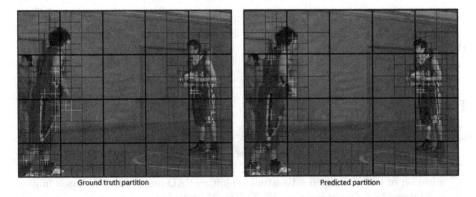

Ground truth partition   Predicted partition

Fig. 3. CU partition result of BasketballPass with QP = 37. frame = 20. LDP configuration.

$$\triangle T = \frac{T_P - T_o}{T_o} \times 100 \ (\%) \tag{6}$$

Table 1 demonstrates the performance comparison of the proposed scheme and three other state-of-the-arts schemes.

From this table, we can conclude that the proposed scheme is better in terms of computational complexity reduction than other state-of-the-arts schemes. Specifically, the time saving of our approach is 60.43% on average, which exceeds the 55.26% obtained by [12], the 44.64% achieved by [13], and 54.70% of [14], respectively. On the other hand, the proposed approach can reduce the PSNR performance by −0.046 dB, which is better than −0.085 dB of [12], −0.092 dB of [13] and −0.048 of [14]. Furthermore, the approaches [12] achieve better value of BitRate 0.350% than ours. Meanwhile, the proposed approach outperforms the existing approaches [13,14] in terms of BitRate by 0.797%.

As shown in this table, the proposed approach can reduce the complexity reduction by 70.85% for class E video sequences, since these sequences have low motion activities and homogeneous regions, where the blocks CU partition is larger. From the overall performance assessment, our proposed fast scheme provides competitive HEVC coding efficiency tradeoffs compared to state-of-the-art approaches.

Figure 3 shows the comparison between the CU partition predicted by the proposed deep learning model and the ground truth partition when QP equals 37 under the LDP configuration.

4   Conclusion

This paper proposed a deep learning approach to predict the CU partition, which combines the CNN-LSTM network to reduce inter-mode HEVC complexity. To train the proposed model, the inter database was built. According to the temporal correlation of the CU partition of neighboring frames, we developed a new LSTM architecture to learn the long and short-term dependencies of the CU partition across frames, instead of brute-force RDO search. In summary, the proposed scheme saves a significant encoding complexity compared to other state-of-the-art works.

References

1. Sullivan, G.J., Ohm, J.R., Han, W.J., Wiegand, T.: Overview of the high efficiency video coding. IEEE Trans. Circuits Syst. Video Technol. **22**, 1649–1668 (2012)
2. Khemiri, R., Kibeya, H., Sayadi, F.E., Bahri, N., Atri, M., Masmoudi, N.: Optimisation of HEVC motion estimation exploiting SAD and SSD GPU-based implementation. IET Image Process. **12**(2), 243–253 (2017)
3. Khemiri, R., et al.: Fast Motion Estimation's Configuration Using Diamond Pattern and ECU, CFM, and ESD, Modes for Reducing HEVC Computational Complexity. IntechOpen, Digital Imaging Book, London, pp. 1–17 (2019)
4. Cho, S., Kim, M.: Fast CU splitting and pruning for suboptimal CU partitioning in HEVC intra coding. IEEE TCSVT **23**, 1555–1564 (2013)
5. Shen, L., Zhang, Z., Liu, Z.: Adaptive inter-mode decision for HEVC jointly utilizing inter-level and spatio-temporal correlations. IEEE Trans. Circuits Syst. Video Technol. **24**, 1709–1722 (2014)
6. Cebrián-Márquez, G., Martinez, J.L., Cuenca, P.: Adaptive inter CU partitioning based on a look-ahead stage for HEVC. Sig. Process. Image Commun. **76**, 97–108 (2019)
7. Xiong, J., Li, H., Wu, Q., Meng, F.: A fast HEVC inter CU selection method based on pyramid motion divergence. IEEE Trans. Multimed. **16**, 559–564 (2014)
8. Shen, X., Yu, L.: CU splitting early termination based on weighted SVM. EURASIP J. Image Video Process. **4**, 1–11 (2013)
9. Zhu, L., Zhang, Y., Kwong, S., Wang, X., Zhao, T.: Fuzzy SVM based coding unit decision in HEVC. IEEE Trans. Broadcast **64**, 681–694 (2017)
10. Kim, K., Ro, W.: Fast CU depth decision for HEVC using neural networks. IEEE Trans. Circuits Syst. Video Technol. (2018). https://doi.org/10.1109/TCSVT.2018.2839113
11. Li, N., Zhang, Y., Zhu, L., Luo, W., Kwong, S.: Reinforcement learning based coding unit early termination algorithm for high efficiency video coding. J. Visual Commun. Image R **60**, 276–286 (2019)
12. Bouaafia, S., Khemiri, R., Sayadi, F.E., Atri, M.: Fast CU partition based machine learning approach for reducing HEVC complexity. J. Real-Time Image Process. **7**, 185–196 (2019)
13. Li, Y., Liu, Z., Ji, X., Wang, D.: CNN based CU partition mode decision algorithm for HEVC inter coding. In: 2018 25th IEEE International Conference on Image Processing (ICIP), pp. 993–997 (2018)
14. Xu, M., Li, T., Wang, Z., Deng, X., Yang, R., Guan, Z.: Reducing complexity of HEVC: a deep learning approach. IEEE Trans. Image Process. **27**(10), 5044–59 (2018)
15. Bossen, F.: Common test conditions and software reference configurations. Document JCTVC-L1100, Joint Collaborative Team on Video Coding (2013)
16. Xiph.org.: Xiph.org Video Test Media (2017). https://media.xiph.org/video/derf

Microcontrollers on the Edge – Is ESP32 with Camera Ready for Machine Learning?

Kristian Dokic[✉] [ID]

Polytechnic in Pozega, 34000 Pozega, Croatia
kdjokic@vup.hr

Abstract. For most machine learning tasks big computing power is needed, but some tasks can be done with microcontrollers. In this paper well-known SoC ESP32 has been analyzed. It is usually used in IoT devices for data measurement, but some authors started to use simple machine learning algorithms with them. Generally, this paper will analyze the possibility of using ESP32 with a built-in camera for machine learning algorithms. Focus of research will be on durations of photographing and photograph processing, because that can be a bottleneck of a machine learning tasks.

For this purpose, logistic regression has been implemented on ESP32 with camera. It has been used to differentiate two handwritten letters on the greyscale pictures ("o" and "x"). Logistic regression weights have been calculated on the cloud, but then they have been transferred to an ESP32. The output results have been analyzed. The duration of photographing and processing were analyzed as well as the impact of implemented PSRAM memory on performances. It can be concluded that ESP32 with camera can be used for some simple machine learning tasks and for camera picture taking and preparing for other more powerful processors. Arduino IDE still does not provide enough level of optimization for implemented PSRAM memory.

Keywords: ESP32 · Logistic regression · Machine learning · IoT

1 Introduction

A microcontroller is a cheap and programmable system that generally includes memory and I/O interfaces on a single chip. They have been developed for decades but the main paradigm hasn't changed all that time until the first decade of 21st century. The ubiquity of the internet has resulted in the appearance of services that offer to send, collecting and analyzing data from microcontrollers on the cloud services. In most cases connection between microcontroller and the Internet has been made through Wi-Fi. Cloud services offer lots of advantages like the reliability of cloud services and data visualization but in the last few years new paradigm has arrived – edge computing. Some authors declared that "edge computing refers to the enabling technologies allowing computation to be performed at the edge of the network, on downstream data on behalf of cloud services and upstream data on behalf of IoT services" [1].

On the other hand, in the last few years, a lot of microcontroller producers have worked on machine learning implementation on microcontrollers. Some of them have

© Springer Nature Switzerland AG 2020
A. El Moataz et al. (Eds.): ICISP 2020, LNCS 12119, pp. 213–220, 2020.
https://doi.org/10.1007/978-3-030-51935-3_23

developed special libraries with machine learning functions [2, 3] but the others have implemented special hardware with enhanced machine learning capabilities [4, 5].

In this paper, a low-cost Chinese SoC ESP32 with camera has been analyzed. The ability to apply simple machine learning algorithms (Logistic Regression) has been tested as well as the impact of PSRAM memory implementation on performances. The focus of the research has been on durations of photographing and photograph processing. ESP32 has been chosen because it has caused great interest from the start of its production.

In the Sect. 2, few papers with ESP32 used for machine learning algorithms are presented. In Sect. 3, Logistic Regression implementation on an ESP32 with camera is presented. ESP32 board with a camera has been used to differentiate two letters on greyscale photos that have been taken from implemented camera. Logistic regression has been used to solve that problem, and the first part has been released with Google Collaboratory service. After that, final coding and microcontroller programming have been done with Arduino IDE. In Sect. 4, camera and picture processing speed have been analyzed, as well as the impact of PSRAM memory implementation on performances. Finally, in Sects. 5 and 6, discussion and conclusion can be found.

2   Related Work

ESP32 SoC is the second generation of Espressif corporation IoT solution and it includes WiFi and Bluetooth. It is based on the 32-bit RISC Tensilica Xtensa LX106 MCU with included FPU and DSP. Clock speed is 240 MHz, and it has 520 KB SRAM. Ivkovic put it in the MCU IoT Ready group because of FPU, DSP and WiFi components integrated [6]. In the ESP32 datasheets there is the application list where ESP32 can be used. Some of them are speech and picture recognition, internet radio players, an energy monitors and smart lighting, etc. [7].

Only a few authors used ESP32 for some ML tasks. Kokoulin et al. used ESP32 to reduce high network traffic and computing load of the central face recognition server. They implemented system based on a microcontroller that processes video stream from a public place and detects the presence of a face or silhouette fragment. Only pictures with faces or silhouette are sent to the main server. Estimated traffic decrease gains up to 80–90% [8].

Espressif System has been developed ESP-WHO framework for face recognition and detection and it is available on the GitHub [9, 10]. They cited that their framework is based on Multi-task Cascaded Convolutional Networks model and new mobile architecture - MobileNetV2 [11, 12].

3   Materials and Methods

Logistic regression is an algorithm for classification purposes. It is used when a model has to return a limited number of values, and the dependent variable is categorical. With only two possible outcomes Logistic Regression is called Binary Logistic Regression [13].

Logistic Regression is similar to Linear Regression, but the only difference is in the fact that output the weighted sum has to pass through a function that can map any value between zero and one. For that purpose, sigmoid function is used.

In this paper, binary logistic regression has been used, and θT values have been defined in the cloud. These values have been transferred to the ESP32 SoC, and they have been used for deciding about the letter in front of the camera connected with the microcontroller.

3.1   Hardware Part

In this paper, ESP32 with camera module is used. There are lots of vendors that produce ESP32 with a camera with different characteristics. In this paper, ESP32-CAM produced by AI-Thinker has been used as well as M5CAMERA by M5STACK. The main difference between them is that M5CAMERA is in a plastic box and it has 4 MB PSRAM memory. ESP32-CAM hasn't PSRAM memory, and it has only 512 kB RAM on the board. Both boards have the same camera, and it is OV2640.

3.2   Software Development

Logistic regression functions are well known, and it is implemented in lots of machine learning libraries. In this paper, logistic regression weights are calculated in the cloud, and they are transferred on the microcontroller board with a camera. The main idea was to use logistic regression to separate handwritten pictures of letters in two groups: pictures with letter "o" and pictures with letter "x". Described solution can be divided into three parts. First, some handwritten pictures with letters "o" and "x" have to be taken, and then conversion to vectors has to be done. After that, weights for output vector with Logistic Regression have to be calculated, and finally, weights have to be transferred to a microcontroller, including code for decision making.

First, on the twenty little papers, ten letters "o" and ten letters "x" have been written by hand. The pictures of these papers have been taken with the smartphone XIAOMI Redmi 5+ in resolution 3000×3000 pixels. These twenty pictures have been converted to 1000×1000 resolution with program IrfanView, and then pictures have been renamed. Pictures with letter "o" have been renamed to 0XX.jpg where XX is a number of picture. Pictures with letter "x" have been renamed to 1XX.jpg where XX is a number of picture.

These converted and renamed pictures have been uploaded to Collaboratory environment. Collaboratory is a free Jupyter notebook environment that runs entirely on the cloud. It can be used to write, save and execute code in Python. It provides access to powerful computing resources for free. After pictures uploading Python code converts color pictures to grayscale pictures and then, 1000×1000 pixels pictures to 3×3 pixel pictures.

After that values of 9 bytes from all twenty pictures have been transferred to an array with nine columns and 20 rows. One row for every picture, and one column for every pixel grey value. In the last tenth column value from the first number for every file, the name has been added (0 or 1). That array has been converted to CSV file named output3.csv. Values are between zero and one. Zero is for black color and one is for white color. It can be seen that the content of the first row of the file are values for the letter "x". It is easy to prove it because the value for the fifth pixel in the center must be almost zero. It is 0.192156863.

Next step is weights calculating with Logistic Regression. Logistic Regression library from scikit-learn has been used. Code with comments is available on the GitHub. Output is file named weights3.csv with weights. It is nine-dimension vector that has to be transferred to a ESP32 with a camera, and it will be used to decide what picture is in the front of the camera.

Program for ESP32-CAM and M5CAMERA has been developed on Arduino IDE and the vector from the previous step has been imported as an array.

ESP32-CAM and M5CAMERA both have OV-2640 camera implemented, and different resolutions can be used, but the highest is 1600×1200. All available resolutions are rectangular and not quadratic. Pictures that have been used for calculating Logistic Regression weights have been quadratic. The solution is to cut parts of the taken picture from the left and the right side. All pictures and code for both SoC boards with comments is available on the address https://github.com/kristian1971/LogRegESP32-CAM/.

The program output from the ESP32-CAM and M5CAMERA is a number that can be received over the serial port. Programmed ESP32-CAM and M5CAMERA have been tested with different pictures and results were similar. Boards have sent numbers around 0,25 when a paper with letter "o" has been in front of the camera, as well as the numbers around value 0,70 when a paper with letter "x" has been in front of the camera. The device is sensitive to light, and light intensity change causes slight changes in the output values. Overview of proposed approach can be seen on the Fig. 1.

Fig. 1. Overview of proposed approach

4   Testing and Results

After Logistic Regression implementation next goal has been to analyze the speed of the camera and photo processing on an ESP32-CAM and M5CAMERA boards. These boards can be used for taking pictures as well as preparing photos for some more powerful processor that can be connected with ESP32 board with serial, WiFi or Bluetooth connection. This second processor can be used for some more intensive calculations. This part of the research has been separated in the two parts. In the first part, ESP32-CAM and M5CAMERA have been used without 4 MB PSRAM and in the second part PSRAM has been enabled on M5CAMERA board.

The measurements were performed by inserting function micros() in the program code. This function returns a number of microseconds after microcontroller reset. When these values are subtracted, results are the execution times between inserted functions.

Duration of photographing presents time between command for photograph taking and the moment after the command execution. An example is here:

```
time1 = micros();
  //taking photograph
time2 = micros();
Total_time = time2-time1;
```

Photograph processing time presents the time needed to downsample picture and calculate logistic regression value. Downsampling is more time consuming than logistic regression calculating because all used pixels have to be taken into account.

4.1 Processing Speed Without PSRAM

Without 4 MB PSRAM memory, ESP32-CAM and M5CAMERA boards can use only one photo buffer. In Table 1, four different resolutions and duration of photographing and processing can be seen.

Table 1. Processing speed without PSRAM

Resolution	Pixels	Used pixels	Duration of photographing (ms)	Photograph processing (ms)	Overall time (ms)
QVGA	76800	57600	70	9	79
VGA	307200	230400	145	54	199
XGA	786432	589824	463	130	593
SXGA	1310720	1048576	540	254	794

It is obvious that the duration of photographing depends on photo resolution as well as photo processing depends on it. It can be seen in Figs. 2 and 3, too. There are two different charts because the camera takes a photo in full resolution but microcontroller processes only quadratic part of the photo. There is a difference between pixel numbers in taken and processed photos.

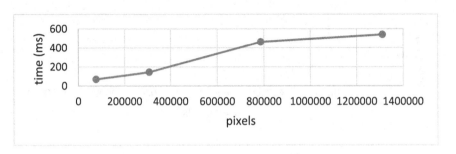

Fig. 2. Duration of photographing without PSRAM

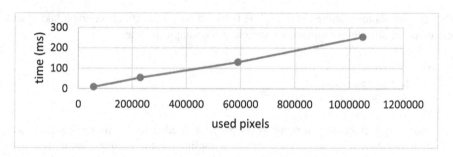

Fig. 3. Photo processing speed without PSRAM

4.2 Processing Speed with PSRAM

With 4 MB PSRAM memory, M5CAMERA board can use more than one photo buffer, so two photo buffers have been enabled in this part of analyze. The results are in Table 2.

Table 2. Processing speed with PSRAM

Resolution	Pixels	Used pixels	Photographing (ms)	Photo processing (ms)	Overall time (ms)
QVGA	76800	57600	<1	12	12
VGA	307200	230400	<1	360	360
XGA	786432	589824	<1	292	292
SXGA	1310720	1048576	<1	1710	1710

It can be seen that the duration of photographing is less than a millisecond despite photo resolution. On the other hand, photo processing times are much higher than in Table 1. Duration of photographing are not presented with the chart, but pho-to processing times are presented on Fig. 4.

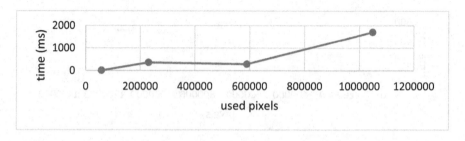

Fig. 4. Photo processing speed with PSRAM

5 Discussion

It is obvious that elapsed time to take a picture depends on picture resolution. It can be concluded that there is a linear connection between number of pixels and elapsed time to take a picture when microcontroller without PSRAM is used. It can be seen in Fig. 2. Values are between 70 ms and 540 ms.

Microcontroller with PSRAM has elapsed time to take a picture less than 1 ms, but as it is mentioned before, a microcontroller with PSRAM use two buffers. While the first picture is being processed, the second is already waiting in the buffer.

Picture processing time, as well as elapsed time to take a picture, depends on picture resolution. Picture processing time consists of downsampling time and logistic regression calculations. Downsampling process consists of two integer multiplications and three integer addition for each pixel in our case. The size of the picture after the downsampling process does not affect the duration of the process because all pixels have to be taken into account.

Microcontroller without PSRAM has downsampling time within the expected range. About 240 ns is needed per pixel for downsampling [14]. Logistic regression calculation time depends on downsampled picture size and includes floating-point multiplication and addition per downsampled picture pixel. Floating-point multiplication and addition take 54 ns with 32bit floating-point numbers [14]. Logistic regression calculation time is very short in our case because our downsampled picture has only nine pixels.

Microcontroller with PSRAM has much higher photo processing times. On producer website can be found that PSRAM uses the same cache as the external flash but when accessing large chunks of data (>32 KB) speeds will fall back to the access speed of the external RAM and it is little over 7 MB/sec [15, 16].

6 Conclusion

In this paper, ESP32 microcontroller has been presented because its quality is proven as well as the boards with cameras have been developed in the last year. Cameras on tested boards (ESP32-CAM and M5CAMERA) have sufficient resolution for most machine learning tasks.

Downsampling process can be time-consuming on ESP32, but machine learning algorithms usually use low-resolution pictures, so it is recommended to set camera capture resolution to the lowest levels. Logistic regression calculation speed depends on number of inputs, but with single-precision floating-point numbers, it lasts about 100 ns per input node. In our case with nine inputs it is negligible.

From our experience, PSRAM usage is not recommended with Arduino core for ESP32 WiFi chip, version 1.0.2. It looks like that Arduino IDE still does not provide enough level of optimization for all boards type. Espressit Systems has developed Espressif IoT Development Framework, and it is the official development framework for the ESP32 chip, and it probably provides the highest optimization level.

We can conclude that ESP32 with camera has enough computing power for simple machine learning tasks and for camera picture taking and preparing for other more powerful processors. In the future analysis it will be interesting to test ESP32 with different neural networks and to try to use both Tensilica Xtensa LX106 cores in calculations because ESP32 has two cores.

References

1. Shi, W., et al.: Edge computing: vision and challenges. IEEE. Int. Things J. **3**(5), 637–646 (2016)
2. ST Microelectronics, STM32Cube.AI: Convert Neural Networks into Optimized Code for STM32, 9 January 2019. https://blog.st.com/stm32cubeai-neural-networks/ Accessed 13 June 2019
3. Lai, L., Suda, N., Chandra, I.V.: CMSIS-NN: efficient neural network kernels for arm cortex-M CPUs. Comput. Res. Repository.svez. abs/1801.06601 (2018)
4. General vision, Presentation of the CurieNeurons on Arduino/Genuino101, 6 June 2016. https://www.general-vision.com/publications/PR_CurieNeuronsPresentation.pdf   Accessed 16 June 2019
5. Allan, A.: Getting Started with the NVIDIA Jetson Nano Developer Kit, 15 April 2019. https://blog.hackster.io/getting-started-with-the-nvidia-jetson-nano-developer-kit-43aa7c298797 Accessed 18 June 2019
6. Ivkovic, I.J., Ivkovic, L.: Analysis of the performance of the new generation of 32-bit microcontrollers for IoT and big data application. In: Proceedings of the International Conference on Information Society and Technology (ICIST 2017), Kopaonik (2017)
7. Espressif Systems, ESP32 Datasheet, 10 April 2019. https://www.espressif.com/sites/default/files/documentation/esp32_datasheet_en.pdf Accessed 11 June 2019
8. Kokoulin, A.N., Tur, A.I., Yuzhakov, A.A., Knyazev, I.A.I.: Hierarchical convolutional neural network architecture in distributed facial recognition system. In: 2019 IEEE Conference of Russian Young Researchers in Electrical and Electronic Engineering (EIConRus), Saint Petersburg and Moscow (2019)
9. Allan, A.: Face Detection and Recognition on the ESP32, 10 December 2018. https://blog.hackster.io/face-detection-and-recognition-on-the-esp32-3b4b9a35c765 Accessed 7 June 2019
10. Espressif System, ESP-WHO. https://github.com/espressif/esp-who Accessed 19 June 2019
11. Sandler, M., Howard, A., Zhu, M., Zhmoginov, I.L.A., Chen, C.: MobileNetV2: inverted residuals and linear bottlenecks. In: Proceedings of the IEEE Conference on computer vision and Pattern Recognition, Salt Lake City (2018)
12. Zhang, K., Zhang, Z., Lii, Z., Qiao, Y.: Joint face detection and alignment using multi-task cascaded convolutional networks. IEEE. Sig. Proc. Lett. **23**(10), 1499–1503 (2016)
13. Cox, D.: The regression analysis of binary sequences. J. Roy. Stat. Soc. Ser. B (Methodol.) **20**(2), 215–242 (1958)
14. Erich11, GitHub, 8 May 2019. https://github.com/espressif/arduino-esp32/issues/2538. Accessed 14 June 2019
15. ESP_Angus, ESP32.com, 13 September 2018. https://esp32.com/viewtopic.php?t=7158. Accessed 14 June 2019
16. Espressif, Support for external RAM, Espressif (2016). https://docs.espressif.com/projects/esp-idf/en/latest/api-guides/external-ram.html. Accessed 15 June 2019

Speech Enhancement Based on Deep AutoEncoder for Remote Arabic Speech Recognition

Bilal Dendani[1,2]([✉]), Halima Bahi[1], and Toufik Sari[1,2]

[1] Computer Science Department, University Badji Mokhtar Annaba,
Annaba, Algeria
{bilal.dendani,halima.bahi,
toufik.sari}@univ-annaba.org
[2] Labged Laboratory, University Badji Mokhtar Annaba, Annaba, Algeria

Abstract. Remote applications that deal with speech need the speech signal to be compressed. First, speech coding transforms the continuous waveform into a numerical form. Then, the digitized signal is compressed with or without loss of information. This transformation affects the original waveform and degrades performances for further recognition of the speech signal. Meanwhile, the transmission is another source of speech degradation. To restore the original "clean" speech, speech enhancement (SE) is widely used, and deep learning algorithms are state-of-the-art, nowadays. In this paper, the target application is a remote Arabic speech recognition system, and the aim of using SE is to improve the accuracy of the speech recognizer. For that purpose, a Deep Auto Encoder (DAE) is used. The effect of the DAE-based SE is studied through different configurations, and the performances are evaluated through accuracy. The results showed an improvement of about 3.17 between the accuracy prior to the SE and that computed with the enhanced speech.

Keywords: Mobile speech recognition · Arabic language · Speech enhancement · Deep learning · Deep AutoEncoder (DAE)

1 Introduction

IN the last few years, there has been a large availability of communication technology, a growing number of people are accessing the internet easily, and digital devices have become increasingly powerful and cheaper; this sparkled the emergence of mobile Automatic Speech Recognition (ASR) applications. Indeed, speech recognition could be used as a core component of several applications such as healthcare [1, 2] or language learning [3, 4]. Two approaches can be used for the deployment of mobile ASR applications, a client-based approach or a remote server-based approach. The remote speech recognition architectures are Network Speech Recognition (NSR) and Distributed Speech Recognition (DSR) [5]; in both of them, the speech signal is captured at the client-side and is transmitted to the server-side for decoding and recognition. The feature extraction stage is located at the client-side for DSR architectures and the server-side for NSR ones.

© Springer Nature Switzerland AG 2020
A. El Moataz et al. (Eds.): ICISP 2020, LNCS 12119, pp. 221–229, 2020.
https://doi.org/10.1007/978-3-030-51935-3_24

In remote speech recognition, the speech signal has to be transmitted, and for that purpose, the speech signal needs to be coded. The coding process aims to minimize bits used to represent a speech signal while preserving the quality of the transmitted speech [6]. An ideal speech coder represents the input speech by a few bits without quality degradation. Obviously, there is a trade-off between the codec bit rate and the quality of the transmitted voice. The transmission process is another source of signal degradation [6]. This paper is about a remote speech recognition system of which performances highly depend on the quality of the coded and transmitted (transcoded) speech signal; to restore the original signal, SE is required.

The SE technique is intended to recover the clean speech signal from its corrupted form. For that purpose, many methods exist such as the minimum mean square error [7] or the Wiener filtering [8], these traditional methods perform well when the speech is corrupted by stationary noises, but are limited in the non-stationary noise environments. Recently, Deep Learning (DL) algorithms have become state-of-the-art in SE, and they deal with stationary and non-stationary noises as well [9–20].

While a great number of researches have addressed the SE of noisy speech, researches that deal with coded SE are rare. However, in [12], the authors have implemented a Convolutional Neural Network (CNN) to enhance the coded speech. The proposed CNN architecture included three kinds of layers: convolutional, max pooling, and up-sampling layers. The authors compared their solution to the G.711 post-processing as a baseline and found that their proposition improved the speech quality in terms of PESQ (Perceptual Evaluation of Speech Quality) for G.711, G.726, and AMR-WB coders. On another side, works focusing on the enhancement of Arabic speech, and based on DL models are quasi-inexistent.

The paper tackles the problem of SE of transcoded speech in a remote context, implemented throughout a network-based ASR architecture. In this paper, we propose the use of a DAE (Deep AutoEncoder) algorithm for SE; the effect of the DAE is studied in different contexts of the input window. The performances of the proposed architecture are assessed in terms of the accuracy of the speech recognizer. The speech recognizer is considered as a black box and is based on Hidden Markov Models (HMMs) [21, 22].

The next section outlines the DL-based SE models. Section 3 presents the target application and the context of the study from speech coding towards SE and recognition. Results and discussion are presented in Sect. 4. Finally, a conclusion is drawn.

2   Related Works

Recently, different DL models, known as data-driven models, have been explored, and they provided good results for SE. Most of the time, the DL-based SE methods model the non-linearity between noisy speech signals and clean speech signals, when the clean speech is accessible, thus they can recover clean speech signals from corrupted versions.

The fully connected-based models exploit the mapping function to enhance speech. These models include the deep denoising auto-encoders (DDAE) [9, 13], and the fully connected deep neural network (DNN) models [14, 15]. Lu et al. [9] have applied a

DDAE algorithm for noise reduction and SE, in the training stage, both input and output were clean speech signals. Thus, the DDAE is expected to only encode statistical information of the clean speech. In their further experiments reported in [13], the DDAE was trained using noisy speech as input and clean speech as output. Therefore, the DDAE is expected to explicitly learn the difference between clean and noisy signals. Furthermore, Xu et al. [14, 15] proposed a regression-based SE DNN approach to estimate the mapping function between noisy and clean utterances. The DNN model was trained on a large data of about 100 h with multiple noise conditions. Other works focused on recurrent neural networks (RNN) using noisy/clean speech pairs as input and output respectively during the training stage [16].

While the fully connected models use a lot of parameters, CNN-based SE architectures use a fewer number of parameters. Indeed, the CNN-based models deal with the local temporal and spectral structure of the speech. Fu et al. [17] investigated the CNN model to restore clean speech from a noisy version using SNR-aware algorithms. Later, they applied a fully convolution network (FCN) to model simultaneously the high and the low-frequency components of a raw waveform [19]. The results of the FCN algorithm outperformed those of the CNN and the DNN when the waveforms are used as inputs. The FCN has been applied also in an embedded system to enhance speech for hearing aids [18]. The authors in [11] used both convolutional and recurrent neural network architectures for SE. The proposed model improves the PESQ and the WER (Word Error Rate) by 0.6 and 1% respectively, the model also generalizes better on unseen noise. Other DL-based models have been tested for SE such as the generative adversarial networks [20].

3 Speech Enhancement Based on Deep AutoEncoder for Remote Arabic Speech Recognition

Our investigation is about the impact of SE on network speech recognizer performances. In the NSR model, the speech signal is captured at the client-side, then, it is transmitted to the server where the recognition is performed [23, 24]. Figure 1 depicts the block diagram of the proposed remote Arabic speech recognition system.

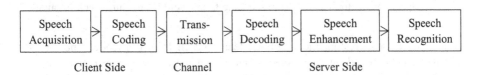

Fig. 1. Block diagram for remote Arabic speech recognition

The received signal is coded at the client-side. For the purpose of this study, the G.711 codec is applied. Then, the coded speech is transmitted to the server-side in real-life conditions. At the server-side, the received signal is enhanced by DAE previously trained. Finally, the HMM-based Arabic speech recognizer is fed with the enhanced speech signal. The computed accuracy serves to assess the performances of the DAE-based SE system.

3.1 G.711 Codec

G.711 is a waveform, lossless compression speech codec, developed by ITU-T [25] and released in 1972. G.711 codec is mainly used for speech transmission over the telephony system, provides a bit rate at 64 kbps and exists in two versions: μ-law and A-law. Both versions code speech sample with 8 bits, provide a reduction of 50% of file size and use a bandwidth compared to the original signal which uses 16 bits as a sample size. In [24], experiments, made in the context of the Arabic language, showed that the narrowband high bit rate codec's G.711 provides good performances.

3.2 Deep AutoEncoder for Arabic Speech Enhancement

DAE is a feedforward neural network with many hidden layers. Given a pattern X, the encoding layers reduce X into a smaller dimension, then, the decoder layers reconstruct the pattern Y (a version close to X). When applied in SE, DAEs aim to recover the clean version of the speech from its corrupted one. For that purpose, the model is trained on noisy/clean pairs of speech. In the training stage, the DAE is fed with a corrupted version of the speech signal; in our case with the transcoded speech, as input and the output is the original signal. During the test stage, the DAE is fed with the transcoded version of the speech and it is intended to provide the clean one (Fig. 2).

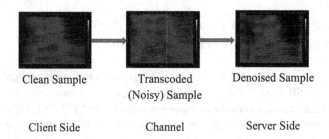

Clean Sample   Transcoded   Denoised Sample
                    (Noisy) Sample

Client Side   Channel   Server Side

Fig. 2. Spectrograms for a sample corresponding to the word "Zero", in several locations

We endeavor to restore the original (clean) speech from transcoded speech. We trained the DAE based on clean speech samples, using an Arabic speech corpus for isolated words. The incoming speech is framed into windows of 512 samples. From each frame are extracted coefficients corresponding to the spectrum power.

3.3 Speech Recognizer

For the recognition purpose, HMMs stand for the acoustic models. The training dataset contains approximately 70% of the utterances in which the list of isolated words is pronounced by 50 native Arabic speakers. Testing data represents 30% of the total number of utterances. The HMM-based speech recognizer is used as a black box without tuning during our experiments related to SE.

4   Experimental Results and Discussion

4.1   Arabic Speech Data Set

The corpus used in experimentations is an Arabic speech corpus for isolated words [26]. The corpus has been developed by the Department of Management Information Systems of King Faisal University. It contains 9992 recorded utterances of 50 speakers pronouncing 20 words. The list of the pronounced words is:

نعم, التسديد, الرصيد, التحويل, التنشيط, تسعه, ثمانيه, سبعه, سته, خمسه, أربعه, ثلاثه, اثنان, واحد, صفر}
.{انهاء, الحساب, البيانات, التمويل, لا

Translated as: {Zero, One, Two, Three, Four, Five, Six, Seven, Eight, Nine, Activation, Transfer, Balance, Payment, Yes, No, Funding, Data, Account, End}.

The recordings are sampled at 44100 Hz (stereo channel) with 16 bits precision. The corpus has been adapted by down-sampling it to 8 kHz sampling rate as if the recordings were done through mobile devices, and a mono channel is considered.

The DAE is trained with approximately 1500 utterances from 50 speakers. The test dataset contains 445 utterances from 15 speakers.

4.2   Effects of Coding, Transmission, and Enhancement

Prior to the coding and transmission tasks, the accuracy is computed at the client-side. Then, to assess the effect of the transmission process, the accuracy is computed at the server-side after the reception of the uncompressed signal. Finally, to assess coding and transmission effects, the accuracy is computed at the server-side after the reception of the compressed signal. The HMM-based recognition takes place without prior enhancement of the incoming signal. Table 1 reports the accuracy of the ASR system in the three situations considering the utterances of the test dataset.

Table 1. Accuracy Offline, Online with and without compression

Mode	Offline	Online Uncompressed	Online G.711
Accuracy	64,44	64,33	62,55

Table 1 shows that transmission slightly decreased the accuracy while the coding/transmission clearly did with a loss of about 1.89.

To overcome the degradation of the accuracy due to the use of coding and transmission, we suggest the use of the DAE algorithm to enhance the corrupted signal. Figure 3 shows the impact of the SE stage on the accuracy computed at the server-side.

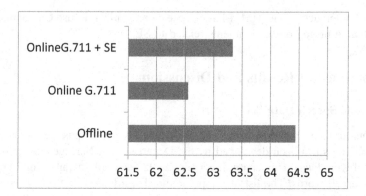

Fig. 3. Accuracy computed in offline mode, after transmission, and after SE

Figure 3 shows an improvement of the accuracy when the speech signal submitted to the decoder was already enhanced.

4.3   Effect of the Context Length

Although the proposed DNN is only intended to enhance the speech, the first tuning concerns the inclusion of the left/right context of the target window. As already said, from each window of 512 samples were extracted the corresponding spectrum power coefficients, and as the obtained vector is symmetric, only 257 coefficients are kept. Thus, the input layer, as well as the output layer, comprises 257 neurons. Alternately, as the context of neighbor frames is important in speech, we make experiments where both sides of a target window are considered. Figure 4 depicts DAE architecture where two frames from the left and two from the right of the target window are considered.

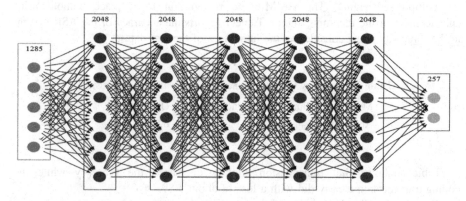

Fig. 4. DAE where the input corresponds to five consecutive frames and the output stands for the middle one

Table 2 reports accuracy obtained after the enhancement of the received signal considering models of DAE trained with several context length of the target window. Considering the DAE depicted in Fig. 4, and the utterances of the test dataset, the results are reported in Table 2.

Table 2. Accuracy of enhanced speech considering several context lengths

#Input frames	01	03	**05**	07	11	13
Accuracy	63.333	64.276	**65.724**	53.153	50.922	48.981

Table 2 shows that an increasing number of windows surrounding the target speech frame impacts accuracy. Indeed, adding one window at each side improves the accuracy, and by enlarging the context to two windows at each side, the accuracy grows to 65.72%. Although, when the context becomes larger and the complexity of the DAE increases (three windows per side or more), the accuracy decreased drastically.

In summary, the DAE-based SE process including one or two windows at each side of the target frame (or even only the target frame) improved the accuracy of the ASR application and provides a robust remote ASR system even when the HMMs were trained on clean speech.

5   Conclusion

Remote speech recognition is a challenging task and is for much importance for many applications such as security, healthcare, education, etc. The presented work aims to improve the communication client-server on mobile networks by reducing the impact of degraded ASR performance which is introduced by speech G.711 codec. A deep auto-encoder algorithm is used to enhance speech degraded by the G.711 codec and the transmission. Results showed that the accuracy of speech recognizer has been increased when enhancement of the speech is performed. Future works will address several configurations of the DAE architecture depth, and the number of neurons in each layer. One of the remaining challenges is to explore other algorithms such as the CNN and to consider larger Arabic speech corpora.

References

1. Hossain, M.S.: Patient sate recognition system for healthcare using speech and facial expressions. J. Med. Syst. **40**, 12 (2016)
2. Necibi, K., Bahi, H., Sari, T.: Automatic speech recognition technology for speech disorders analysis. In: Speech, Image and Language Processing for Human-Computer Interaction: Multi-modal Advancements. IGI-global (2012)
3. O'Brien, M.G., et al.: Directions for the future of technology in pronunciation research and teaching. J. Second Lang. Pronunciation **4**(2), 182–207 (2018)
4. Lee, J., Lee, C.H., Kim, D.-W., Kang, B.-Y.: Smartphone-assisted pronunciation learning technique for ambient intelligence. IEEE Access **5**, 312–325 (2016)

5. Schmitt, A., Zaykovskiy, D., Minker, W.: Speech recognition for mobile devices. Int. J. Speech Technol. **11**(2), 63–72 (2008)
6. Chu, W.C.: Speech Coding Algorithms: Foundation and Evolution of Standardized Coders. Wiley, Hoboken (2004)
7. Ephraim, Y., Malah, D.: Speech enhancement using a minimum-mean square error short-time spectral amplitude estimator. IEEE Trans. Acoust. Speech Sig. Process. **32**, 1109–1121 (1984)
8. Sreenivas, T.V., Pradeep, K.: Codebook constrained Wiener filtering for speech enhancement. IEEE Trans. Speech Audio Process. **4**(5), 383–389 (1996)
9. Lu, X., Matsuda, S., Hori, C., Kashioka, H.: Speech restoration based on deep learning autoencoder with layer-wised pretraining. In: Proceedings of INTERSPEECH, Portland, USA, pp. 1504–1507 (2012)
10. Kumar, A., Florencio, D.: Speech enhancement in multiple-noise conditions using deep neural networks. In: Proceedings of the Annual Conference of the International Speech Communication Association, INTERSPEECH, San Francisco, USA (2016)
11. Zhao, H., Zarar, S., Tashev, I., Lee, C.H.: Convolutional-recurrent neural networks for speech enhancement. In: Proceedings of ICASSP, pp. 2401–2405 (2018)
12. Zhao, Z., Liu, H., Fingscheidt, T.: Convolutional neural networks to enhance coded speech. IEEE/ACM Trans. Audio Speech Lang. Process. **27**(4), 663–678 (2018)
13. Lu, X., Tsao, Y., Matsuda, S., Hori, C.: Speech enhancement based on deep denoising autoencoder. In: Annual Conference of the International Speech Communication Association, INTERSPEECH, pp. 436–440 (2013)
14. Xu, Y.: An experimental study on speech enhancement based on deep neural networks. IEEE/ACM Trans. Audio Speech Lang. Process. **23**(1), 7–19 (2015)
15. Xu, Y., Du, J., Dai, L.R., Lee, C.H.: A regression approach to speech enhancement based on deep neural networks. IEEE/ACM Trans. Audio Speech Lang. Process. **23**(1), 7–19 (2015)
16. Weninger, F., et al.: Speech enhancement with LSTM recurrent neural networks and its application to noise-robust ASR. In: Vincent, E., Yeredor, A., Koldovský, Z., Tichavský, P. (eds.) LVA/ICA 2015. LNCS, vol. 9237, pp. 91–99. Springer, Heidelberg (2015). https://doi.org/10.1007/978-3-319-22482-4_11
17. Fu, S.W., Tsao, Y., Lu, X.: SNR-aware convolutional neural network modeling for speech enhancement. In: Proceedings of Interspeech (2016)
18. Park, S.R., Lee, J.W.: A fully convolutional neural network for speech enhancement. In: Annual Conference of the International Speech Communication Association INTERSPEECH, vol. 2017, no. 2, pp. 1993–1997 (2017)
19. Fu, S.W., Tsao, Y., Lu, X., Kawai, H.: Raw waveform-based speech enhancement by fully convolutional networks. In: Asia-Pacific Signal & Information Processing Association Annual Summit Conference APSIPA ASC 2017, vol. 2018, no. December, pp. 6–12 (2018)
20. Pascual, S., Bonafonte, A., Serra, J.: SEGAN: speech enhancement generative adversarial network. In: Proceedings of the Annual Conference of the International Speech Communication Association, INTERSPEECH, vol. 2017, pp. 3642–3646 (2017)
21. Rabiner, L.R.: A tutorial on hidden Markov models and selected applications in speech recognition. Proc. IEEE **77**–2, 257–286 (1989)
22. Bahi, H., Sellami, M.: Combination of vector quantization and hidden Markov models for Arabic speech recognition. In: Proceedings ACS/IEEE International Conference on Computer Systems and Applications, Beirut, Lebanon (2001)
23. Tan, Z.H., Varga, I.: Network, distributed and embedded speech recognition: an overview. In: Tan, Z.-H., Lindberg, B. (eds.) Automatic Speech Recognition on Mobile Devices and over Communication Networks. Advances in Pattern Recognition. Springer, London (2008). https://doi.org/10.1007/978-1-84800-143-5_1

24. Dendani, B., Bahi, H., Sari, T.: A ubiquitous application for Arabic speech recognition. In: Proceedings of the International Conference on Artificial Intelligence and Information Technology, Ouargla, Algeria, pp. 281–284 (2019)

25. Digital cellular telecommunications system (Phase 2+); Half rate speech; ANSI-C code for the GSM half rate speech codec (GSM 06.06 version 7.0.1 Release 1998) GLOBAL SYSTEM FOR MOBILE COMMUNICATIONS (2000)

26. Alalshekmubarak, A., Smith, L.S.: On improving the classification capability of reservoir computing for Arabic speech recognition. In: Wermter, S., et al. (eds.) ICANN 2014. LNCS, vol. 8681, pp. 225–232. Springer, Cham (2014). https://doi.org/10.1007/978-3-319-11179-7_29

Pattern Recognition

Handwriting Based Gender Classification Using COLD and Hinge Features

Abdeljalil Gattal[1(✉)], Chawki Djeddi[1,2], Ameur Bensefia[3], and Abdellatif Ennaji[2]

[1] Department of Mathematics and Computer Science, Larbi Tebessi University, Tebessa, Algeria
{abdeljalil.gattal,c.djeddi}@univ-tebessa.dz
[2] LITIS Laboratory, University of Rouen, Rouen, France
abdel.ennaji@univ-rouen.fr
[3] Higher Colleges of Technology, CIS Division, Abu Dhabi, UAE
abensefia@hct.ac.ae

Abstract. Gender Classification from handwriting is still considered to be challenging due to homogeneous vision comparing male and female handwritten documents. This paper presents a new method based on Cloud of Line Distribution (COLD) and Hinge feature for distinguishing the gender from handwriting. The SVM classifier combination decides the assigned class based on the maximum of the two decisions values resulting from COLD and Hinge feature. The proposed approach is evaluated on the standard QUWI dataset and following the framework protocol described in the ICFHR 2016 competition. Obtained results are promising regarding the classification rates announced in the literature.

Keywords: Handwriting · COLD feature · Hinge feature · Gender classification · SVM classifier

1 Introduction

The automated off-line handwriting analysis is one of the most researched problems in pattern recognition, due to its wide variety of applications, such as text recognition, writer identification [1], script identification [2], gender classification [3–8] and much more.

The automatic classification of gender-based on handwritten samples is the task that allows distinguishing between the writer's samples produced by males from those produced by females. Some psychological studies [9–11] confirmed that we could distinguish between the two genders writings thanks to some differences; where in general female handwritings tend to be a more uniform and regular, while male handwritings tend to be spikier and slanted. This task has attracted a lot of interest due to its application in forensic document examination, where it allows the examiners to focus on a particular category of writers. Thus, a handwriting based gender recognition system can be used in combination with a writer recognition system to process only part of the dataset, which may improve the processing rate in terms of performance and

© Springer Nature Switzerland AG 2020
A. El Moataz et al. (Eds.): ICISP 2020, LNCS 12119, pp. 233–242, 2020.
https://doi.org/10.1007/978-3-030-51935-3_25

time. From handwriting variability point of view, the gender classification is in the middle, between handwriting recognition; where the variability needs to be blurred to emerge the common characteristic across all writers, and the writer identification, where the variability needs to be highlighted to arise differences between all writers. Therefore, in a gender recognition task, we need to emerge the handwriting variability, but not as much as in the writer identification, and not too low as in the handwriting recognition, since we are looking to find a common feature between a particular group of writers (males vs females), which makes this task delicate. Despite the fact, that the gender classification has received a lot of attention in the document analysis community [3–8] most of the research has been conducted by enhancing the feature extraction or/and classification process. In general, the gender classification problem makes a challenging task for a two-class problem. The study aims to focus on the enhancement of the feature extraction step to improve the classification rates on off-line handwritten documents.

In this paper, we present an original set of global features designed for gender recognition task: Cloud of Line Distribution (COLD) feature and Hinge features, coupled with two dedicated SVM classifiers for each set. We demonstrate through this study, how a maximum value of both SVM decisions; for COLD and Hinge features, can achieve high accuracy rates on handwriting-based gender recognition. Furthermore, the paper analyses how current state-of-the-art methods in gender classification perform on handwritten document dataset of ICFHR 2016 competition [12]. The best approach proposed in this competition was the "MCS-NUST", which uses Histogram of Local Binary Patterns (LBP), Histogram of Oriented Gradients (HoG) and Gray-level Co-occurrence Matrices (GLCM) to extract features from binarized images. These features trained using the SVMs classifier as well as SVM ensembles with bagging. Thus, we settled the ICFHR 2016 competition as a framework of comparison between our approach and those proposed in this event [12–14] where different evaluations mode have been performed (single and muti-scripts).

The paper is organized as follows: first, we present a literature review about some of the significant works conducted in gender recognition based on handwriting. Second, we detail our methodology where the features and the classifiers employed in our study are discussed. Section 4, details the experiments conducted along with a comparative analysis and discussion on the realized results. Finally, we conclude the paper with a discussion on future perspectives on the subject.

2   Literature Review

In this section, we present an exhaustive list of works that dealt with automatic gender classification based on handwriting. Since the task of gender classification is a recognition task, we focused on the elementary modules of such systems: features and classifier, in addition to the dataset used to evaluate the proposed system. In general, most of the proposed works articulate their methodologies either on local features and/or global features.

The authors in [6] proposed an approach that focuses on individual classification based on their age, nationality or gender. The formulation of this problem has been considered the same, whatever is the classification criterion. In other words, the same set of features and classification have been used in all cases. To this end, the authors proposed a set of features made of: direction, curvature, tortuosity, chain code and edge-based direction; combined with a Random Forest classifier (RF) and a kernel discriminant analysis (KDA). A rate of 74.05% in gender prediction was achieved when all the features where used with in QUWI dataset.

Another approach, in gender recognition, where local features were preferred to global ones, is presented in [5]. The Histogram Oriented Gradient (HOG) and the Local Binary Pattern (LBP) and grid features were used. A similar approach proposed in [15], where the same set of features was used, in addition to a feature resulting from a segmentation-based fractal texture analysis (SFTA) and features extracted from gray-level co-occurrence matrices (GLCM). The approach was evaluated on the QUWI dataset with a different combination of classifiers (SVM, ANN, KNN) and features. The best rate obtained by the authors is 85% when using all features with the three classifiers.

In [7] the gender classification has been tackled with the usage of global features; wavelet transform and symbol dynamic filtering (SDF). Firstly, the handwritten document is decomposed into sub-bands using the discrete wavelet transforms (DWT). These sub-bands are then extended into data-sequences; where a maximum entropy partitioning is applied. Finally, a features vector is built by concatenating all the SDF features obtained by the algorithm described in [16]. Two different datasets were used alternatively, for training and test, the QUWI and the MSHD dataset [17], combined with two different classifiers: ANN and SVM. The best gender recognition rate announced by the authors is 80%. Liwicki et al. [4] also stated that gender classification could be successful with global features: Gaussian Mixture Models (GMM). In their methodology, offline and online features were used, since they processed the two types of handwritten documents. Even though the online gender classification is out of the scope of our paper, we consider that it is interesting to mention the impact of additional features in the performances rates. The system was evaluated with the IAM dataset, where 100 samples were used for training and validation and 50 samples for testing (equally distributed between males and females). Among the different combination of features used, the authors reached 67.75% as the best classification rate where both online and offline features were used and only 55.39% with the offline features.

Recently, gender classification from offline multi-script handwriting images using Oriented Basic Image Features (oBIFs) has been applied on handwritten document dataset of ICFHR 2016 competition [18]. The authors focused on the textural information in handwriting using combinations of different configurations of oBIFs histograms and oBIFs columns histograms. These features were used to train an SVM classifier and have been evaluated on the QUWI dataset, where 76% of good recognition rate was reached.

Previous works have shown that the texture features can improve the performance of gender classification system. We assume that the shape information and the curvature-based information provided by Hinge and COLDs features can characterize the writing style.

3   Proposed Methodology

Most of the works conducted on gender classification are continuity of works undertaken in the writer's identification, where almost the same features were used. We consider that this assumption can be valid if we use global features rather than local features. Indeed, global features, even if they have discriminant power, they tend to emerge similarities between writings, which make them suitable for the gender recognition task. Therefore, in this section, we aim to design a new method based on Cloud of Line Distribution (COLD) and Hinge feature to capture female or male writing styles of handwritten documents (See Fig. 1).

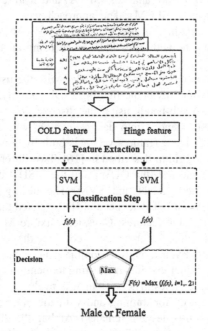

Fig. 1. Overview of the proposed system

3.1   Features Extraction

In our proposed methodology of gender classification, we have selected to use two sets of features: COLD feature and Hinge feature. This set of features has been successfully applied to the writer identification problem [19, 20]. In our study, we will take advantage of their strong discriminatory representation captured through the curvature information and contour information embedded in the handwritten document. The subsections below provide a detailed description of these features.

COLD Feature
The COLD feature [19] is inspired by the shape context descriptor, with respect to the extracted information into a log-polar histogram, which is more sensitive to regions of nearby the center than to those farther away [21]. The feature extracts unique shapes of

handwritten text components by analyzing the relationship between dominant points; such as straight, angle-oriented features and curvature over contours of handwritten text components. Although, the COLD captures the writing style and the variations between different handwriting samples from the same gender by comparing the COLDs using the dominant points.

The log-polar space is based on three parameters: the distance between two consecutive rings in the log space Dc, number of angular intervals Np, and the number of distance intervals Nq. In our experimentations, we have empirically set these parameters as: Dc = 5, Np = 12, and Nq = 7 [19]. Besides, the combination of COLDs with different k (level of connection between dominant points) achieves the best performance. Therefore, the ultimate COLD feature is generated by concatenating the COLDs with different k together.

Hinge Feature

The contour-based feature was designed to capture the curvature of the ink trace of the document images, which is considered to be very discriminatory between handwritings. The best contour-based features reported in the literature are the Hinge [22], Quill-Hinge [23], and Delta-n Hinge features [20].

The Hinge feature [22] is the probability distribution of orientations of two legs of the obtained "hinge" based on edges or contours together attached at a current pixel. There are two parameters in the Hinge feature: the number of angle bins p and the leg length r. In our case, we set p = 40, as suggested in [23]. Therefore, it has been extended to the Delta-n Hinge [23] to achieve the rotation-invariant property. There are four parameters in the proposed method, the number of angle bins p, leg length r, Manhattan distance Δl, and the number of derivative n. In our experimentations we set p = 40, the Manhattan distance $\Delta l = 7$, and n = 2.

The two sets of features COLD and Hinge are combined only at the decision level. In other words, we proceed with the gender classification by submitting each set of features to a separate classifier, as discussed in the following section.

3.2 Decision Strategy

In an attempt to enhance the reliability of the accuracy rates of the proposed system, a decision module is designed to produce the final decision. This module is proposed with two different SVM classifiers [24], where each classifier is trained by one set of features: COLD or Hinge. The training of each SVM requires the selection of two parameters, which are the regularization parameter C and the Radial Basis Function (RBF) kernel parameter (σ) is selected in the range $\sigma = [1:50]$, with the soft margin parameter C fixed to 10, which allows improvement in the accuracy rate. The choice of hyperparameters values selected by experimentation to achieve better performance while overcoming over fitting to avoid identity or isometric identity kernel matrix. However, there is no robust rule available to guide how to choose the hyper parameters values appropriately. The final decision of our module is taken by selecting the maximum value of the decision function produced by the two SVM classifiers. Experimentally, the maximum (Max) decision resulting from COLD and Hinge features achieved the best performance compared to other standard statistical reasoning measures such as average (Avr) and minimum (Min). In the next section, we present the experimental settings and the corresponding results.

4   Experimentations and Results

We carried out a series of experiments to evaluate the effectiveness of the proposed system for gender classification from handwriting on competition datasets.

A subset of QUWI dataset, used in ICFHR 2016 competition [12, 25], comprised 1000 handwritten documents samples were selected: 500 samples were provided as a training set, 250 as the validation set and 250 as a test set. The most interesting aspect of this competition is the usage of a dataset with handwritten documents of the same writer in multiple scripts, and also the opportunity to study the performance of script-dependency (Sub-task 2A and Sub-task 2B) and script-independency (Sub-task 2C and Sub-task 2D) in a multi-script experimental setup. Our experiments aim to study the effect of different k combination of COLDs, and various leg length r from Hinge and Delta-n Hinge features, while the number of angle bins $p = 40$. The realized classification rates are illustrated in Fig. 2 and Fig. 3.

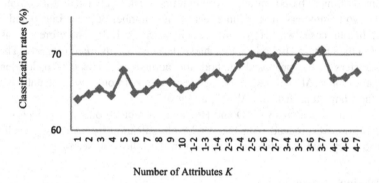

Fig. 2. Classification rates on both scripts QUWI dataset used in ICFHR 2016 competition using COLD feature and its combination.

It can be seen that the combination of COLDs features while k = {3,4,5,6,7} outperforms the other COLDs features. Therefore, the Hinge feature with r = 25 while p = 40 does better than others Hinge and Delta-n Hinge features configurations. The combination of COLDs features is extracted from the complete handwriting image as well as they are obtained by applying the Uniform Grid Sampling (UGS) [26] to the handwriting; this allows extracting features from different regions of the image separately.

According to the previous results, we evaluated the proposed method using the decision strategy with optimal features on both scripts (all samples in Arabic and English). Table 1 reports the different features and the proposed method.

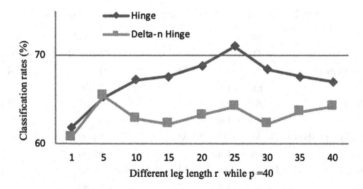

Fig. 3. Classification rates on both scripts QUWI dataset used in ICFHR 2016 competition using hinge feature and Delta-n Hinge feature.

Table 1. Gender classification rates with different features the proposed method

Features description		Feature parameters	UGS	Dim	Classification rates (%)
F1	Hinge	r = 25, p = 40	No	780	71.00
F2	Deltahinge	r = 5, p = 40	No	780	65.40
F3	COLD feature	k = 5	No	84	68.00
F4	COLDs features Combination	k = 3,4,5,6,7	No	420	70.60
F5		k = 3,4,5,6,7	1x2	840	71.00
Proposed method					73.60

It can be seen that the maximum value on the decision functions produced by COLDs features extracted from the 1×2 regions of the image while k = {3,4,5,6,7} and the Hinge feature with r = 25 while p = 40 outperforms all other features reporting classification rates of 73.60% on both scripts. Furthermore, we use our proposed methodology according to the competition protocol in script-dependent and script-independent evaluations. According to sub-task 2A and sub-task 2B for Script-dependent evaluation described in the ICFHR2016 competition, the protocol is performed using 500 script samples in the training set and 250 same script samples (English (En) or Arabic (Ar)) in the test set. The results of our proposed approach with different ones in Script-dependent evaluation are summarized in Table 2.

It can be seen from Table 2 that our approach realizes an average rate of 74.20%, which is comparable to performances of the top 2 systems, according to the competition protocol. However, it should be noted that the curvature and contour information used in our method can compete for the textural information [18]. These results validate the effectiveness of the proposed method in Script-dependent evaluation mode.

According to sub-task 2C and sub-task 2D in Script-independent evaluation, the experiments is performed using 500 script samples in the training set and 250 other script samples in the test set. Table 3 shows that the classification rates of our proposed approach in script-independent evaluation compared with different approaches.

Table 2. Classification rates of script-dependent evaluations.

Method	Script		Classifier	Classification rates (%)	
	Train	Test		Script dependent	Average
Proposed method	Ar	Ar	SVM	74.80	74.20
	En	En		73.60	
Gattal et al. [18]	Ar	Ar	SVM	74.80	75.00
	En	En		75.20	
MCS-NUST1 method [12]	Ar	Ar	SVM	61.60	58.80
	En	En		56.00	
MCS-NUST2 method [12]	Ar	Ar	SVM with bagging	60.80	59.20
	En	En		57.60	
Nuremberg1 method [12]	Ar	Ar	SVM	58.00	56.00
	En	En		54.00	
Nuremberg2 method [12]	Ar	Ar	SVM	46.40	60.20
	En	En		74.00	

Table 3. Classification rates of script-independent evaluations.

Method	Script		Classifier	Classification rates (%)	
	Train	Test		Script independent	Average
Proposed method	Ar	En	SVM	64.00	64.40
	En	Ar		64.80	
Gattal et al. [18]	Ar	En	SVM	66.00	68.00
	En	Ar		70.00	
MCS-NUST1 method [12]	Ar	En	SVM	57.60	58.60
	En	Ar		59.60	
MCS-NUST2 method [12]	Ar	En	SVM with bagging	58.40	58.80
	En	Ar		59.20	
Nuremberg1 method [12]	Ar	En	SVM	56.00	58.60
	En	Ar		61.20	
Nuremberg2 method [12]	Ar	En	SVM	72.40	58.80
	En	Ar		45.20	

We notice that the proposed methodology achieves higher average rate in script-independent evaluations than other methods in the literature, except Gattal et al. [18] method. The average classification rate is 64.40%. We clearly note that the proposed method is in the top 2 methods, according to the competition protocol in script-dependent and script-independent evaluations. Besides, in some cases, our proposed method can provide unsuccessful classification results. As example, male handwritten documents found to have a homogeneous vision comparing to female handwritten documents and vice versa.

5   Conclusions and Future Works

An effective approach for characterizing gender based on their handwriting is presented, which that exploits COLD and Hinge as features. Different combinations of the selected features are investigated using an SVM classifier. The system is evaluated using the same experimental protocol described in the ICFHR2016 Competition on Multi-script Writer Demographics Classification. The results showed that our approach outperforms the existing methods reported in the mentioned competition, except the new method proposed by Gattal et al. [18].

In our further study on this problem, we intend to investigate other kinds of features to characterize the gender from handwriting by exploring feature selection techniques to identify the most appropriate descriptors for this problem. Finally, we also plan to determine different demographic traits, such as age and handedness from handwriting.

References

1. Srihari, S., Cha, S.-H., Arora, H., Lee, S.: Individuality of handwriting. J. Forensic Sci. **47**, 856–872 (2002)
2. Singh, P.K., Sarkar, R., Nasipuri, M.: Offline script identification from multilingual Indic-script documents: a state-of-the-art. Comput. Sci. Rev. **15–16**, 1–28 (2015)
3. Siddiqi, I., Djeddi, C., Raza, A., Souici-meslati, L.: Automatic analysis of handwriting for gender classification. Pattern Anal. Appl. **18**(4), 887–899 (2015)
4. Liwicki, M., Schlapbach, A., Bunke, H.: Automatic gender detection using on-line and off-line information. Pattern Anal. Appl. **14**(1), 87–92 (2011)
5. Bouadjenek N., Nemmour, H., Chibani Y.: Local descriptors to improve off-line handwriting-based gender prediction. In: 2014 6th International Conference of Soft Computing and Pattern Recognition (SoCPaR), pp. 43–47 (2014)
6. Al Maadeed, S., Hassaine, A.: Automatic prediction of age, gender, and nationality in offline handwriting. EURASIP J. Image Video Process. **2014**(1), 1–10 (2014). https://doi.org/10.1186/1687-5281-2014-10
7. Akbari, Y., Nouri, K., Sadri, J., Djeddi, C., Siddiqi, I.: Wavelet-based gender detection on off-line handwritten documents using probabilistic finite state automata. Image Vis. Comput. **59**, 17–30 (2017)
8. Mirza, A., Moetesum, M., Siddiqi, I., Djeddi, C.: Gender classification from offline handwriting images using textural features. In: 2016 15th International Conference on Frontiers in Handwriting Recognition (ICFHR), pp. 395–398 (2016)
9. Goodenough, F.L.: Sex differences in judging the sex of handwriting. J. Soc. Psychol. **22**(1), 61–68 (1945)
10. Hartley, J.: Sex differences in handwriting: a comment on spear. Br. Educ. Res. J. **17**(2), 141–145 (1991)
11. Burr, V.: Judging gender from samples of adult handwriting: accuracy and use of cues. J. Soc. Psychol. **142**(6), 691–700 (2002)
12. Djeddi, C., Al-Maadeed, S., Gattal, A., Siddiqi, I., Ennaji, A., Abed, H.E.: ICFHR2016 competition on multi-script writer demographics classification using 'QUWI' database. In: 2016 15th International Conference on Frontiers in Handwriting Recognition (ICFHR), pp. 602–606 (2016)

13. Hassaïne, A., Al Maadeed, S., Aljaam, J., Jaoua, A.: ICDAR 2013 competition on gender prediction from handwriting. In: 2013 12th International Conference on Document Analysis and Recognition, pp. 1417–1421 (2013)
14. Djeddi C., Al-Maadeed S., Gattal A., Siddiqi I., Souici-Meslati L., El Abed H.: ICDAR2015 competition on multi-script writer identification and gender classification using 'QUWI' database. In: 2015 13th International Conference on Document Analysis and Recognition (ICDAR), pp. 1191–1195 (2015)
15. Ahmed, M., Rasool, A.G., Afzal, H., Siddiqi, I.: Improving handwriting based gender classification using ensemble classifiers. Expert Syst. Appl. **85**, 158–168 (2017)
16. Bahrampour, S., Ray, A., Sarkar, S., Damarla, T., Nasrabadi, N.M.: Performance comparison of feature extraction algorithms for target detection and classification. Pattern Recognit. Lett. **34**(16), 2126–2134 (2013)
17. Djeddi, C., Gattal, A., Souici-Meslati, L., Siddiqi, I., Chibani, Y., El Abed, H.: LAMIS-MSHD: a multi-script offline handwriting database. In: 2014 14th International Conference on Frontiers in Handwriting Recognition, pp. 93–97 (2014)
18. Gattal, A., Djeddi, C., Siddiqi, I., Chibani, Y.: Gender classification from offline multi-script handwriting images using oriented Basic Image Features (oBIFs). Expert Syst. Appl. **99**, 155–167 (2018)
19. He, S., Schomaker, L.: Writer identification using curvature-free features. Pattern Recognit. **63**, 451–464 (2017)
20. He S., Schomaker L.: Delta-n Hinge: rotation-invariant features for writer identification. In: 2014 22nd International Conference on Pattern Recognition, pp. 2023–2028 (2014)
21. Belongie, S., Malik, J., Puzicha, J.: Shape matching and object recognition using shape contexts. IEEE Trans. Pattern Anal. Mach. Intell. **24**(4), 509–522 (2002)
22. Bulacu, M., Schomaker, L.: Text-independent writer identification and verification using textural and allographic features. IEEE Trans. Pattern Anal. Mach. Intell. **29**(4), 701–717 (2007)
23. Brink, A.A., Smit, J., Bulacu, M.L., Schomaker, L.R.B.: Writer identification using directional ink-trace width measurements. Pattern Recogn. **45**(1), 162–171 (2012)
24. Apnik, V.N.: The Nature of Statistical Learning Theory. Springer, Heidelberg (1995). https://doi.org/10.1007/978-1-4757-2440-0
25. Maadeed, S.A., Ayouby, W., Hassaïne, A., Aljaam, J.M.: QUWI: an Arabic and English handwriting dataset for offline writer identification. In: 2012 International Conference on Frontiers in Handwriting Recognition, pp. 746–751 (2012)
26. Gattal, A., Chibani, Y., Djeddi, C., Siddiqi, I.: Improving isolated digit recognition using a combination of multiple features. In: 2014 14th International Conference on Frontiers in Handwriting Recognition, pp. 446–451 (2014)

Extraction and Recognition of Bangla Texts from Natural Scene Images Using CNN

Rashedul Islam$^{(\boxtimes)}$(iD), Md Rafiqul Islam, and Kamrul Hasan Talukder

Computer Science and Engineering Discipline, Khulna University,
Khulna 9208, Bangladesh
rashedcse98@yahoo.com, dmri1978@yahoo.com, k.h.t@alumni.nus.edu.sg

Abstract. The semantic information presents in the scene images may be the useful information for the viewers who is searching for a specific location or any specific shop and address. This type of information can also be useful in licenseplate detection, controlling the vehicle on the road, robot navigation, and assisting visually impaired persons. An efficient method is presented in this paper to detect and extract Bangla texts from scene images based on a connected component approach along with rule-based filtering and vertical scanning scheme. Next, extracted characters are recognized by using Convolutional Neural Network (CNN). The method consists of the four basic consecutive steps such as detection and extraction of the Region of Interest (ROI), segmentation of the words, extraction of characters, and recognition of the extracted characters. After extracting the ROI from the input image, connected component(CC) analysis and bounding box technology are used for segmentation of Bangla words. To separate and extract Bangla characters from the segmented Bangla words, vertical scanning based method along with a dynamic threshold value has been applied. Finally, character recognition is carried out using CNN. The proposed algorithm is applied to 600 scene images of different writing styles and colors, and we have obtained 89.25% accuracy in text detection and 94.50% accuracy in the extraction of characters. We have achieved an accuracy of 99.30% and 95.76% in recognition of Bangla digits and characters respectively. By combining both the digits and characters, obtained recognition accuracy is 95.39%.

Keywords: Bangla text · Bounding box · Connected component · CNN · Vertical projection

1 Introduction

It is always challenging as well as an important task to extract and recognize texts from natural scene images. These types of images include banners, posters, billboards, license plates, etc. which may contain valuable information. This type of information can be used in many applications like the text to speech

© Springer Nature Switzerland AG 2020
A. El Moataz et al. (Eds.): ICISP 2020, LNCS 12119, pp. 243–253, 2020.
https://doi.org/10.1007/978-3-030-51935-3_26

conversion, text based image indexing, text mining [1], robot navigation, license plate recognition [2,3] etc. The variation in font size, color, style, alignment, light intensity, blurry image, noise, etc. makes it a difficult issue to design a standard Text Information Extraction (TIE) system. The extraction of Bangla text is another challenging issue as headline or 'matra' presents in this type of text. A 'matra' is a horizontal line located at the upper portion of a character. A Bangla text may be partitioned into three zones as shown in Fig. 1.

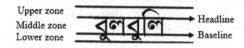

Fig. 1. Different zones of Bangla text.

As Bangla characters are connected by a headline or 'matra', we have proposed and applied a new algorithm to separate characters from each of the Bangla words by the method of vertical projection along with dynamic threshold values. The whole process of character detection, extraction, and recognition has been described in Sect. 3.

There is no benchmark database of scene images containing Bangla texts to perform research on extraction and recognition of Bangla characters. From this point of view, we have contributed to this field by providing a database of scene images consisting of 600 images. Another contribution of this paper is that we have a rich collection of Bangla characters which can be used by other researchers in developing a system of searching and recognition of office documents, text in scene images, etc.

2   Related Work

Text detection is a very challenging task for researchers who work with natural scene images. Various methods have been introduced earlier for the detection and localization of texts from scene images. In [2–4], text detection and localization techniques have been discussed based on the edge, texture, CC, stroke, and different combination of these methods. An edge detector is used in edge based method [5–7] for detecting the edges followed by morphological operation.

Bangla text extraction from the natural scene images is still now an ongoing research [8]. In the early stage, most of the researchers were concerned only with the images of printed documents, where the text was written in black color with white background [9]. Another method proposed by A. Asaduzzaman et al. [10] to detect and recognize Bangla text from printed documents using the heuristic method and Artificial Neural Network (ANN).

U. Bhattacharya et al. [11], proposed a method for the recognition of Bangla characters from scene images. The method can separate the CCs from scene images using morphological operation by calculating height and standard deviation of the CCs. Their achieved precision and recall values were 68.8% and 71.2% respectively considering a set of 100 images. R. Ghoshal et al. [12] proposed a morphological approach for Bangla text extraction from images. Their approach was limited to highlighted texts only. The algorithm can perform detection of text area and segmentation of CCs.

In [13], a texture based method was proposed to detect text at gray level natural scene images. A probabilistic model with ANN based classifier is used here to separate text from non-text objects. They achieved text detection and false alarm rate of 64% and 25% respectively.

The detail description of the originality and other contributions of the work are given below.

1. The system architecture has been proposed for the extraction and recognition of Bangla characters from scene images. This proposed architecture is a new one and got very good results both in extraction and recognition of Bangla characters.
2. The proposed method of character extraction is a new one and can effectively perform the task and gives better results than the existing methods.
3. The experimental results of character extraction and recognition have been compared with other related methods. The results show that our proposed method gives better results both in extraction and recognition of Bangla characters.

3   Proposed Method

In this paper, the proposed method is executed in two phases. Such as: Character Extraction and Character Recognition. In the first phase, the main emphasis is given on text localization and extraction that lead to better accuracy of character extraction. In the second phase, character recognition is performed by using CNN. At first, the text area is selected then each of the text regions is marked by a rectangular bounding box and finally, individual characters are extracted from each text region using the newly proposed vertical scanning algorithm.

3.1   Character Extraction

In this phase, a database of scene image is prepared. Then some pre-processing measures are taken to resize the images into 500×500 pixels. Then the images are converted to Binary image. Some other necessary steps are taken to extract the characters from the scene images. The detail description of each of the steps of this phase is stated below.

Preparing a Database of Scene Images: Since there is no benchmark database of scene images with Bangla text, we have collected scene images from different locations of Bangladesh using the digital camera and the camera of the smartphone. Then the captured images were renamed as 1.jpg, 2.jpg, 3.jpg, etc. and resized to a resolution of 500 × 500 and stored in a folder. The images were collected from different types of sources like banners, posters, billboards, vehicle license plates, etc. In this way we have created a database of scene images.

Pre-processing: This step involves two subsections as mentioned below.

Convert to the Grayscale Image: The captured images are the RGB image. So, to prepare them for the next step, we have to convert them into grayscale images. We have done it by using the National Television Standard Committee (NTSC) standard as shown in (1).

$$Gray\_Image = 0.2999 * R + 0.587 * G + 0.114 * B \qquad (1)$$

Convert to the Binary Image: To convert the grayscale image to a binary image, a threshold value is selected and all the gray level pixels below the threshold value are classified as 0 (black or background) and all the gray level pixels, equal to or greater than the threshold value are classified as 1(white or foreground) as shown in (2).

$$g(x,y) = \begin{cases} 1, & if f(x,y) >= T \\ 0, & otherwise \end{cases} \qquad (2)$$

Here, g(x, y) represents the threshold image pixel at (x, y) and f(x, y) represents grayscale image pixel at (x, y).

Detection and Extraction of ROI: In this process, the best possible regions are selected as ROI by the users. It helps to decreases false positive and also helps to collect more Bangla characters for preparing the training and test set.

Word Segmentation: CC based approach along with bounding box technology is applied to select each of the Bangla words as CCs. For this purpose, we have used the labeling of CCs of the binary image. Here all the CCs are marked by the red color rectangular bounding boxes as shown in Fig. 2(c).

Character Extraction: Bangla character extraction is one of the challenging tasks of the character recognition system. As the words are connected by a headline or 'matra', it is difficult to segment out individual characters. The technique to remove headlines to separate characters from Bangla words has been followed by the existing methods. But the main problem of removing 'matra' is that after removing the 'matra' some characters will be changed to another character. Some examples of such characters are shown in Fig. 3. We can solve this problem by the following way. At first, we take all the CCs as input and

Fig. 2. (a) Original color image (b) Binary image (c) Localization of Bangla words.

count the number of white pixels in every column to determine minimum value among all the columns. The column that contains minimum value will be treated as a separating zone among the characters of a word. To separate two characters vertically, we have set the pixel values of all the pixels of a specific column to 0 where the number of white pixels of the column is less than (minimum+5). Then the separated characters are resized to 16 × 16 pixels and store them into a specific folder as Bangla characters.

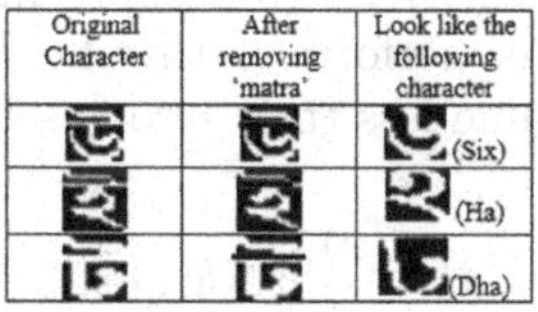

Original Character	After removing 'matra'	Look like the following character
ড	ড	(Six)
হ	হ	(Ha)
ঢ	ঢ	(Dha)

Fig. 3. Effects of removing matra.

3.2   Character Recognition

This is the final stage of the proposed method. In this stage, experiment is performed based on the two consecutive phases such as the training and the testing phase. The brief description of each of the phases is stated below:

Training Phase

Prepare Training and Test Dataset: To prepare the data sets, at first it is required to load the database named 'banglacharacter' as an imagedatastore. The main function of the imagedatastore object is to automatically labels the images based on folder names. Finally data are stored as an imagedatastore object. To prepare the training data set, the system will randomly select a fixed number of images as mentioned by the user from each of the folders containing Bangla characters. In this experiment, we have assigned 250 as the number of images to be selected from each folder for training. The remaining characters of the folder will be treated as a test data set.

Initialize the CNN Layers: CNN is designed with many layers. To work with CNN, at first we have to define each of the layers by specific parameter values. Brief description of the layers of our designed CNN is stated below:

- *Image Input Layer:* In this layer, the image size is specified for our database. We have specified the said size as 16-by-16-by-1. Here, height and width of the image is 16 and the channel size is 1. The 'banglacharacter' data consists of binary images, so the channel size is 1. For a color image, the channel size 3 is recommended.
- *Convolutional Layer:* This layer contains three parameters. The first parameter is the filter size. The second parameter is the number of filters, which represents the number of neurons that connect to the same region of the input. 'Padding' name-value pair is used to add padding to the input feature map. We have used the following hyperparameters for the function convolution2dLayer(3, 16, 'Padding', 1).
FilterSize: [3 3], NumChannels: 'auto', NumFilters: 16, Stride: [1 1], Padding-Mode: 'manual', and PaddingSize: [1 1 1 1].
- *Batch Normalization Layer:* This layer of CNN helps to speed up network training and reduces the sensitivity to network initialization.
- *ReLU Layer:* A ReLU layer performs a threshold operation to each element of the input, where any value less than zero is set to zero as shown in 3.

$$f(x) = \begin{cases} x, & x > 0 \\ 0, & x < 0 \end{cases} \tag{3}$$

- *Max Pooling Layer:* The max pooling layer returns the maximum values of rectangular regions of inputs. the hyperparameters that we have used for this layer are as follows:
PoolSize: [2 2], Stride: [2 2], PaddingMode: 'manual', PaddingSize: [0 0 0 0]
- Fully Connected Layer: The last fully connected layer combines the features to classify the images. This layer is equal to the number of classes in the target data. For the classification of Bangla digits, we have used the function fullyConnectedLayer(10) with the two following hyperparameters.
InputSize: 'auto' and OutputSize: 10

Figure 4 shows the working process of a CNN with input image 'zero'.

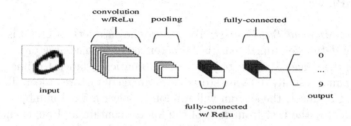

Fig. 4. Classification process of CNN.

Set the Training Options: Before train the CNN classifier, it is required to specify the training options for classification of Bangla digits and characters. The Following training options have been used in this experiment.

options = trainingOptions('sgdm', 'InitialLearnRate', 0.01, 'MaxEpochs', 4, 'ValidationData', imdsValidation, 'ValidationFrequency', 30, 'Verbose', false, 'Plots', 'training-progress').

Train the Network: The Main purpose of training is to perform the task of recognition successfully. For this, the training data set is used along with predefined values of CNN layers and training options. These three parameters help to train the CNN successfully.

Testing Phase

Classify Using the Trained CNN: In this step, all the characters under the test data set are classified using the trained CNN. In this process labels of test data set are matched with the labels of the training data and obtained result is stored as predicted_data.

Calculation of Accuracy: At first, labels of test data set are stored as test_validation. Then Recognition accuracy is calculated by making a one-to-one comparison between the predicted_data and the test_validation. Figure 5 shows the system architecture of the proposed method.

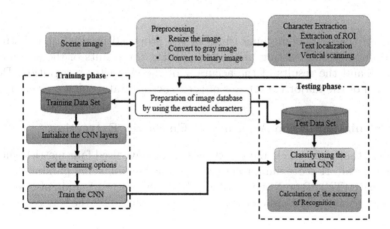

Fig. 5. System architecture of the proposed method.

4   Experimental Results

The experiments were conducted in the following two phases. Such as

a) Character extraction and
b) Calculation of the accuracy of recognition.

In the first phase, Bangla characters were extracted from the natural scene images and in the second phase training and testing were performed on the extracted characters and the accuracy of recognition was calculated. All the experiments were performed in MATLAB environment using the images of our image database. The proposed method was applied to 600 scene images. The algorithm will not work properly or fail in the case of character extraction if all the characters are connected with each other by any way other than "matra". The Algorithm will fail in another case where the texts are written in a curved or round shape. A few such images are shown in Fig. 6 where the proposed algorithm will fail to extract Bangla characters.

Fig. 6. Images where the proposed algorithm will fail.

Though there are some limitations of the proposed method, it is better in comparison with the existing methods regarding the results of the accuracy of extraction and the results of the accuracy of the character recognition. Detail description of the major two phases of the experimental results is given below.

4.1   Calculation of the Accuracy of Character Extraction

To analyze the results of character extraction, we have used four metrics, such as precision, recall, f1-score, and accuracy based on the following parameters [14]. True Positive (TP), True Negative (TN), False Positive (FP), and False Negative (FN). The accuracy of character extraction is calculated by the way as shown in (4).

$$Accuracy = \frac{TP + TN}{TP + TN + FP + FN} \tag{4}$$

Table 1 shows the percentage of precision, recall, f1-score, and the accuracy of character extractions from different types of scene images like banners, posters and license plates.

Table 1. Results of character extraction.

Image type	No. of images	Pr(%)	RR(%)	F1 score(%)	Accuracy(%)
Banner	240	92.73	94.08	93.00	92.92
Poster	260	95.30	94.13	94.45	95.09
License plate	100	95.43	97.58	96.32	95.48

4.2 Calculation of the Accuracy of Recognition

To calculate the accuracy of recognition, CNN predicts the labels of the test data using the trained network, and calculate the final validation accuracy. Accuracy is the fraction of labels that the network predicts correctly. Recognition accuracy is calculated by the following equation as shown in (5).

$$Accuracy = \frac{total\ number\ of\ matching\ labels}{total\ number\ of\ elements\ in\ the\ test\ data} \times 100\% \qquad (5)$$

The comparison of recognition accuracy is shown in Table 2. The cited approaches mentioned in Table 2 do not use the same database as ours. In the table, '–' indicates that the result was not found in the respective paper. From Table 2, it is clear that the proposed method outperforms the existing methods.

Table 2. Comparison of the recognition accuracy with existing methods.

Methods	No. of images	No. of characters	Recognition accuracy(%)
Proposed	600	21108	95.39
Moyeen, M.A., et al. [8]	400	—	73.25
Ghoshal, R., et al. [15]	100	7500	85.93
Ghoshal, R., et al. [16]	250	7100	92.00

5 Conclusions

The proposed method of Bangla character recognition has been tested on Bangla digits and letters extracted from the varied sorts of scene images and achieved smart ends up in comparison with the present strategies. To separate Bangla characters from the words, we have applied the vertical scanning algorithm. In the case of the extraction of Bangla characters, we've achieved 94.50% accuracy from 600 natural scene images. Within the recognition phase, character recognition is performed exploitation CNN classifier. We've used the CNN for the popularity due to its high accuracy. A hierarchical model is followed in the CNN that works on building a network, like a funnel, and at last offers out a fully-connected layer wherever all the neurons are connected and the output is processed. The achieved recognition accuracy for Bangla digits is 99.30% and for Bangla characters, it is 95.76% and their combined result is 95.39% that is best than the results of the present strategies. Our future set up is to counterpoint our information with all the essential characters and joined letters of the Bangla alphabet and to represent the recognized characters in the editable form.

References

1. Bouakkaz, M., Ouinten, Y., Loudcher, S., Fournier-Viger, P.: Efficiently mining frequent itemsets applied for textual aggregation. Appl. Intell. **48**(4), 1013–1019 (2017). https://doi.org/10.1007/s10489-017-1050-9
2. Zhu, Y., Yao, C., Bai, X.: Scene text detection and recognition: recent advances and future trends. Front. Comput. Sci. **10**(1), 19–36 (2016)
3. Zhang, H., Zhao, K., Song, YZ., Guo, J.: Text extraction from natural scene image: a survey. Neurocomputing **122**, 310–323 (2013)
4. Unar, S., Hussain, A., Shaikh, M., Memon, K.H., Ansari, M.A., Memon, Z.: A study on text detection and localization techniques for natural scene images. IJCSNS **18**(1) (2018)
5. Yu, C., Song, Y., Zhang, Y.: Scene text localization using edge analysis and feature pool. Neurocomputing **175**, 652–661 (2016)
6. Silva, B.L.S., Ciarelli, P.M.: Edge detection and confidence map applied to identify textual elements in the image (2016)
7. Lee, S., Cho, M.S., Jung, K., Kim, J.H.: Scene text extraction with edge constraint and text collinearity. In: 20th International Conference on Pattern Recognition (ICPR), pp. 3983–3986. IEEE (2010)
8. Moyeen, M.A., Alam, K.M.R., Awal, M.A.: Bangla text extraction from natural scene images for mobile applications. J. Electr. Eng. Inst. Eng. **EE 39**(I & II) (2013)
9. Aurich, V., Weule, J.: Non linear Gaussian filters performing edge preserving diffusion. In: 17 DAGM Symposium, pp. 538–545 (1995)
10. Asaduzzaman, A., Molla, M.K.I., Ali, M.G.: Printed Bangla text recognition using artificial neural network with heuristic method. In: Proceedings of ICCIT, Dhaka, Bangladesh (2002)
11. Bhattacharya, U., Parui, S.K., Mondal, S.: Devanagari and Bangla text extraction from natural scene images. In: Proceedings of International Conference on Document Analysis and Recognition, pp. 26–29 (2009)
12. Ghoshal, R., Roy, A., Bhowmik, T.K., Parui, S.K.: Headline based text extraction from outdoor images. In: Kuznetsov, S.O., Mandal, D.P., Kundu, M.K., Pal, S.K. (eds.) PReMI 2011. LNCS, vol. 6744, pp. 446–451. Springer, Heidelberg (2011). https://doi.org/10.1007/978-3-642-21786-9_72
13. Hanif, S.M., Prevost, L.: Texture based text detection in natural scene images: a help to blind and visually impaired persons. In: Conference Workshop on Assistive Technologies for People with Vision Hearing Impairments Assistive Technology for All Ages CVHI
14. Dalal, N., Triggs, B.: Histograms of oriented gradients for human detection. In: IEEE Computer Society Conference on Computer Vision and Pattern Recognition, vol. 1, pp. 886–893. IEEE (2005)

15. Ghoshal, R., Roy, A., Parui, S.K.: Recognition of Bangla text from scene images through perspective correction. In: 2011 International Conference on Image Information Processing (ICIIP), pp. 1–6 (2011)
16. Ghoshal, R., Roy, A., Dhara, B.C., Parui, S.K.: Recognition of Bangla text from outdoor images using decision tree model. Int. J. Knowl.-Based Intell. Eng. Syst. **21**(1), 29–38 (2017)

Detection of Elliptical Traffic Signs

Manal El Baz(✉) ⓘ, Taher Zaki ⓘ, and Hassan Douzi ⓘ

Image and Pattern Recognition - Intelligent and Communicating Systems Laboratory,
Faculty of Sciences, Ibn Zohr University, Agadir, Morocco
elbazmanal@gmail.com

Abstract. Detection of elliptical features is a challenging and important task in computer vision. In fact, ellipses can describe many objects in real images like manufactured objects, cells, ball, or traffic signs. Furthermore, an ellipse is defined by five parameters: the center coordinates, the semi major axe, the semi minor axe and the orientation, which require more computational power to estimate them. Some non-ideal ellipses cause also difficulties for detection such as occlusion, appearance of multiple ellipses at same time and non-parallel ellipse. In this paper, we are interested in detecting elliptical traffic signs. We present a method for detecting different cases of ellipses: simple, partially occluded, non-parallel, multiple ellipses in images. The method selects three lines to find the ellipse center, and then it calculates the value of the semi minor and major axes of the ellipse. Experiments show that the proposed method performs well on real images in the presence or not of noises.

Keywords: Detection of traffic signs · Ellipse detection · Binary image

1 Introduction

Detecting ellipse in an image has received attention by many researchers in recent decades [2–10]. Many methods are developed to detect an ellipse in an image. The most method used is the Hough transform. Firstly, Hough transform is invented in 1962 by Paul Hough [1] to detect straight lines in an image. Later, Hough transform is developed to detect other shapes such as ellipses [2–4]. An ellipse is a conic that is defined by five parameters: the semi major axe, the semi minor axe, the center and the orientation. The standard Hough transform and its derivatives calculate those parameters to extract an ellipse in an image. Teng et al. [2] use two-level voting algorithms to estimate the five ellipse parameters. They developed the randomized Hough transform to eliminate noisy edge points dissimilar to ellipse patterns by analyzing the variation of the angles of tangential lines for edge points. Latour et al. [3] adopt the elliptical Hough transform to extract ellipses in an image using two edge points and their associated image gradients. Their method is based on algebraic framework that is expressed in the point form or the tangential form. Xie et al. [4] accumulate the length of ellipse minor axis by using one dimensional accumulator array. If the vote of

© Springer Nature Switzerland AG 2020
A. El Moataz et al. (Eds.): ICISP 2020, LNCS 12119, pp. 254–261, 2020.
https://doi.org/10.1007/978-3-030-51935-3_27

minor axis is greater than the threshold, then the ellipse and its parameters are extracted.

In other hand, other methods for detecting ellipse; which are not based on Hough transform; are proposed: Ouellet et al.'s method [5] is based on the dual conic model. It exploits directly the raw gradient information in the neighborhood of an ellipse's boundary, and minimize the algebraic distance to constrain the dual conic to a dual ellipse. It exists also methods based on tracking arc segments and connectivity between edge pixels to detect ellipses. Kim et al. [6] extract arc segments and then merge them into an ellipse. Ellipses can be detected from the extended elliptical arc segments. Jia et al. [7] used a projective invariance of characteristic number to prune line segments and pick arc segments lying on an ellipse. The picked arc segments are used to estimate ellipse parameters. They determine the ellipse center by finding the algebraic average of coordinates of intersections of bisectors of 16 parallel chords of each arc segments lying on an ellipse. Libuda et al. [8] firstly extract short straight lines, then they merge these lines to small arcs which will be combined to extended arcs. Then, they create an ellipse by merging extended arcs. Mai et al.'s method [9] is also based on arc segments extraction. They use the hierarchical algorithm to detect ellipse in four stages: firstly, they extract line segments that will be linked to synthetic arc segments. Then, they fit ellipse in each arc group by using RANSAC.

In this paper, we present a new method for detecting elliptical traffic signs in an image. We use three lines to find the ellipse center. Two of those lines are parallels; therefore, the third goes through the middles of the two lines. The middle of the third line is the ellipse center. Hence, we calculate the semi major and minor of the detected ellipse. The result of the proposed method shows that it works better than the state-of-the-art ellipse detectors on detecting elliptical traffic signs: simple ellipse, partially occluded ellipse, non-parallel ellipse, multiple ellipses, and noisy image. This paper is organized as follow: firstly, we present our method in Sect. 2, then in Sect. 3, we discuss the result of our method and finally we make some conclusions in Sect. 4.

2   Method Description

2.1   Circle Detection

At first of our work in shape detection development, we were interested in finding circles in images. We take 500 points in an image and then we relied on pairs of points to determine the circle center. We choose randomly two points: $P_1(x_1, y_1)$ and $P_2(x_2, y_2)$ in the image. The coordinates of circle center $\Omega(a, b)$ is:

$$a = \frac{x_1 + x_2}{2}. \tag{1}$$

$$b = \frac{y_1 + y_2}{2}. \tag{2}$$

and the circle radius is:

$$r = \frac{\sqrt{(x_1 - x_2)^2 + (y_1 - y_2)^2}}{2}. \tag{3}$$

We calculate the number of points that satisfy the Cartesian equation of each circle:

$$(x - a)^2 + (y - b)^2 = r^2. \tag{4}$$

We consider NbMax as the maximum of points that satisfy a Cartesian equation of a circle. If more than half of NbMax of points are belonging to one circle in the image, we extract the present circle. The result of this method shows that it does not detect circles correctly and the false positive rate is higher (we got $29,56\%$ as F-measure). Therefore, we decide to change our purpose: detecting ellipses rather than circles and defining lines to extract the center.

2.2   Ellipse Detection

Our method is about finding the ellipse center by three lines (D_1), (D_2) and (D_3) that intersect the ellipse. Two of those lines (D_1) and (D_2) are parallels. However, the third line (D_3) goes through the middles of intersections of the two lines and the ellipse. The middle of intersections of (D_3) and the ellipse is the ellipse center (see Fig. 1).

We consider two parallel lines (D_1) and (D_2), and the points $P_1(x_1,y_1), P_2(x_2,y_2)$ and $P_3(x_3,y_3), P_4(x_4,y_4)$ are respectively intersections of (D_1) and (D_2) and the ellipse. Equations of (D_1) and (D_2) are:

$$\forall m_1, k_1 \in \mathbb{R} \quad D_1 : y = m_1 x + k_1. \tag{5}$$

$$\forall m_2, k_2 \in \mathbb{R} \quad D_2 : y = m_2 x + k_2. \tag{6}$$

We have:

$$D_1 \parallel D_2 \quad \Leftrightarrow \quad m_1 = m_2. \tag{7}$$

and we have $P_1, P_2 \in D_1$ and $P_3, P_4 \in D_2$ so:

$$\begin{cases} y_1 = m_1 x_1 + k_1 \\ y_2 = m_1 x_2 + k_1 \end{cases} \Rightarrow \begin{cases} m_1 = \frac{y_2 - y_1}{x_2 - x_1} \\ k_1 = y_1 - m_1 x_1 \end{cases} \tag{8}$$

$$\begin{cases} y_3 = m_2 x_3 + k_2 \\ y_4 = m_2 x_4 + k_2 \end{cases} \Rightarrow \begin{cases} m_2 = \frac{y_4 - y_3}{x_4 - x_3} \\ k_2 = y_3 - m_2 x_3 \end{cases} \tag{9}$$

The middle of $[P_1 P_2]$ is the point $M_1(x_{M_1}, y_{M_1})$ where:

$$x_{M_1} = \frac{x_1 + x_2}{2}. \tag{10}$$

$$y_{M_1} = \frac{y_1 + y_2}{2}. \tag{11}$$

And the middle of $[P_3P_4]$ is the point $M_2(x_{M_2}, y_{M_2})$ where:

$$x_{M_2} = \frac{x_3 + x_4}{2}. \tag{12}$$

$$y_{M_2} = \frac{y_3 + y_4}{2}. \tag{13}$$

We consider (D_3) goes through M_1 and M_2 so the equation of (D_3) is as follow:

$$\forall m_3, k_3 \in \mathbb{R} \quad D_3 : y = m_3 x + k_3. \tag{14}$$

where:

$$m_3 = \frac{y_{M_2} - y_{M_1}}{x_{M_2} - x_{M_1}} \quad and \quad k_3 = y_{M_1} - m_3 x_{M_1}. \tag{15}$$

We consider the points $P_5(x_5, y_5)$ and $P_6(x_6, y_6)$ are intersections of (D_3) and the ellipse. The middle of $[P_5P_6]$ is the center of the ellipse $C(x_C, y_C)$ so:

$$x_C = \frac{x_5 + x_6}{2}. \tag{16}$$

$$y_C = \frac{y_5 + y_6}{2}. \tag{17}$$

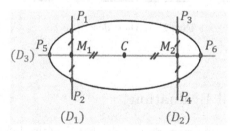

Fig. 1. The ellipse center

For finding semi major and minor axes of ellipse, we choose randomly four points $P_1'(x_{P_1'}, y_{P_1'})$, $P_2'(x_{P_2'}, y_{P_2'})$, $P_3'(x_{P_3'}, y_{P_3'})$ and $P_4'(x_{P_4'}, y_{P_4'})$ from the image where the ellipse center $C(x_C, y_C)$ is the middle of segments $[P_1'P_2']$ and $[P_3'P_4']$.

The values of a and b are respectively distance of $[P_1'P_2']$ and $[P_3'P_4']$:

$$a = \frac{\sqrt{(x_{P_1'} - x_{P_2'})^2 + (y_{P_1'} - y_{P_2'})^2}}{2}. \tag{18}$$

$$b = \frac{\sqrt{(x_{P_3'} - x_{P_4'})^2 + (y_{P_3'} - y_{P_4'})^2}}{2}. \tag{19}$$

a and b are respectively the semi major and minor axe of the ellipse if they satisfy the following conditions:

$$\begin{cases} 0 < b < a \\ 0 < \frac{\sqrt{a^2-b^2}}{a} < 1 \\ (P_1'P_2') \perp (P_3'P_4') \end{cases} \tag{20}$$

The proposed method can be described in steps showed in Fig. 2.

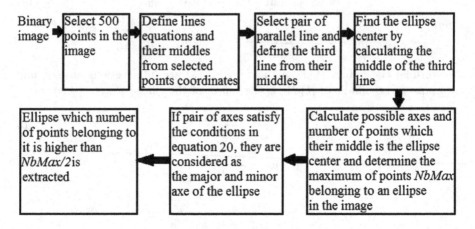

Fig. 2. Method description.

3   Experimental Evaluation

We experiment the methods on 69 images that contain different cases possible of ellipse: simple ellipse, partially occluded ellipse, non-parallel ellipse, multiple ellipses, noisy image; and its different size and position.

We use the following metrics to measure the performance of the methods:

$$Precision = \frac{TP}{TP + FP}. \tag{21}$$

$$Recall = \frac{TP}{TP + FN}. \tag{22}$$

$$F - measure = 2 \times \frac{Precision \times Recall}{Precision + Recall}. \tag{23}$$

where:

TP: True positive.
FP: False positive.
FN: False negative.

Table 1. Result of methods.

	Precision	Recall	F-measure	Time
Proposed method	73.27%	**79.57%**	**76.29%**	**3,07 s**
ELSDc [10]	**73.56%**	68.82%	71.11%	17,97 s
Jia et al.'s method [7]	57.55%	65.59%	61.31%	4,92 s
Xie et al.'s method [4]	46.79%	54.84%	50.50%	15.34 s

We compare our method to three popular state-of-the-art ellipse detectors: joint ellipse and line segment detector ELSDc proposed by Patraucean et al. [10], Jia et al.'s method [7] and Xie et al.'s method [4]. ELSDc [10] is a line segment and ellipse detector based on the contrario theory. This detector [10] performs accurately on synthetic and real images. ELSDc is available online at dev.ipol.im/~jirafa/elsdc/. Jia et al. method [7] is based on projective invariant to prune lines and pick out ellipses. It [7] proved its high performance in public datasets. This detector is shared online at https://github.com/dlut-dimt/ellipse-detector. Xie et al.'s method [4] is an ellipse hough detection method. This method [4] is based on a one-dimensional accumulator array to vote on the minor axis. We reimplement Xie et al.'s method [4] according to the original paper. In experiments, we are interested on detecting elliptical traffic signs in images. Table 1 gives the result of the methods. It shows that our method has achieved higher recall and F-measure rate than the other methods. However, ELSDc performs better than our method in terms of precision. Furthermore, our method consumes lower computation time compared to ELSDc, Jia et al. method [7] and Xie et al.'s method [4]. ELSDc [10] suffers from large computation time because of the detection of line segments and ellipses all together.

The result of the proposed method on each case of ellipse is depicted in Fig. 3. The proposed method works well on all ellipse cases except multiple ellipses case. Its performance is poor if the image contains multiple ellipses. That happens because the false negative rate increases when more ellipses appear at the same time in an image (see Fig. 3 (f)). Moreover, our method has achieved good result on non-parallel ellipses. In contrast, the result that we obtained for ELSDc [10] and Xie et al.'s method [4] on non-parallel ellipses is lower than our method and Jia et al.'s method [7] in terms of F-measure: 78.26% for our method, 68.97% for Jia et al.'s method [7], 60.00% for Xie et al.'s method [4] and 50.00% for ELSDc [10].

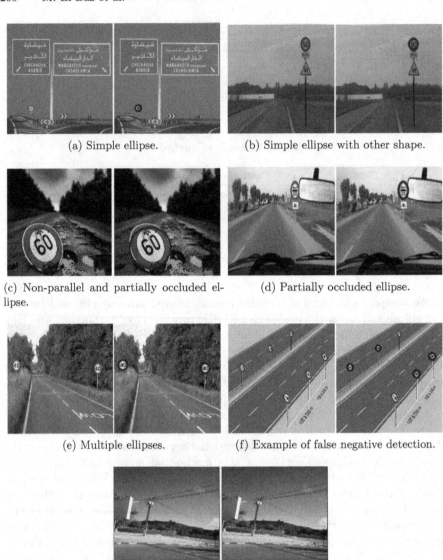

(a) Simple ellipse.

(b) Simple ellipse with other shape.

(c) Non-parallel and partially occluded ellipse.

(d) Partially occluded ellipse.

(e) Multiple ellipses.

(f) Example of false negative detection.

(g) Image without traffic sign.

Fig. 3. Elliptical traffic signs detection.

4   Conclusion

This paper has presented a method for detecting elliptical traffic signs in images. This method selects firstly two parallel lines, and then it determines the midpoints of segments limited by intersections of those two lines and the ellipse. The third line; which goes through the middles; is defined and its middle is the center of the researched ellipse. It has been shown that the proposed method can detect simple, partially occluded, non-parallel and multiple ellipses and it performs accurately on real images in presence or not of noises. Currently, we are developing the present method in order to increase its performance on images in which appear multiple ellipses.

References

1. Hough, P.V.C.: Method and means for recognizing complex patterns. U.S. Patent 3069654, Ser. No. 177156 Claims (1962)
2. Teng, Z., Kim, J.-H., Kang, D.-J.: Ellipse detection: a simple and precise method based on randomized Hough transform. Opt. Eng. **51**(5), 057203-1-057203-14 (2012). https://doi.org/10.1117/1.OE.51.5.057203
3. Latour, P., Van Droogenbroeck, M.: Dual approaches for elliptic Hough transform: eccentricity/orientation vs center based. In: Couprie M., Cousty J., Kenmochi Y., Mustafa N. (eds.) Discrete Geometry for Computer Imagery. DGCI 2019. LNCS, vol. 11414, pp. 367–379. Springer, Charm (2019). https://doi.org/10.1007/978-3-030-14085-4_29
4. Xie, Y., Ji, Q.: A new efficient ellipse detection method. In: International Conference on Pattern Recognition, vol. 2, pp. 957–960. IEEE, Quebec (2002). https://doi.org/10.1109/ICPR.2002.1048464
5. Ouellet, J.-N., Hebert, P.: A simple operator for very precise estimation of ellipses. In: Fourth Canadian Conference on Computer and Robot Vision (CRV 2007), pp. 21–28. IEEE, Montreal (2007). https://doi.org/10.1109/CRV.2007.8
6. Kim, E., Haseyama, M., Kitajima, H.: Fast and robust ellipse extraction from complicated images. In: Proceedings of the IEEE International Conference on Information Technology and Applications (2002)
7. Jia, Q., Fan, X., Luo, Z., Song, L., Qiu, T.: A fast ellipse detector using projective invariant pruning. IEEE Trans. Image Process. **26**(8), 3665–3679 (2017). https://doi.org/10.1109/TIP.2017.2704660
8. Libuda, L., Grothues, I., Kraiss, K.-F.: Ellipse detection in digital image data using geometric features. In: Ranchordas, A., Araujo, H., Encarnaçao, B. (eds.) Proceedings of the First International Conference on Computer Vision Theory and Applications, vol. 1, pp. 175–180. SciTePress, Setubal (2006). https://doi.org/10.5220/0001362301750180
9. Mai, F., Hung, Y.S., Zhong, H., Sze, W.F.: A hierarchical approach for fast and robust ellipse extraction. In: 2007 IEEE International Conference on Image Processing. ICIP 2007, vol. 5, pp. V-345–V-348. IEEE, San Antonio (2007). https://doi.org/10.1109/ICIP.2007.4379836
10. Patraucean, V., Gurdjos, P., Von Gioi, R.G.: Joint a contrario ellipse and line detection. IEEE Trans. Pattern Anal. Mach. Intell. **39**(4), 788–802 (2017). https://doi.org/10.1109/TPAMI.2016.2558150

Image-Based Place Recognition Using Semantic Segmentation and Inpainting to Remove Dynamic Objects

Linrunjia Liu[✉], Cindy Cappelle, and Yassine Ruichek

CIAD, Univ. Bourgogne Franche-Comté, UTBM, Belfort, France
{linrunjia.liu,cindy.cappelle, yassine.ruichek}@utbm.fr

Abstract. Place recognition is an important step in intelligent driving, allowing the vehicle recognize where it is to plan its route. Obtaining distinguishable features can ensure the success of image-based place recognition. However, generating robust features across drastically appearance changing images is still a challenging problem. Deep features are frequently chosen instead of local features in the tasks of place recognition following the development of convolutional neural networks. But even the deep features generated by powerful neural models can cause unsatisfactory recognition results. This is perhaps due to a lack of information selecting process. The technology of semantic segmentation allows recognizing and classifying image information. Semantic segmentation followed by image inpainting provide a possibility of detecting, deleting and reconstructing annoying information.

This paper proves that dynamic information present in images such as vehicles and pedestrians damages the performance of place recognition and proposes a feature extraction system that includes a step to decrease the presence of dynamic information of an image. This system is composed of two stages: 1) dynamic objects detection and removing, 2) image inpainting to reconstruct the background of removed regions. Objects detection and removing consists of deleting unstable objects recognized by semantic segmentation method from images. Image inpainting and reconstructing deals with generating inpaint-images by repairing missing regions through image inpainting method. The robustness of the proposed approach is evaluated by comparing to the non-selecting deep feature based place recognition approaches over three datasets.

Keywords: Place recognition · Image inpainting · Semantic segmentation

1 Introduction

Place recognition for vehicle localization aims to select the image in a long-term robotcar dataset, which represents the same place as a query image. Extracting

Supported by CSC.

(a) Original-image   (b) Mask   (c) Inpaint-image

Fig. 1. (a) Original-image. (b) Mask generating by FCN based semantic segmentation. Regions depicted in grey are dynamic objects recognized by FCN and are seen as missing regions by EdgeConnect system based inpainting. (c) Inpaint-image getting from EdgeConnect system.

robust features to represent an image is then a critical step of place recognition. Since feature generation technique of Convolutional Neural Networks (CNNs) has been gradually applied into the tasks of intelligent vehicle [18], place recognition has also progressed significantly. Nevertheless, due to the severe appearance changes in long-term robot navigation, there are still many challenges and problems to achieve efficient image based place recognition. In addition to dramatic changes in weather conditions, illumination and viewpoint, dynamic objects, such as vehicles and pedestrians, also contribute to the appearance changes and make it difficult to distinguish the same place at different time.

Since learning invariant features of an image can improve the performance of place recognition, what about deleting all the dynamic objects in an image and extract features by the remaining stable background? The development of semantic segmentation and image inpainting approaches makes it possible.

The main goal of this paper is to provide a new method that addresses the challenges in traffic variations of image-based place recognition. Inpaint-images are generated from original images from which dynamic objects–objects classified as cars, buses, trucks, and pedestrians–are removed and the stable background is completed by a two-stages process (Fig. 1): 1) objects detection and removing and 2) image inpainting and reconstructing. Firstly, we use FCN (Fully Convolutional Networks) [8] to detect the dynamic objects in the original images and label them into masks. Then, the parts labeled as masks, which are then seen as missing regions, are filled by fine details using EdgeConnect [11] based inpainting approach. The outputs of Edge-Connect are called inpaint-images in this paper. A CNN network is then used to extract global features from these inpaint-images. Image retrieval process finally performed using Euclidean distance as a metric.

In this paper, the source of images for experiments are selected from three publicly available datasets: St.lucia, CMU, and Oxford RobotCar [10]. Three sequences of different routes are chosen in each dataset and their inpaint-images are generated independently. Then, four different CNNs [6,7,15,17] are used to learn the features of both inpaint-images and original-images (sequences without dynamic detection and inpainting). In each test, comparing by using CNN features of original-images in the process of image retrieval, a noticeable improvement is obtained on the task of place recognition performed using the inpaint-images CNN features.

2   Related Work

Generating distinguishable features to represent images with severe appearance variation determines the results of place recognition in changing environments.

Traditional methods design hand-craft local features that are illumination-, viewpoint-, and scale-invariant, such as speeded-up robust features (SURF) [1] and scale-invariant feature transforms (SIFT) [9].

Because of the generalization of features learned by representative Convolutional Neural Networks (CNNs), a recent trend is to exploit CNN features into place recognition. Papers [2,14] use holistic features of images. [2] concludes that different layers of CNN features have different effect in changing environments. In the work of [14], different layers of CNN features are connected together to represent an image. This kind of holistic features contains more information but also increases computation complexity.

Though holistic CNN features outperform classical features, they sometimes fail to recognize correctly the same places when facing the realistic street scenes with significant appearance changes. Increasing the features complexity to learn as much as possible different appearance representations was the goal of many place recognition approaches. [4] observes the environmental changes over the course of day and night and combines the different features together. [12] learns systematic appearance changes so that to predict the changes under different environmental conditions. These works require lots of training data and didn't show generality abilities in the different conditions as with training data.

Some approaches choose to make the features invariable such as using landmarks features. [16] applies Edge Boxes to extract landmark proposals and then learns the features of these proposals by ConvNet. Regions extracted by a landmark detector in [5] are deep local features and robust to changes in viewpoint. These works need the environmental conditions with distinguished landmarks which is not always happen in real world.

Our approach follows the idea of using invariable features, but instead of selecting useful information, we delete the dynamic information, as cars, vehicles, pedestrians, without damaging context information. This is achieved through semantic segmentation to detect objects to remove and image inpainting to reconstruct the removed regions. To our best knowledge, this is the first work that tackles dynamic information removal and background information reconstruction in place recognition.

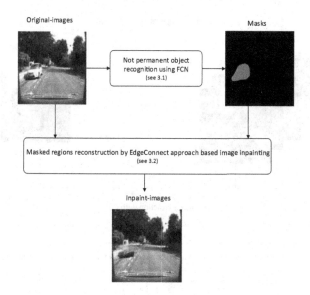

Fig. 2. Proposed system to decrease the impact of dynamic information of the image and reconstruction of its background.

3 Proposed System

Image-based localization aims to recognize the image in a local navigation dataset that represents the same place as a query image (the current image) based on their appearance.

To describe each image, a CNN model is used to extract image features. After comparing the similarity of a query image with each images in the reference sequence, the image which is the most similar to the query image is selected.

The proposed method that tries to decrease the impact of the dynamic objects in images is based on two-stages (Fig. 2): 1) objects detection and removing, and 2) image inpainting and reconstructing.

3.1 Proposed Dynamic Objects Removing

The first step of the proposed approach is to eliminate the objects of the image that are not permanent in the scene (as vehicles, pedestrians, ⋯). For that, FCN [8] is used to recognize these objects in image and make local predictions of these objects. Given an image of any size as input, this end-to-end fully convolutional network (FCN) provides an output image of the same size with dense prediction of each pixel. With this prediction, the position of each object can be easily marked. Without considering the position of the other objects, only the positions of pedestrians and vehicles are recognized and marked in this paper. These marked dynamic objects are not directly removed from the image, but are seen as missing regions in the image inpainting method that is chosen. The 3.2 part will explain this step in detail.

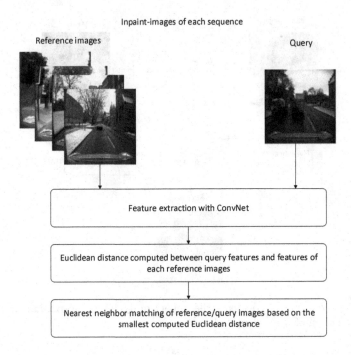

Fig. 3. The process of place recognition.

To get the output of FCN, a fine-tuned classification model is needed. In our work, the resnet101 Model which is trained on PASCAL VOC dataset is picked. The output image of FCN will be considered as an input mask image in image inpainting step.

3.2 Images Repairing with Image Inpainting Approach

Image inpainting means reproducing the missing regions of a ground truth image as if it has not being corrupted. In our paper, we choose a state-of-art image inpainting method: EdgeConnect [11]. Given an original image as the ground truth image and a mask image generated by FCN, the output of EdgeConnect is our expected inpainting-image. In the process of inpainting, EdgeConnect will see the dark part of mask image as background and recognize the rest part as missing regions which will be reconstructed later.

The reconstructing of EdgeConnect consists of two parts. First, it trains an edge generator to hallucinate edges in the missing regions. Then, it needs an image completion network to combine the color and texture information of the existing parts with the edges in the missing parts of the image. Combining the two parts of work together, an end-to-end trainable network is proposed to reconstruct missing regions with fine details. By solving two computer vision problems: Image-to-Edges and Edges-to-Image, EdgeConnect improves the inpainting quality of the corrupted images.

Table 1. Testing datasets description.

Datasets	Sequences	No. of frames	Variations in			Resolution	
			Traffic	Weather	Season	Original	Resized
St.Lucia	different time in a day	227	Mild	None	None	640 × 480	224 × 224
CMU	summer V.S fall	207	Severe	Moderate	Moderate	1024 × 768	224 × 224
Oxford Robotcar	summer V.S winter	282	Severe	Moderate	Severe	1280 × 960	224 × 224

3.3 Place Recognition Approach

By comparing the Euclidean distances between the feature vector of query image and of every reference images, a nearest matching reference image is found per query image. The place of this nearest neighbor reference image is then considered as the place of the query image. And the featurization process in this paper is to learn the deep feature of each image by a ConvNet. The whole place recognition process is shown in Fig. 3.

4 Experimental Setup

4.1 Datasets

In order to evaluate the proposed approach, three sequences are selected from three datasets which deal with mild, moderate and severe variations in environmental conditions, including illumination, traffic and pedestrians, weather and seasonal changes. Details about the datasets are described below and summarized in Table 1.

- St. Lucia Dataset[1]: The St.Lucia Dataset was captured in a suburb. The car pass through some routes several times.
- CMU Seasons Dataset[2]: The CMU Seasons dataset is a part of CMU Visual localization dataset created by *Badino et al.* [3]. We use the left mono images that represent the scenes of the same route in summer and fall. The summer images are used as references and fall images are used as queries.
- Oxford RobotCar Dataset: The Oxford RobotCar Dataset [10] is composed of over 100 different sequences of Oxford. We choose a fixed route and compare it between summer and winter.

4.2 Data Pre-processing

One fixed route in each dataset is selected. The sequence of each route is a subset of two traversals which were taken over different time. After considering

---

[1] https://wiki.qut.edu.au/display/cyphy/OpenRatSLAM+datasets.
[2] https://www.visuallocalization.net/datasets/.

Table 2. Place recognition precision comparison when using original-images and inpaint-images with four featurization ConvNets.

Precision (%)	VGG19		NIN_ImageNet		AlexNet		bvlc_GoogleNet	
	Original	Inpaint	Original	Inpaint	Original	Inpaint	Original	Inpaint
St.Lucia	98.24	**99.56**	97.80	**99.12**	98.68	**99.56**	99.56	99.56
CMU	84.54	**86.96**	92.75	**96.14**	87.92	**92.75**	87.44	**92.27**
Oxford RobotCar	90.78	**98.58**	86.88	**90.43**	88.65	**91.84**	96.10	**98.23**

one traversal as the query traversal, the images in the other traversal are selected one by one to make sure that each query image has a corresponding image, which was captured at the same place. Then, each query image and its corresponding image are renamed with the same number from 1 to the length of the query sequence according to their positions. Since the same number represents the same place and adjacent numbers represent adjacent places, it is easy to find out whether a query image is correctly matched or not by selecting the best matching images according to their numbers.

4.3   Evaluation Criteria and Baselines

Based on the evaluation of [13], a precision parameter to benchmark visual place recognition is defined as the percentage of query images whose predicted place is approximately same as its real place: the query image and its nearest reference image are seperated by no more than 3 images (about 5 m of their ground truth positions).

Table 2 shows the results obtained using deep feature matching on the original query and reference images to benchmark the proposed method.

The method to generate the deep feature vector of an image is applied using four different ConvNets: VGG19, NIN_ImageNet, AlexNet, bvlc_GoogleNet. The best results obtained using the proposed approach are based on layer con5_1 in VGG19, conv3 in NIN_ImageNet, conv4 in AlexNet and pool3/3×3_s2 in bvlc_GoogleNet.

The four ConvNets are detailed as below:

– VGG19. The model used here [15] is an improved version of VGG19 in the ILSVRC-2014 competition.
– NIN_ImageNet [7]. It is a small model for ImageNet, yet fast to train and has slightly better performance than AlexNet.
– AlexNet. AlexNet model is proposed in [6] and is trained on images from ImageNet.
– bvlc_GoogleNet. This model was trained by Sergio Guadarrama[3] and it is a replication of the work in [17].

---

[3] https://github.com/BVLC/caffe/tree/master/models/bvlc_googlenet.

5   Results

Table 2 shows the place recognition precision obtained using original-images and their corresponding inpaint-images for three different datasets through four ConvNets.

Results of Table 2 shows that no matter the traffic conditions and weather conditions are moderate or severe changing, deleting the cars and pedestrians and then repairing the background permits to improve the results whatever the network used (VGG19, NIN_ImageNet, AlexNet). When using bvlc_GoogleNet to generate features of images from St.Lucia dataset, removing the dynamic objects have no impact on the performance of place recognition.

As illustrated in Fig. 1, the FCN is effective in recognizing objects and Edge-Connect can repairing images, but there are also some challenges. FCN is not sensitive to details, so it can not recognize cars which are far away. For some objects, FCN outputs are too coarse and parts of the objects remain, that will result in those objects being reconstructed by EdgeConnect unexpectly. As for EdgeConnect, the inpainting quality may be not good when faces to the images with a large missing portion, and the reconstructed regions may suffer from artifacts.

Even though, the performance of place recognition can be improved without perfect inpainted images.

6   Conclusion and Future Work

This paper proposed a new feature extraction system for visual place recognition. Before putting the whole images into a ConvNet, they are submitted to semantic segmentation and image inpainting method to decrease the presence of dynamic information of images. This method improves the performance of place recognition in changing environments, especially in changing traffic conditions.

The robustness of features extracted by the proposed system relies on the performance of semantic segmentation and image inpainting. It is a hard task for image inpainting methods to inpaint the image with a large missing portion, so improving the image inpainting performance may further improve the performance of place recognition. This is the research direction of our following work.

Acknowledgment. The authors gratefully acknowledge financial support from China Scholarship Council, and useful discussions with Tao Yang.

References

1. Bay, H., Tuytelaars, T., Van Gool, L.: SURF: speeded up robust features. In: Leonardis, A., Bischof, H., Pinz, A. (eds.) ECCV 2006. LNCS, vol. 3951, pp. 404–417. Springer, Heidelberg (2006). https://doi.org/10.1007/11744023_32

2. Chen, Z., Lam, O., Jacobson, A., Milford, M.: Convolutional neural network-based place recognition (2014). arXiv:1411.1509

3. Hernan Badino, D.H., Kanade, T.: The CMU visual localization data set (2011). http://3dvis.ri.cmu.edu/data-sets/localization

4. Johns, E., Yang, G.: Feature co-occurrence maps: appearance-based localisation throughout the day. In: IEEE International Conference on Robotics and Automation, pp. 3212–3218, May 2013

5. Kanji, T.: Self-localization from images with small overlap. In: IEEE/RSJ International Conference on Intelligent Robots and Systems (IROS), pp. 4497–4504 (2016)

6. Krizhevsky, A., Sutskever, I., Hinton, G.E.: ImageNet classification with deep convolutional neural networks. Adv. Neural Inf. Process. Syst. **25**, 1097–1105 (2012)

7. Lin, M., Chen, Q., Yan, S.: Network in network (2013). arXiv:1312.4400

8. Long, J., Shelhamer, E., Darrell, T.: Fully convolutional networks for semantic segmentation. In: The IEEE Conference on Computer Vision and Pattern Recognition (CVPR), June 2015

9. Lowe, D.G.: Object recognition from local scale-invariant features. In: Proceedings of the International Conference on Computer Vision (ICCV), p. 1150 (1999)

10. Maddern, W., Pascoe, G., Linegar, C., Newman, P.: 1 year, 1000 km: the Oxford robotcar dataset. Int. J. Robot. Res. **36**(1), 3–15 (2017). https://doi.org/10.1177/0278364916679498

11. Nazeri, K., Ng, E., Joseph, T., Qureshi, F.Z., Ebrahimi, M.: EdgeConnect: Generative image inpainting with adversarial edge learning (2019). arXiv:1901.00212

12. Neubert, P., Sünderhauf, N., Protzel, P.: Superpixel-based appearance change prediction for long-term navigation across seasons. Robot. Auton. Syst. **69**, 15–27 (2015)

13. Olid, D., Fácil, J.M., Civera, J.: Single-view place recognition under seasonal changes (2018). arXiv:1808.06516

14. Qiao, Y., Cappelle, C., Ruichek, Y., Yang, T.: ConvNet and LSH-based visual localization using localized sequence matching. Sensors **19**(11) (2019)

15. Simonyan, K., Zisserman, A.: Very deep convolutional networks for large-scale image recognition (2014). arXiv:1409.1556

16. Sünderhauf, N., et al.: Place recognition with ConvNet landmarks: viewpoint-robust, condition-robust, training-free. In: Robotics: Science and Systems, July 2015

17. Szegedy, C., et al.: Going deeper with convolutions. In: The IEEE Conference on Computer Vision and Pattern Recognition (CVPR), June 2015

18. Yang, T., Cappelle, C., Ruichek, Y., Bagdouri, M.: Multi-object tracking with discriminant correlation filter based deep learning tracker. Integr. Comput.-Aided Eng. **26**, 1–12 (2019). https://doi.org/10.3233/ICA-180596

CNN-SVM Learning Approach Based Human Activity Recognition

Hend Basly[1]([✉]), Wael Ouarda[2], Fatma Ezahra Sayadi[3], Bouraoui Ouni[1], and Adel M. Alimi[2]

[1] NOCCS-Lab.: Networked Objects Control and Communication Systems Laboratory, National Engineering School of Sousse (ENISO), University of Sousse, BP 264, 4023 Erriadh, Sousse, Tunisia
basly.hend@gmail.com, ouni_bouraoui@yahoo.fr
[2] REGIM-Lab.: REsearch Groups in Intelligent Machines, National Engineering School of Sfax (ENIS), University of Sfax, BP 1173, 3038 Sfax, Tunisia
{wael.ouarda,adel.alimi}@ieee.org
[3] EμE-Lab.: Electronics and Microelectronics Laboratory, Faculty of Sciences of Monastir (FSM), University of Monastir, Environment Avenue, 5019 Monastir, Tunisia
sayadi_fatma@yahoo.fr

Abstract. Although it has been encountered for a long time, the human activity recognition remains a big challenge to tackle. Recently, several deep learning approaches have been proposed to enhance the recognition performance with different areas of application. In this paper, we aim to combine a recent deep learning-based method and a traditional classifier based hand-crafted feature extractors in order to replace the artisanal feature extraction method with a new one. To this end, we used a deep convolutional neural network that offers the possibility of having more powerful extracted features from sequence video frames. The resulting feature vector is then fed as an input to the support vector machine (SVM) classifier to assign each instance to the corresponding label and bythere, recognize the performed activity. The proposed architecture was trained and evaluated on MSR Daily activity 3D dataset. Compared to state of art methods, our proposed technique proves that it has performed better.

Keywords: Convolutional Neural Network (CNN) · Human action recognition · Support Vector Machines (SVM)

1 Introduction

Human Activity Recognition remains a very important research field of numerous computer science organizations because of its potency to provide adapted support for various applications such as human-computer interaction, eHealth

Supported by organization x.

applications and surveillance. Nowadays, according to the method of feature extraction, the recognition of the human activity system can be classified as a classical or a deep model. A classical model is based on hand-crafted feature descriptors which can be categorized in three types; local features, global features or a combination between them to tackle the human activity recognition problem. The global features designate the image as a whole to describe the entire human body motions. However, the local features are extracted from a set of spatio-temporal interest points (STIPs) to describe the image patches of a human action. Although global methods are able to represent more visual informations by maintaining spatio-temporal structures of the occured actions in the video, they are very sensitive to background variations and partial occlusions. The Local features considers the image as small regions, which is practically computationally expensive. On another side, deep models using deep neural networks are a promising alternative in the image analysis applications areas. Convolutional Neural Network (CNN) is considered as one of the successful deep models for image classification tasks. Traditionally, to deal with such problem of recognition, researcher are obliged to anticipate their algorithms of Human activity recognition by prior data training preprocessing in order to extract a set of features using different types of descriptors such as HOG3D [1], extended SURF [2] and Space Time Interest Points (STIPs) [3] before inputting them to the specific classification algorithm such as HMM, SVM, Random Forest [4–6]. It has been proven that the previous approaches are not very robust due to their poor performance and their requirement in time and memory space. Recently, deep learning architectures are employed in order to change the engineering feature extraction phase by an automatic processing where deep neural networks have been directly applied to the raw data without human intervention to extract deep features. Since the training of a new CNN from scratch requires to load huge amount of data and expensive computational resources, we used the concept of transfer learning and fine tune the parameters of a pretrained model. The initial CNN model was trained on a subset of the ILSVRC-2015 of the large scale ImageNet [7] dataset. Consequently, we decreased the training time, and avoid over fitting by insuring the suitable weight initialization given the quite small used data set. In this study, we proposed an advanced human activity recognition method from video sequence using CNN, where the large scale dataset ImageNet pretrains the network. In fact, a pretrained CNN extracts feature vectors that characterize frames from the raw data. The resulting deep sparse representation of features vectors are fed as input to a multi class support vector machines algorithm to be classified. Since the deep neural networks are more difficult to train, the residual learning approach based ResNet model was proposed to facilitate the training phase. The main contribution of the present work is to propose a learning approach for human activity recognition based CNN and SVM able to classify activities from one shot. The proposed framework is trained and tested on a publicly available dataset, i.e., MSRDailyActivity 3D dataset [8]. Obtained results show that the proposed method outperforms the state-of-the-art methods. The rest of this paper is organized as follows: Sect. 2 highlights some related works, in Sect. 3, we describe our proposed approach. We present the experimental evaluation in Sect. 4. Finally, in Sect. 5, we conclude the paper.

2 Related Works

For Human Activity recognition challenge, an activity has to be represented by a set of features. To represent complex activities, authors in [9] have combined the histogram of oriented gradient (HOG), the motion history image (MHI) and the foreground image (FI). The HOG feature represents the magnitude and the direction of corners and edges, MHI feature is extracted to characterize motion direction and the FI is obtained by background subtraction. Finally, all the resulting features have been merged to be fed as input to a simulated annealing multiple instance learning support vector machine (SMILE-SVM) classifier for human activity recognition. The work of [10] extracted a motion space-time feature descriptor characterizing the video frames by combining the histogram of silhouette and the optical flow values. The first feature is obtained by background subtraction and the second is calculated using the algorithm of Lucas-Kanade [11] inside a normalized bounding box. A multi class SVM classifier has been used to classify the activities. This system was set up to face the restraints of long training time and high dimension of the feature vector. [12] investigates a two distinct stream convNets architecture that includes spatial and temporal networks. In the spatial stream, the action recognition is performed from RGB video frames, whereas in the temporal stream, the recognition of action was made from motion information obtained by stacking dense optical flow between consecutive frames. Both streams are employed as ConvNets and are finally combined by late fusion. Two fusion methods have been considered; a fusion by averaging and a fusion by multi-class linear SVM on softmax scores. The purpose in [13] is to classify the human actions from videos into different classes. The process is performed by extracting interest points from each video, segmenting images and constructing motion history images. After selecting discriminating features and representing images by visual words, a histogram of visual words is elaborated based on features extracted from the motion history images. Finally, the extracted features vectors are used to train a support vector machine for action classification. [14] proposed a system to recognize abnormal comportment providing an alert to the accurate user on his android mobile phone. The task is to extract features using Scale Invariant Feature Transform (SIFT) descriptor for each video after dividing them into number of frames. The extracted features are then exploited as input to two different types of classifiers, i.e; the K Nearest Neighbor (KNN) and the Support Vector Machine (SVM) to classify the actions.

3 Proposed Approach

3.1 Convolutional Neural Networks (CNN)

As recent written works [12,24,27] has proven, the deep hierarchical visual feature extractors are currently outperforming traditional hand-crafted descriptor, and are more generalizable and accurate when dealing with important levels of immanent noise problems. To describe the activities in a frame-wise way, we chose to use the CNN approach based on RGB data because of its widespread application in different areas. CNNs are also advantageous by their reduction of the number of parameters and connections used on artificial neural model to facilitate their training phase. In this step, the question now is how to represent the human actions in each extracted frame of the video. To extract the most pertinent and significant features from the raw RGB video frame, we employed a pre-trained deep CNN architecture with pre-trained parameters based on ImageNet. The original CNN was trained on the 1.2M high-resolution images of the ILSVRC2015 classification training subset of the ImageNet dataset. Though, in the proposed method, we used a deep CNN network architecture to generate a probability vector for each input frame which represents the probability of the presence of the different objects present in each individual frame. A ResNet model is used with pre-trained parameters from ImageNet database and applied to extract sparse and pertinent residual representations of features from video frames of each sequence video. The architecture is composed of several ResNet blocks with three layer deep, composed of five composite convolutional layers including small kernels sizing by 7×7, 1×1 and 3×3. The network takes an input of size 224×224 which was reduced five times in the network by a stride of 2. The output obtained from the average pooling operation is applied to the final feature map of the network followed by the fully connected layer. The resulting vector from the last pooling layer is considered as the features representation generated from the reused pretrained model in a feedforward pass. After each convolution, a batch normalization and an ReLU are achieved. The residual units are represented as:

$$x_{l+1} = f(x_l + F(x_l; W_l)) \tag{1}$$

where x_l and x_{l+1} correspond to the input and the output of the $l^t h$ layer, F denotes a nonlinear residual mapping characterized by convolutional filter weights Wl and f corresponds to the ReLU function. The main advantage of handling residual units in such types of networks, is that their skip connections or "shortcuts" allow the direct propagation of signals over all the network' layers. This design is very advantageous mainly during the backpropagation phase; in fact, gradients are directly propagated from the loss layer to all the other preceding layers while skipping some intermediate layers which have the potential to provoke the deterioration or the disappearance of the gradient signal. This strategy helped the network to appreciate the accuracy gained from deeper architectures. Since training a new deep CNN model from scratch requires important loads of data and elevated resources of computation, we have implemented a transfer learning procedure to fine-tune the parameters of a pre-trained model. We adopted an original CNN model that was pretrained on a subset of the large-scale image classification dataset such as the ImageNet. Proceeding in this way, we succeed to reduce the required time for training and to avoid our dataset from overfitting by assuring a good initialization of weights, given the quiet small available dataset. In fact, the dataset was artificially augmented by using three techniques. First random reflect frames in the left direction, second a random horizontal translation that consists of moving frames along the horizontal direction, and finally, a random vertical translation is applied by moving frames on the vertical direction. In reality, the last layer of the adopted CNN model is a classification layer; though, in the present study, we removed this layer and exploited the output of the preceding layer as frame features for the classification step. Instead of the eliminated layer, the SVM classifier has been employed to predict the human activity label. Figure 1 summarizes the architecture of the proposed action recognition model.

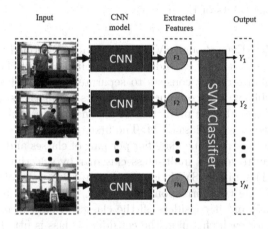

Fig. 1. Architecture of the proposed action recognition model.

3.2 Support Vector Machines (SVM)

SVM is supposed as machine learning classifier method that gives good results in comparison with other types of classifier. We decided to use it in this study because of its effectiveness when dealing with quiet small datasets and its performance in high dimensional spaces [15,16,25–29]. The principal idea behind the use of SVM is to applicate a supervised learning algorithm facilitating to find the optimal hyperplane that separates the feature space. During training, the SVM generates hyperplanes in a high dimensional space to separate the training dataset into different classes. If the training data subset are not linearly separable, a kernel function SVM is used to transmit the data to a new vector space. SVM performs well with large scale training datasets and yields to accurate and effective results. For a given training dataset; $D(x_1, y_1), (x_2, y_2), ...(x_N, y_N)$ where $x_i \in \mathbb{R}^n$ and memberships $y_i \in \pm 1$ classes; i represents the label corresponding to each action in the defined dataset. To determine a decision function for a linear classification, the hyperplane separation is represented by:

$$y_i = sng((w \Delta x_i) + b) \tag{2}$$

A generic hyperplane is defined by satisfying the condition:

$$w \cdot x_i + b = 0 \tag{3}$$

When delimited by margins, the set of hyperplanes can be written as:

$$y_i \cdot ((w \cdot x_i) + b) \geq 1 \tag{4}$$

To formulate the optimal hyperplane that separates the data, we should minimize:

$$1/2\|w\|^2 \tag{5}$$

Subject to the constraints of Eq. 4).

Multi-class SVM. Even though SVM were initially developed for binary classification, it can be successfully extended to be applied to multiclass classification problems. The main strategy consists to separate the multiclass problem into many biclass problems and combine the outputs of all the sub-binary classifiers to provide the final class prediction of a sample. Fundamentally, there are two main methods for multiclass SVM. The first type is called "oneagainstone" [17], it consists to construct one classifier per pair of classes and combine binary classifiers in a way to form a multi-class classifier by selecting the most voted class. So, $N(N-1)/2$ binary SVM classifiers are needed, each of them is trained on the samples of the two corresponding classes. The second method is called "oneagainstall" [27] and it considers all the classes of data in one optimization problem. In fact, for each classifier, the considered class is fitted against all the other classes, so, N number of classes use N SVM classifiers. When using the latter technique, the training process takes a long time.

4   Dataset and Evaluation Procedure

4.1   MSRDailyActivity 3D Dataset

The "MSRDailyActivity 3D" dataset [12] is an RGB sequences dataset that contains sixteen daily human activities. The database was captured by a kinect camera around various objects, and the humans in question are located at different distances from the camera. Activities are accomplished by ten different subjects, the most of them are categorized as "human object interactions". Activities were performed twice by each person in two different positions; i.e; the "standing" and the "sitting" situation.

4.2   Implementation Details

The deep CNN model was trained using Matlab 2018. Our approach based CNN model was performed on a machine equipped with a NVIDIA GeForce 960M GPU, 64 GB memory and an Intel Core i7-6700 HQ (2.60 GHz) processor. Our dataset was artificially augmented. This technique allows to avoid the problem of dataset overfitting. Each video from our dataset were split into frames which serve as input to the pre-trained CNN model. In the training stage, a 224 × 224 frame is randomly reflected from the selected frame; it then undergoes a random horizontal and vertical translation. These operations are applied in such a way that the training dataset is augmented at each iteration. The 2048 dimensional vector resulting from the last pooling layer of the ResNet model were used to activate the training and testing subsets. The resulting vectors were used as training and test data for the multi-class SVM classifier. The training process is performed using a mini-batch stochastic gradient descent with a momentum set to 0.9 to learn the network weights. At each iteration, a mini-batch size of 50 samples is constructed by sampling the training videos by 50, from which a single frame is selected randomly. During our experimentation, the learning rate is initially set to $1e^{-4}$ and the network is trained for 6 epochs. We also tried to increment the number of epochs but we got always overfitting. For our used multi–class SVM classifier, we chose to employ the linear function kernel to project the original linear or nonlinear dataset into a higher dimensional space in order to make it linearly separable and to give a better performance for the SVM. The Linear kernel is a simple kernel function based on the penalty parameter C described by the following format:

$$K(x, y) = x^T y + c \tag{6}$$

4.3 Evaluation Methodology

During experimentation, we evaluated our method on the dataset described above: 70% used for the training stage and 30% from data are used for testing. Firstly, each frame is resized to 224 × 224 resolution. We have determined the confusion matrix of our proposed system in order to demonstrate the correspondence between the predicted labels along the x-axis and the true labels along the y-axis and to represent the recognition performance for each action class in the MSRDailyActivity 3D dataset. Generally, a confusion matrix involves four groupings: TP (True positive) mean the instances that are correctly identified as positives, FP (False positive) refers to the negative examples incorrectly identified as positive, TN (True negative) refers to the negative instances that are correctly predicted as negative, and FN (false negative) represents the positive instances incorrectly predicted as negative. We also evaluate different performance metrics of our proposed approach by calculating the precision, recall and f-measure values as shown in Table 1.

Table 1. Performance metrics results obtained with our proposed approach.

Accuracy (%)	Precision (%)	Recall	F-measure
99.92	98.77	99.79	99.28

Figure 2 demonstrates that the most confusion is between sit down and stand up labels. This misclassification can be explained by the similarity in a few steps when carrying out both of actions which contain a person in a half-sitting position. İn fact, the middle frames of the two classes sit down and stand up presenting a person in a half setting position are making the confusion, because of their repetition in the two cases. Whereas more than half of the classes have been correctly classified at 100%.

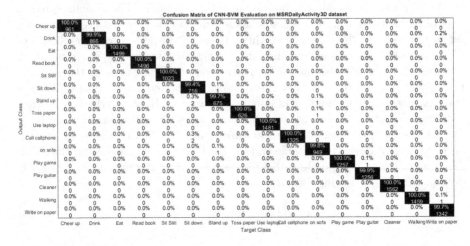

Fig. 2. Confusion matrix of the learning CNN–SVM approach evaluation on MSRDailyActivity3D dataset.

Table 2 notices that our approach has achieved a good recognition performance and outperforms other state-of-the-art methods on MSRDailyActivity 3D dataset. Achieved performance confirms the generalization competence of our learned representations across domains. The work of [18] has obtained bad results in this dataset despite it was based on the combination of two deep neural network models which are the CNN and LSTM. Whereas the implemented CNN model for feature extraction is not based transfer learning concept. Based on all these observations, we can deduce that pretraining a model on a large-scale dataset and fine tune his hyper-parameters on a small one is very efficient to obtain good performance rate. We have also combined the same pretrained ResNet model which was used to extract features, once with a Multi Layer Perception (MLP) classifier and another time with a Long Short Term Memory (LSTM) network. The obtained results show that using a multi-class SVM classifier, gives the best result.

In order to investigate on the effect of the choice of the SVM kernel, we have performed a classification using Radial Basis Function (RBF) kernel. The results were not interesting due to the relevance of the feature representation obtained from convolutional neural network.

Table 2. Comparison of different HAR approaches on MSRDAily Activity 3D dataset.

Reference	Method	Accuracy (%)
Sial et al. [19]	D-STIP + D-DESC+ RGB-DESC +SVM	92.00
Nunez et al. [18]	CNN + LSTM	63.10
Lu Xia et al. [20]	DSTIP + DCSF + SVM	96.70
Wang et al. [21]	LOP + FTP + AEM +SVM	85.75
Shahroudy et al. [22]	Dense trajectories with HOG, HOF and MBH + skeleton joints + SVM	91.25
Wang et al. [23]	WHDMM + Three channel 3D ConvNet	85.00
LAHAR-CNN (Ours)	Pretrained CNN + MLP	99.40
DTR-HAR (Ours)	Pretrained CNN + LSTM	91.56
Our approach	Pretrained CNN + SVM	99.92

5 Conclusion

In this study we presented the support vector machines approach for human activity recognition task. We proposed to use a pre-trained CNN approach based ResNet model in order to extract spatial and temporal features from consecutive video frames. Our proposed architecture was trained and tested on MSRDaily-Activity 3D dataset and it achieved a good recognition performance. For our future works, we propose to use a combination of a genetic algorithm with support vector machines in order to optimize the weights of the used CNN model leading to automatically improve the performance. Likewise, we would like to expend the proposed model for more large-scale dataset such as NTU RGB+D because the used dataset is small and the used pretrained CNN model can be more effective when applied to a big one.

References

1. Klaser, A., Marszałek, M., Schmid, C.: A spatio-temporal descriptor based on 3D-gradients. In: 19th British Machine Vision Conference, BMVC 2008 on Proceedings, pp. 275:1–10. BMVA Press, Leeds (2008)
2. Willems, G., Tuytelaars, T., Van Gool, L.: An efficient dense and scale-invariant spatio-temporal interest point detector. In: Forsyth, D., Torr, P., Zisserman, A. (eds.) ECCV 2008. LNCS, vol. 5303, pp. 650–663. Springer, Heidelberg (2008). https://doi.org/10.1007/978-3-540-88688-4_48
3. Dollár, P., Rabaud, V., Cottrell, G., Belongie, S.: Behavior recognition via sparse spatio-temporal features. In: Proceedings. 2nd Joint IEEE International Workshop on Visual Surveillance and Performance Evaluation of Tracking and Surveillance (VS-PETS), pp. 1:65–72. IEEE, Beijing (2005)
4. Zhu, C., Sheng, W.: Multi-sensor fusion for human daily activity recognition in robot-assisted living. In: Proceedings of the 4th ACM/IEEE International Conference on Human Robot Interaction, pp. 303–304. ACM (2009)
5. Ghosh, A., Riccardi, G.: Recognizing human activities from smartphone sensor signals. In: Proceedings of the 22nd ACM International Conference on Multimedia, pp. 865–868. ACM (2014)

6. Rahman, S.A., Merck, C., Huang, Y., Kleinberg, S.: Unintrusive eating recognition using Google Glass. In: Proceedings of the 9th International Conference on Pervasive Computing Technologies for Healthcare, PervasiveHealth 2015, pp. 108–111. Institute of Electrical and Electronics Engineers Inc, United States (2015). https://doi.org/10.4108/icst.pervasivehealth.2015.259044

7. Russakovsky, O., et al.: Imagenet large scale visual recognition challenge. Int. J. Comput. Vis. **115**(3), 211–252 (2015)

8. Wang, J., Liu, Z., Wu, Y., Yuan, J.: Mining actionlet ensemble for action recognition with depth cameras. In: Proceedings of the 2012 IEEE Conference on Computer Vision and Pattern Recognition (CVPR), pp. 1290–1297. IEEE, Providence (2012)

9. Hu, Y., Cao, L., Lv, F., Yan, S., Gong, Y., Huang, T.S.: Action detection in complex scenes with spatial and temporal ambiguities. In: Proceedings of 12th IEEE International Conference on Computer Vision, pp. 128–135. IEEE, Kyoto (2009)

10. Chathuramali, K.M., Rodrigo, R.: Faster human activity recognition with SVM. In: International Conference on Advances in ICT for Emerging Regions (ICTer), pp. 197–203. IEEE, Colombo (2012)

11. Lucas, B.D., Kanade, T.: An iterative image registration technique with an application to stereo vision. In: Proceedings of the 7th International Joint Conference on Artificial Intelligence (IJCAI 1981), pp. 674–679. Morgan Kaufmann Publishers Inc., Vancouver (1981)

12. Simonyan, K., Zisserman, A.: Two-stream convolutional networks for action recognition in videos. In: Advances in Neural Information Processing Systems, pp. 568–576. MIT Press, Montreal (2014)

13. Siddiqui, S., Khan, M.A., Bashir, K., Sharif, M., Azam, F., Javed, M.Y.: Human action recognition: a construction of codebook by discriminative features selection approach. Int. J. Appl. Pattern Recogn. **5**(3), 206–228 (2018)

14. Kale, G.: Human activity recognition on real time and offline dataset. Int. J. Intell. Syst. Appl. Eng. **7**(1), 60–65 (2019)

15. Cortes, C., Vapnik, V.: Support-vector networks. Mach. Learn. **20**(3), 273–297 (1995)

16. Ikram, S.T., Cherukuri, A.K.: Improving accuracy of intrusion detection model using PCA and optimized SVM. J. Comput. Inf. Technol. **24**(2), 133–148 (2016)

17. Xu, Y., Zomer, S., Brereton, R.G.: Support vector machines: a recent method for classification in chemometrics. Critical Rev. Anal. Chem. **36**(3–4), 177–188 (2006)

18. Nunez, J.C., Cabido, R., Pantrigo, J.J., Montemayor, A.S., Velez, J.F.: Convolutional neural networks and long short-term memory for skeleton-based human activity and hand gesture recognition. Pattern Recogn. **76**, 80–94 (2018)

19. Sial, H.A., Yousaf, M.H., Hussain, F.: Spatio-temporal RGBD cuboids feature for human activity recognition. Nucleus **55**(3), 139–149 (2018)

20. Xia, L., Aggarwal, J.K.: Spatio-temporal depth cuboid similarity feature for activity recognition using depth camera. In: Proceedings of the IEEE Conference on Computer Vision and Pattern Recognition, pp. 2834–2841. IEEE Computer Society, USA (2013)

21. Wang, J., Liu, Z., Wu, Y., Yuan, J.: Learning actionlet ensemble for 3D human action recognition. IEEE Trans. Pattern Anal. Mach. Intell. **36**(5), 914–927 (2013)

22. Shahroudy, A., Ng, T.T., Yang, Q., Wang, G.: Multimodal multipart learning for action recognition in depth videos. IEEE Trans. Pattern Anal. Mach. Intell. **38**(10), 2123–2129 (2015)

23. Wang, P., Li, W., Gao, Z., Zhang, J., Tang, C., Ogunbona, P.O.: Action recognition from depth maps using deep convolutional neural networks. IEEE Trans. Hum.-Mach. Syst. **46**(4), 498–509 (2015)
24. Jarraya, I., Ouarda, W., Alimi, A.M.: Deep neural network features for horses identity recognition using multiview horses' face pattern. In: Ninth International Conference on Machine Vision (ICMV), pp. 103410B. International Society for Optics and Photonics (2016)
25. Sassi, A., Ouarda, W., Ben Amar, C., Miguet, S.: Neural approach for context scene image classification based on geometric, texture and color information. In: Chen, L., Ben Amor, B., Ghorbel, F. (eds.) RFMI 2017. CCIS, vol. 842, pp. 110–120. Springer, Cham (2019). https://doi.org/10.1007/978-3-030-19816-9_9
26. Nasri, H., Ouarda, W., Alimi, A.M.: ReLiDSS: novel lie detection system from speech signal. In: IEEE/ACS 13th International Conference of Computer Systems and Applications (AICCSA), pp. 1–8. IEEE (2016)
27. Larbi, K., Ouarda, W., Drira, H., Amor, B.B., Amar, C.B.: DeepColorFASD: face anti spoofing solution using a multi channeled color spaces CNN. In: IEEE International Conference on Systems, Man, and Cybernetics (SMC), pp. 4011–4016. IEEE, Miyazaki (2018)
28. Ehaimir, M.E., Jarraya, I., Ouarda, W., Alimi, A.M.: Human gait identity recognition system based on gait pal and pal entropy (GPPE) and distances features fusion. In: Sudan Conference on Computer Science and Information Technology (SCCSIT), pp. 1–5. IEEE, Elnihood (2017)
29. Ghabri, S., Ouarda, W., Alimi, A.M.: Towards human behavior recognition based on spatio temporal features and support vector machines. In: Ninth International Conference on Machine Vision (ICMV), pp. 103410E. International Society for Optics and Photonics (2016)

Convolutional Neural Networks Backbones for Object Detection

Ayoub Benali Amjoud$^{(\boxtimes)}$ ⓘ and Mustapha Amrouch

IRF-SIC Laboratory, Faculty of Sciences, Ibn Zohr University, Agadir, Morocco
a05.benali@gmail.com, m.amrouch@uiz.ac.ma

Abstract. Detecting objects in images is an extremely important step in many image and video analysis applications. Object detection is considered as one of the main challenges in the field of computer vision, which focuses on identifying and locating objects of different classes in an image. In this paper, we aim to highlight the important role of deep learning and convolutional neural networks in particular in the object detection task. We analyze and focus on the various state-of-the-art convolutional neural networks serving as a backbone in object detection models. We test and evaluate them in the common datasets and benchmarks up-to-date. We Also outline the main features of each architecture. We demonstrate that the application of some convolutional neural network architectures has yielded very promising state-of-the-art results in image classification in the first place and then in the object detection task. The results have surpassed all the traditional methods, and in some cases, outperformed the human being's performance.

Keywords: Object detection · Convolutional neural networks · Review

1 Introduction

Detecting objects in a scene proved to be a very difficult task, which has been investigated for a variety of applications in recent years, such as face detection, self-driving cars, medical disease detection, video surveillance, and for natural disaster protection. The convolutional neural networks (CNNs) represent the heart of state-of-the-art object detection methods. They are used for extracting features. Several CNNs are available, for instance, AlexNet, VGGNet, and ResNet. These networks are mainly used for object classification task and have evaluated on some widely used benchmarks and datasets such as ImageNet (Fig. 1). In image classification or image recognition, the classifier classifies a single object in the image, outputs a single category per image, and gives the probability of matching a class. Whereas in object detection, the model must be able to recognize several objects in a single image and provides the coordinates that identify the location of the objects. This shows that the detection of objects can be more difficult than the classification of images.

Traditional object detection models tend to use methods such as Haar-Like features [1], HOG [2], and Scale-Invariant Feature Transform [3] for extracting the features in the image. Those approaches have been based on the way we could manually design the features or the model according to our understanding. Recently it has been proven

© Springer Nature Switzerland AG 2020
A. El Moataz et al. (Eds.): ICISP 2020, LNCS 12119, pp. 282–289, 2020.
https://doi.org/10.1007/978-3-030-51935-3_30

that it is more efficient to let the machine handle these tasks. And this is when the convolutional neural networks came to take control, achieving impressive successes [4, 5]. The present paper will be structured as follows. First, we review the leading state-of-the-art convolutional neural network architectures used in object detection. We then introduce the image datasets we have used to compare the networks along with the experiments. We further report the results of each architecture when used with state-of-the-art object detection models.

2   Convolutional Neural Network Backbones

The selection of CNN architectures to be covered in this article is not made randomly, but according to their popularity and performance in different state of the art object detection models.

2.1   AlexNet

Krizhevsky et al. [4] in 2012, developed a convolutional neural network composed of 8 layers, where 5 are convolutional and 3 are fully connected. The network is called AlexNet. In comparison to LeNet-5, AlexNet [6] has more layers and contains around 60 million parameters. Rectified Linear Units (ReLUs) are used for the first time as activations in AlexNet instead of sigmoid and tanh activations to add non-linearity. AlexNet is used in object detection models such as R-CNN [7], and HyperNet [8].

2.2   VGG-16

In 2014 a network called VGG-16 [9] was released, composed of 13 convolutional and 3 fully connected layers with ReLU activation. VGG-16 provides more layers compared to AlexNet and uses smaller filters of 2×2 and 3×3. It includes 138 million parameters. A deeper version of VGG called VGG-19 is available. VGG-16 is one of the most used architectures in object detection and achieved interesting performances; it's used for instance in algorithms like Fast R-CNN [10], Faster R-CNN [11], HyperNet [8], RON384 [12], SSD [13] and RefineDet [14].

2.3   GoogLeNet

Also called Inception V1, GoogLeNet [15] is a small network developed by Szegedy et al. in 2014. Their method is different from that of VGGNet and AlexNet. They came up with a new notion known as blocks of inception, where it embeds multi-scale convolutional transformations. The inception block includes filters of varying sizes 1×1, 3×3 and 5×5. It employs a 1×1 convolution in the middle of the network to reduce dimensionality and they opted to use global average pooling instead of fully connected layers. The network is made of 22 layers with 5 million parameters. GoogLeNet mainly is used in YOLO [16] object detection model.

2.4 ResNets

Convolutional neural networks have become more and more deeper with the addition of layers, but once the accuracy gets saturated, it quickly drops off. To solve this issue, He et al. in 2015 developed ResNets [5] which are based on residuals or skip connections. They also use Batch Normalization [17]. ResNets are mainly consisting of convolutional and identity blocks. There are many variants of ResNets, for instance, ResNet-34, ResNet-50 which is composed of 26 million parameters, ResNet-101 with 44 million parameters and ResNet-152 which is deeper with 152 layers. ResNet-50 and ResNet-101 are used widely in object detection models. While ResNet-50 is used in some object detection frameworks such as BlitzNet [18] and RetinaNet [19]. ResNet-101 is used in Faster R-CNN [5], R-FCN [20], and CoupleNet [21], etc.

2.5 Inception-ResNet-V2

Szegedy et al. published in 2016, Inception-ResNet-V2 [22], a CNN inspired by the ResNet and based on a hybrid approach by combining Inceptions and ResNet architectures, which use residual connections as an alternative to concatenation filters. Inception-ResNet-V2 is composed of 164 deep layers and about 55 million parameters. The Inception-ResNet models have led to better accuracy performance at shorter epochs. Inception-ResNet-V2 is used in Faster R-CNN G-RMI [23], and Faster R-CNN with TDM [24] object detection models.

2.6 DarkNet-19

A network developed to be small and efficient at the same time. It is based on many previous ideas like the Darknet reference, Network In Network [25], Inception [15, 26] and Batch Normalization [17]. Darknet-19 [27] uses convolutional layers instead of fully connected layers. It is composed of 19 convolutional and 5 max-pooling layers. It uses only 3×3 convolutional kernels and several 1×1 convolutional kernel to reduce the number of parameters. DarkNet-19 is used in YOLOv2 [27].

3 Data, Experiments and Results

3.1 Data

To assess the different CNNs mentioned above, we used several common data sets in the field of classification and object detection. First, we used the ImageNet database [28], one of the largest databases available today, it contains more than 14 million images from different categories. We used ImageNet to calculate Top-1 and Top-5 accuracy rates. Afterward, we used Pascal VOC [29] (2007 and 2012), and the Common Object in Context (COCO) [30] dataset for the object detection purposes.

Fig. 1. Examples of images from the ImageNet 2012 dataset.

3.2 Experiments and Results

In this section, we experiment with the CNNs mentioned in this paper along with the object detection models based on these networks under the different datasets and benchmarks. In Table 1, with the exception of the DarkNet-19, all the experiments are carried out with PyTorch[1], an open-source machine learning framework and Nvidia T4 GPU. The input size resolution is 224 × 224 for all networks except for Inception-ResNet-V2 where the input size is 299 × 299. To evaluate the computational complexity of each network we use the Multiply-And-Accumulate (MAC) operation that could be considered as two separate floating-point operations (FLOPs) [31]. In Table 2 and Table 3, the detectors are trained on Pascal VOC07 trainval and Pascal VOC12 trainval. The +S suffix means that the model is trained also for segmentation and extra annotations. For the Table 4, the models are trained on MS COCO trainval35k set.

Table 1. Network's performance on the ImageNet 1-crop accuracy rates.

Network	Params(M)	MACs(G)	Top-1 accuracy	Top-5 accuracy
AlexNet	61.1	**0.72**	56.55	79.09
VGG-16	138.36	15.5	71.59	90.38
GoogLeNet	6.62	1.52	69.78	89.53
ResNet-50	25.56	4.12	76.15	92.87
ResNet-101	44.55	7.85	77.37	93.56
Inception-ResNet-V2	55.84	13.22	**80.3**	**95.1**
DarkNet-19[a]	–	–	72.9	91.2

[a]https://pjreddie.com/darknet/imagenet/#darknet19

Table 2. Comparative results on Pascal VOC 2007 test set (%).

Detector	Backbone	Data	mAP
HyperNet [8]	AlexNet	07++12	65.9
PFPNet-R512 [32]	VGG-16	07++12	82.3
YOLOv1 [16]	GoogLeNet	07++12	63.4
BlitzNet512 [18]	ResNet-50	07++12+S	81.5
CoupleNet [21]	**ResNet-101**	07++12	**82.7**

---

[1] https://pytorch.org/.

Table 3. Comparative results on Pascal VOC 2012 test set (%).

Detector	Backbone	Data	mAP
PFPNet-R512 [32]	VGG-16	07++12	80.3
YOLOv1 [16]	GoogLeNet	07++12	57.9
CoupleNet [21]	**ResNet-101**	07++12	**80.4**

Table 4. MS COCO test-dev 2015 detection results (%).

Detector	Backbone	Data	mAP@.5	mAP@ [.5,.95]
PFPNet-R512 [32]	VGG-16	trainval35k	**57.6**	35.2
BlitzNet512 [18]	ResNet-50	trainval35k	50.9	32.5
CoupleNet [21]	**ResNet-101**	trainval35k	57.5	**36.4**
Faster R-CNN G-RMI [23]	Inception-ResNet-V2	trainval	55.5	34.7
YOLOv2 [27]	DarkNet-19	trainval35k	44.0	21.6

4   Discussion

In Table 1, we note that the Inception-ResNet-V2 network achieved a Top-1 Accuracy of 80.3% and 95.1% in the Top-5, higher than all other networks. Both ResNet-50 and ResNet-101 perform almost as well as Inception-ResNet-V2. Whereas AlexNet had 56.5% and 79.09% in Top-1 and Top-5 respectively, the remaining networks achieved nearly similar results. From Table 1, architectures based on the residual concept achieve better accuracy using a very reduced number of parameters compared to other architectures. For example, ResNet-50 has about 25 million parameters, ResNet-101 has around 44 million parameters and Inception-ResNet-v2 contains almost 55 million parameters, whereas VGG-16 has more than 138 million parameters. Although Alex-Net and Inception-Resnet-V2 have a very similar number of parameters, the accuracy and number of MACs are much lower in AlexNet compared to Inception-Resnet-V2. Table 2 clearly shows that in object detection the networks with the best performance are VGG and ResNets. ResNet-101 with CoupleNet, ResNet-50 with BlitzNet512 and PFPNet-R512 with VGG-16 performed an accuracy of 82.7%, 81.5%, and 82.3% respectively in the Pascal VOC 2007 test set. In Pascal VOC 2012, Table 3 indicates that PFPNet-R512 with VGG-16 and CoupleNet with ResNet-101 achieved an accuracy of 80.3% and 80.4% respectively, while YOLOv1 with GoogLeNet achieved only an accuracy of 57.9%. For MS COCO we notice that the models based on VGG-16, ResNet-101 and Inception-ResNet-V2 achieved interesting results which are 57.6%, 57.5% and 55.5% respectively for mAP@.5, and 35.2%, 36.4% and 34.7% for the mAP@ [.5,.95]. While YOLOv2 with DarkNet-19 produced a mAP@.5 of 44% and a mAP@ [.5,.95] of 21.6%. According to the results obtained, we could mention the networks contributing to the highest performance in object detection are VGG-16, the ResNets family and also Inception-ResNets-v2, which combines the Inception and ResNets networks. This explains the wide use of these architectures in different object-detector models.

The following Table 5 shows the main features added for each architecture to improve the performance.

Table 5. Main added features in the CNNs.

Network	What's Novel?
AlexNet	- Apply Rectified Linear Units (ReLU) to add non-linearity
VGG-16	- Deep network, approximately twice as deep as AlexNet
GoogLeNet	- Using dense modules as opposed to stacking convolutional layers
ResNets	- Using batch normalization and skip connections
Inception-ResNet-V2	- Using residual inception blocks instead of inception modules - Adding the inception module (Inception-A) after the stem module - Using more inception modules
DarkNet-19	- Combine Darknet extraction, Network In Network, Inception and Batch Normalization in a single model

5 Conclusion

In this paper, we studied the state-of-the-art CNNs for object detection. We devoted the study to networks that have achieved remarkable performance. We outlined the datasets used for testing the CNNs as well as the object detection models. We compared those networks and models on multiple benchmarks and datasets. We report that the application of convolutional neural networks in object detection has given impressive state-of-the-art results.

Acknowledgements. This work is carried out with the support of the 'Centre National pour la Recherche Scientifique et Technique (CNRST)' of Morocco, through the Research Excellence Awards Program 2019.

References

1. Lienhart, R., Maydt, J.: An extended set of Haar-like features for rapid object detection. In: Proceedings of the International Conference on Image Processing, pp. I-900–I-903. IEEE, Rochester (2002). https://doi.org/10.1109/ICIP.2002.1038171
2. Dalal, N., Triggs, B.: Histograms of oriented gradients for human detection. In: 2005 IEEE Computer Society Conference on Computer Vision and Pattern Recognition (CVPR 2005), pp. 886–893. IEEE, San Diego (2005). https://doi.org/10.1109/CVPR.2005.177
3. Lowe, D.G.: Distinctive image features from scale-invariant keypoints. Int. J. Comput. Vis. **60**, 91–110 (2004). https://doi.org/10.1023/B:VISI.0000029664.99615.94
4. Krizhevsky, A., Sutskever, I., Hinton, G.E.: ImageNet classification with deep convolutional neural networks. In: Pereira, F., Burges, C.J.C., Bottou, L., Weinberger, K.Q. (eds.) Advances in Neural Information Processing Systems 25, pp. 1097–1105. Curran Associates, Inc. (2012)

5. He, K., Zhang, X., Ren, S., Sun, J.: Deep residual learning for image recognition. In: 2016 IEEE Conference on Computer Vision and Pattern Recognition (CVPR), pp. 770–778. IEEE, Las Vegas (2016). https://doi.org/10.1109/CVPR.2016.90

6. Krizhevsky, A., Sutskever, I., Hinton, G.E.: ImageNet classification with deep convolutional neural networks. Commun. ACM **60**, 84–90 (2017). https://doi.org/10.1145/3065386

7. Girshick, R., Donahue, J., Darrell, T., Malik, J.: Rich feature hierarchies for accurate object detection and semantic segmentation. arXiv:1311.2524 [cs]. (2013)

8. Kong, T., Yao, A., Chen, Y., Sun, F.: HyperNet: towards accurate region proposal generation and joint object detection. In: 2016 IEEE Conference on Computer Vision and Pattern Recognition (CVPR), pp. 845–853. IEEE, Las Vegas (2016). https://doi.org/10.1109/CVPR.2016.98

9. Simonyan, K., Zisserman, A.: Very Deep Convolutional Networks for Large-Scale Image Recognition. arXiv:1409.1556 [cs] (2014)

10. Girshick, R.: Fast R-CNN. arXiv:1504.08083 [cs] (2015)

11. Ren, S., He, K., Girshick, R., Sun, J.: Faster R-CNN: towards real-time object detection with region proposal networks. IEEE Trans. Pattern Anal. Mach. Intell. **39**, 1137–1149 (2017). https://doi.org/10.1109/TPAMI.2016.2577031

12. Kong, T., Sun, F., Yao, A., Liu, H., Lu, M., Chen, Y.: RON: reverse connection with objectness prior networks for object detection. In: 2017 IEEE Conference on Computer Vision and Pattern Recognition (CVPR), pp. 5244–5252. IEEE, Honolulu (2017). https://doi.org/10.1109/CVPR.2017.557

13. Liu, W., Anguelov, D., Erhan, D., Szegedy, C., Reed, S., Fu, C.-Y., Berg, A.C.: SSD: Single Shot MultiBox Detector. arXiv:1512.02325 [cs]. 9905, 21–37 (2016). https://doi.org/10.1007/978-3-319-46448-0_2

14. Zhang, S., Wen, L., Bian, X., Lei, Z., Li, S.Z.: Single-shot refinement neural network for object detection. In: 2018 IEEE/CVF Conference on Computer Vision and Pattern Recognition, pp. 4203–4212. IEEE, Salt Lake City (2018). https://doi.org/10.1109/CVPR.2018.00442

15. Szegedy, C., Liu, W., Jia, Y., Sermanet, P., Reed, S., Anguelov, D., Erhan, D., Vanhoucke, V., Rabinovich, A.: Going deeper with convolutions. In: 2015 IEEE Conference on Computer Vision and Pattern Recognition (CVPR), pp. 1–9. IEEE, Boston (2015). https://doi.org/10.1109/CVPR.2015.7298594

16. Redmon, J., Divvala, S., Girshick, R., Farhadi, A.: You only look once: unified, real-time object detection. In: 2016 IEEE Conference on Computer Vision and Pattern Recognition (CVPR), pp. 779–788. IEEE, Las Vegas (2016). https://doi.org/10.1109/CVPR.2016.91

17. Ioffe, S., Szegedy, C.: Batch Normalization: Accelerating Deep Network Training by Reducing Internal Covariate Shift. arXiv:1502.03167 [cs] (2015)

18. Dvornik, N., Shmelkov, K., Mairal, J., Schmid, C.: BlitzNet: A Real-Time Deep Network for Scene Understanding. arXiv:1708.02813 [cs] (2017)

19. Lin, T.-Y., Goyal, P., Girshick, R., He, K., Dollár, P.: Focal Loss for Dense Object Detection. arXiv:1708.02002 [cs] (2018)

20. Dai, J., Li, Y., He, K., Sun, J.: R-FCN: object detection via region-based fully convolutional networks. In: Lee, D.D., Sugiyama, M., Luxburg, U.V., Guyon, I., Garnett, R. (eds.) Advances in Neural Information Processing Systems 29, pp. 379–387. Curran Associates, Inc. (2016)

21. Zhu, Y., Zhao, C., Wang, J., Zhao, X., Wu, Y., Lu, H.: CoupleNet: coupling global structure with local parts for object detection. In: 2017 IEEE International Conference on Computer Vision (ICCV), pp. 4146–4154. IEEE, Venice (2017). https://doi.org/10.1109/ICCV.2017.444

22. Szegedy, C., Ioffe, S., Vanhoucke, V., Alemi, A.: Inception-v4, Inception-ResNet and the Impact of Residual Connections on Learning. arXiv:1602.07261 [cs] (2016)
23. Huang, J., Rathod, V., Sun, C., Zhu, M., Korattikara, A., Fathi, A., Fischer, I., Wojna, Z., Song, Y., Guadarrama, S., Murphy, K.: Speed/accuracy trade-offs for modern convolutional object detectors. In: 2017 IEEE Conference on Computer Vision and Pattern Recognition (CVPR), pp. 3296–3297. IEEE, Honolulu (2017). https://doi.org/10.1109/CVPR.2017.351
24. Shrivastava, A., Sukthankar, R., Malik, J., Gupta, A.: Beyond Skip Connections: Top-Down Modulation for Object Detection. arXiv:1612.06851 [cs] (2016)
25. Lin, M., Chen, Q., Yan, S.: Network In Network. arXiv:1312.4400 [cs] (2014)
26. Szegedy, C., Vanhoucke, V., Ioffe, S., Shlens, J., Wojna, Z.: Rethinking the Inception Architecture for Computer Vision. arXiv:1512.00567 [cs] (2015)
27. Redmon, J., Farhadi, A.: YOLO9000: Better, Faster, Stronger. arXiv:1612.08242 [cs] (2016)
28. Russakovsky, O., Deng, J., Su, H., Krause, J., Satheesh, S., Ma, S., Huang, Z., Karpathy, A., Khosla, A., Bernstein, M., Berg, A.C., Fei-Fei, L.: ImageNet Large Scale Visual Recognition Challenge. arXiv:1409.0575 [cs] (2015)
29. Everingham, M., Eslami, S.M.A., Van Gool, L., Williams, C.K.I., Winn, J., Zisserman, A.: The Pascal visual object classes challenge: a retrospective. Int J Comput Vis. 111, 98–136 (2015). https://doi.org/10.1007/s11263-014-0733-5
30. Lin, T.-Y., Maire, M., Belongie, S., Bourdev, L., Girshick, R., Hays, J., Perona, P., Ramanan, D., Zitnick, C.L., Dollár, P.: Microsoft COCO: Common Objects in Context. arXiv:1405.0312 [cs] (2015)
31. Xie, S., Girshick, R., Dollár, P., Tu, Z., He, K.: Aggregated Residual Transformations for Deep Neural Networks. arXiv:1611.05431 [cs] (2017)
32. Kim, S.-W., Kook, H.-K., Sun, J.-Y., Kang, M.-C., Ko, S.-J.: Parallel feature pyramid network for object detection. In: Ferrari, V., Hebert, M., Sminchisescu, C., Weiss, Y. (eds.) Computer Vision – ECCV 2018, pp. 239–256. Springer, Cham (2018). https://doi.org/10.1007/978-3-030-01228-1_15

Object Detector Combination
for Increasing Accuracy and Detecting
More Overlapping Objects

Khaoula Drid[1]([✉])[iD], Mebarka Allaoui[2][iD], and Mohammed Lamine Kherfi[1,3][iD]

[1] Kasdi Merbah University Ouargla, Ouargla, Algeria
khaouladrid22@gmail.com
[2] LAGE Laboratory, Kasdi Merbah University Ouargla, Ouargla, Algeria
[3] LAMIA Laboratory, Université du Québec à Trois-Rivières, Trois-Rivières, Canada

Abstract. Object detection is considered as the cornerstone of many modern applications such as Drone vision and Self-driven cars. Object detectors, mainly those which are based on Convolutional Neural Networks (CNNs) have received great attention from many researchers because they were able to yield remarkable results. However, most of them fail when it comes to detecting overlapping and small objects in images. There are two families of detectors: the first family detects more objects but with imprecise bounding boxes, while those of the second family do the opposite. In this paper, we propose a solution to this problem by combining the two families, in a way similar to classifier combination. Our solution has been validated through the combination of two famous detectors, Faster R-CNN which detects more objects and YOLO which produces accurate bounding boxes. However, it is more general and it can be applied to other detectors. The evaluation of our method has been applied to the PASCAL VOC dataset and it gave promising results.

Keywords: Object detector combination · Detecting overlapping objects · Convolutional Neural Networks (CNNs) · YOLO · Faster R-CNN

1 Introduction

Detecting and identifying the different objects in an image is very useful. Object detection is an active research field in computer vision. It deals with two main issues: object classification and object localization. Humans and many other animals use a process called visual perception to quickly decide which locations of an image should be processed in detail and which can be ignored. This allows us to deal with the huge amount of visual information and to employ the capacities of our visual system efficiently. Concerning computer vision, researchers have to deal with the same problems. Therefore, learning from human's behavior provides a promising way to improve existing algorithms. Object detection has

© Springer Nature Switzerland AG 2020
A. El Moataz et al. (Eds.): ICISP 2020, LNCS 12119, pp. 290–296, 2020.
https://doi.org/10.1007/978-3-030-51935-3_31

received great attention from researchers in several areas and many interesting applications such as Drone vision systems, Self-driven cars, Video surveillance, robot navigation, and many other applications which require object and scene recognition.

Recently, CNN-based detectors achieved good results in object detection. They can broadly be categorized into two categories: two-stage family such as Faster R-CNN [16], and one-stage family such as YOLO [14]. However, those detectors suffer from several limitations. The two-stage family has two main problems: it has an expensive computational cost, and yields bounding boxes that are not precise and which may contain more than one object. This latter problem occurs especially in images with overlapping and small objects. As for the second family, it usually detects fewer objects than those of the first family. In this work, we have developed an algorithmic solution for detecting overlapping and small objects by combining the advantages of the two families. Our solution combines Faster R-CNN detector which belongs to the first family and YOLO which belongs to the second, but can be generalized to other detectors. The paper is organized as follows: In Sect. 2, we present the related work briefly. After that, we describe our algorithm in Sect. 3. In Sect. 4, we give details about the experiments we conducted to validate the proposed solution and compare it against YOLO and Faster R-CNN. We finish the paper with a short conclusion.

2   Related Work

Object detection aims at locating and classifying existing objects in any given image and surrounds them with bounding boxes. Object detection methods can roughly be subdivided into two categories: (i) Two-stage detectors: In the first step, a Region Proposal Network is used to generate regions of interest that have high probability of being an object. In the following step, they perform the final classification and bounding-box regression of objects by taking these regions as input. Examples of this family include R-CNN [7], SPP-net [9], Fast R-CNN [6], Faster R-CNN [16], R-FCN [1], FPN [11] and Mask R-CNN [8]. (ii) One-stage detectors: they process object detection as a simple regression problem by taking an input image and learning the class probabilities and bounding box coordinates simultaneously. This family includes detectors like MultiBox [2], AttentionNet [18], G-CNN [13], YOLO [14], SSD [12], YOLOv2 [15], DSSD [5] and DSOD [17]. Both families have drawbacks. The two-stage detectors are composed of several correlated stages, and therefore they need to obtain shared convolution parameters between RPN and detection network. As a result, the time spent in handling different components becomes the bottleneck in real-time application. As for one-stage detectors, they have difficulty in dealing with small objects. To address these limitations, we focus on two relatively successful solutions, namely, (YOLO) [14] and Faster region-based convolutional network (Faster R-CNN) [16]. We propose an algorithm for combining them, which aims to increase both the accuracy and the number of the detected objects.

3   Methodology

Images with overlapping and small objects are a real challenge for most object detectors. As we mentioned before, object detectors belong to two families: one-stage and two-stage detectors. Each family deals with overlapping and small objects differently. The one-stage family predicts less bounding boxes compared to the two-stage family. However, the one-stage family's bounding boxes are more accurate than the two-stage family ones. To benefit from the advantages of both families and at the same time limit their shortcoming, we propose to combine them. Our combination guaranteed at least to preserve the accuracy of the bounding boxes and augments the number of predictions. In this work, we focus on two of the most efficient models: YOLO and Faster R-CNN. YOLO belongs to one-stage family detectors whereas Faster R-CNN is a two-stage detector. In Fig. 1, we give an example of an image with overlapping objects, and the objects predicted by YOLO and Faster R-CNN.

Our Algorithm. As mentioned above that YOLO's bounding boxes are more accurate and tight. Moreover, its false prediction is less than the false prediction of Faster R-CNN. The main idea of our algorithm is as follows: for every bounding box predicted by Faster R-CNN, we check to see if YOLO predicts a similar box. If it does, we retain that prediction based on the probability predicted by YOLO. In the opposite case, the objects predicted by Faster R-CNN are retained. Our idea is illustrated in Fig. 2.

In more technical details:

1. We preserve the entire YOLO's bounding boxes,
2. For every Faster R-CNN's bounding box, we compute the intersection over union between it and YOLO's bounding boxes.
   (a) If the result is larger than a certain threshold so that is mean YOLO detects this object and we ignore the Faster R-CNN's bounding box.
   (b) Otherwise, we accept it.

Our solution is summarized in Algorithm 1.

4   Experimental Evaluation

To validate our idea, we compare our combined model with YOLO and Faster R-CNN in terms of accuracy on a benchmark dataset. In this section, we will present the used materials and metrics to get the desired results.

4.1   The Used Dataset

PASCAL VOC is a collection of datasets with 20 classes for object detection. The most common combination for benchmarking is using 16551 samples from both PASCAL VOC 2007 [4] and 2012 [3] for training, and 4952 sample from PASCAL VOC 2007 for the testing.

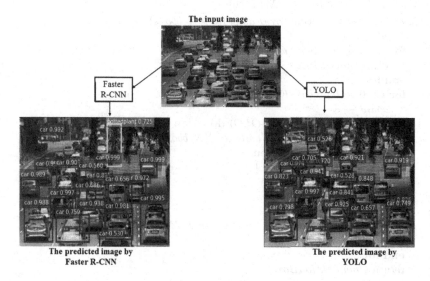

Fig. 1. YOLO and Faster R-CNN predictions.

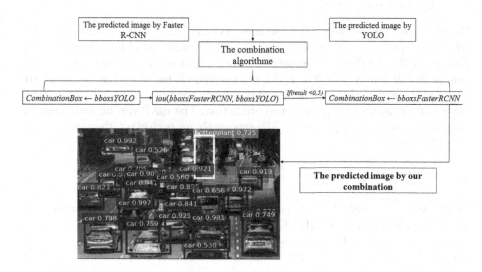

Fig. 2. Our combination method [10], the figure illustrates how our method can combine the predicted bounding boxes from both considered object detectors. In our combination, we preserve YOLO's bounding boxes as they are because they are more accurate than those of Faster R-CNN's. Then we compute the intersection over union (IOU) between the YOLO's boxes and Faster R-CNN's boxes. If the IOU result greater than a certain threshold, we ignore this box, elsewhere we accept it, because this means YOLO does not predict a bounding box similar to this one.

Algorithm 1. Combination Algorithm

1: **function** COMBINATION(*bboxsFasterRCNN, bboxsYOLO*)
2:     *CombinationBox* ← []
3:     **for** $i = 0$ **to** *len(bboxsYOLO)* **do**
4:         *CombinationBox* ← *bboxsYOLO[i]*
5:     **end for**
6:     **for** $i = 0$ **to** *len(bboxsFasterRCNN)* **do**
7:         *count* ← 0
8:         **for** $j = 0$ **to** *len(bboxsYOLO)* **do**
9:             *result* ← *iou(bboxsFasterRCNN, bboxsYOLO)*
10:            **if** *result* >= 0.5 **then**
11:                *count* ← *count* + 1
12:            **end if**
13:        **end for**
14:        **if** *count* < 1 **then**
15:            *CombinationBox* ← *bboxsFasterRCNN[i]*
16:        **end if**
17:    **end for**
18:    **return** *CombinationBox*
19: **end function**

4.2   Evaluation Metric

We evaluated our methods using average precision (AP) [1] and its mean (mAP), which is a popular metric in measuring the accuracy of object detectors like Faster R-CNN, SSD, etc. Average precision computes the average Precision values for Recall values from 0 to 1. For the precision it measures how accurate are the predictions, i.e. the percentages of how much the predictions are correct, and the recall measures how good is the prediction. The mAP can be computed by calculating the average precision (AP) separately for each class, then the average over the classes.

4.3   Experiments Results

In this work, the used YOLO and Faster R-CNN detectors are pre-trained on 16551 samples from PASCAL VOC 2007/2012. And the experiment conducted between our combination method against YOLO and Faster R-CNN on test samples from PASCAL VOC 2007. The obtained results are presented in Table 1 and Table 2.

4.4   Discussion

The proposed combination focuses on images with high overlapping and small objects. As we see in Table 1, our method has the highest precision for most classes of the dataset except for motorbike and bird. Indeed YOLO, For the motorbike class it has been able to give 86% and for the bird class, it gave

Table 1. Comparison between Faster R-CNN, YOLO and our combination method on PASCAL VOC2007/2012 dataset according to mAP metric. The table represents the achieved score for the three models in the 11 classes from the dataset in term of average precision.

	Areo	bike	Bird	Boat	bottle	Bus	Car	Cat	Chair	Cow	table
Faster R-CNN	0.78	0.75	0.47	0.53	0.40	0.90	0.77	0.88	0.34	0.78	0.58
YOLO	**0.90**	**0.81**	**0.70**	0.63	0.42	**0.95**	**0.87**	0.81	0.53	0.79	0.60
Combination	**0.90**	**0.81**	0.69	**0.65**	**0.54**	**0.95**	**0.87**	**0.90**	**0.59**	**0.88**	**0.65**

Table 2. Comparison between Faster R-CNN, YOLO and our combination method on PASCAL VOC2007/2012 dataset according to mAP metric. The table represents the achieved score for the three models in the 9 rest classes from the dataset in term of average precision, and the mAP column represents the achieved score for each model.

	dog	Horse	mbike	person	Plant	sheep	Sofa	Train	TV	mAP
Faster R-CNN	0.75	0.77	0.74	0.75	0.42	0.79	0.61	0.78	0.79	0.68
YOLO	0.79	**0.81**	**0.86**	0.78	0.57	**0.88**	0.65	**0.80**	**0.84**	0.75
Combination	**0.87**	**0.81**	0.85	**0.83**	**0.64**	**0.88**	**0.67**	**0.80**	**0.84**	**0.78**

70%. Our method gave the best results because it combines the advantages of the two models with some improvements since we filter the bounding boxes predicted by both YOLO and Faster R-CNN and only keep bounding boxes with a high probability of existence. As a consequence, the combination accuracy obtained is 78% mAP. As expected, our proposed method succeeds in detecting the maximum instances of the existing objects, and that is a good achievement. However, we should notice that in term of speed, our method speed is less than YOLO. Since we run each model separately and then combine the results. Since YOLO is so fast it does not add any significant computational time compared to Faster R-CNN.

5   Conclusion

In this paper, we proposed a combination framework to deal with the shortcoming of object detectors when we detect overlapping and small objects in an image. We tested our model by combining two of the most famous detectors which are Faster R-CNN and YOLO. Our method succeeded to increase the number of detected objects with accurate bounding boxes. In general, the obtained results were good and promising as seen in the experimental section. Our method was able to improve the performance of existing detectors, which makes it useful for several applications where overlapping objects are omnipresent like in self-driven cars. However, our method still needs to be improved in terms of computation time. In the future, this method could be extended to video object detection.

References

1. Dai, J., Li, Y., He, K., Sun, J.: R-FCN: object detection via region-based fully convolutional networks. In: Advances in Neural Information Processing Systems, pp. 379–387 (2016)
2. Erhan, D., Szegedy, C., Toshev, A., Anguelov, D.: Scalable object detection using deep neural networks. In: Proceedings of the IEEE Conference on Computer Vision and Pattern Recognition, pp. 2147–2154 (2014)
3. Everingham, M., Van Gool, L., Williams, C.K.I., Winn, J., Zisserman, A.: The Pascal visual object classes challenge 2012 (VOC2012) results (2012). http://www.pascal-network.org/challenges/VOC/voc2011/workshop/index.html
4. Everingham, M., Van Gool, L., Williams, C.K.I., Winn, J., Zisserman, A.: The Pascal visual object classes challenge 2007 (VOC2007) results (2007)
5. Fu, C.-Y., Liu, W., Ranga, A., Tyagi, A., Berg, A.C.: DSSD: Deconvolutional single shot detector. arXiv preprint arXiv:1701.06659 (2017)
6. Girshick, R.: Fast R-CNN. In: The IEEE International Conference on Computer Vision (ICCV) (2015)
7. Girshick, R., Donahue, J., Darrell, T., Malik, J.: Rich feature hierarchies for accurate object detection and semantic segmentation. In: Proceedings of the IEEE Conference on Computer Vision and Pattern Recognition, pp. 580–587 (2014)
8. He, K., Gkioxari, G., Dollar, P., Girshick, R.: Mask R-CNN. In: The IEEE International Conference on Computer Vision (ICCV) (2017)
9. He, K., Zhang, X., Ren, S., Sun, J.: Spatial pyramid pooling in deep convolutional networks for visual recognition. IEEE Trans. Pattern Anal. Mach. Intell. **37**(9), 1904–1916 (2015)
10. Kherfi, M.L., Drid, K., Zouaoui, O.K.: Object detection and recognition fom images. Master thesis, University Kasdi Merbah Ouargla, Algeria (2019)
11. Lin, T.-Y., Dollár, P., Girshick, R., He, K., Hariharan, B., Belongie, S.: Feature pyramid networks for object detection. In: Proceedings of the IEEE Conference on Computer Vision and Pattern Recognition, pp. 2117–2125 (2017)
12. Liu, W., et al.: SSD: single shot multibox detector. In: Leibe, B., Matas, J., Sebe, N., Welling, M. (eds.) ECCV 2016. LNCS, vol. 9905, pp. 21–37. Springer, Cham (2016). https://doi.org/10.1007/978-3-319-46448-0_2
13. Najibi, M., Rastegari, M., Davis, L.S.: G-CNN: an iterative grid based object detector. In: Proceedings of the IEEE Conference on Computer Vision and Pattern Recognition, pp. 2369–2377 (2016)
14. Redmon, J., Divvala, S., Girshick, R., Farhadi, A.: You only look once: unified, real-time object detection. In: Proceedings of the IEEE Conference on Computer Vision and Pattern Recognition, pp. 779–788 (2016)
15. Redmon, J., Farhadi, A.: Yolo9000: better, faster, stronger. In: Proceedings of the IEEE Conference on Computer Vision and Pattern Recognition, pp. 7263–7271 (2017)
16. Ren, S., He, K., Girshick, R., Sun, J.: Faster R-CNN: towards real-time object detection with region proposal networks. In: Advances in Neural Information Processing Systems, pp. 91–99 (2015)
17. Shen, Z., Liu, Z., Li, J., Jiang, Y.-G., Chen, Y., Xue, X.: DSOD: learning deeply supervised object detectors from scratch. In: Proceedings of the IEEE International Conference on Computer Vision, pp. 1919–1927 (2017)
18. Yoo, D., Park, S., Lee, J.-Y., Paek, A.S., Kweon, I.S.: Attentionnet: aggregating weak directions for accurate object detection. In: Proceedings of the IEEE International Conference on Computer Vision, pp. 2659–2667 (2015)

Segmentation and Retrieval

Graph-Based Image Retrieval: State of the Art

Imane Belahyane$^{(\boxtimes)}$ (iD), Mouad Mammass$^{(\boxtimes)}$ (iD), Hasna Abioui$^{(\boxtimes)}$,
and Ali Idarrou$^{(\boxtimes)}$ (iD)

IRF-SIC Laboratory, Ibn Zohr University, Agadir, Morocco
imane.belahyane@gmail.com, {m.mammass,h.abioui,ali.idarrou}@uiz.ac.ma

Abstract. The paper deals with the problem of semantic Image Retrieval. Indeed, the image has recently gained popularity in several domains such as medical domain, marketing, etc. Image plays a very vital role in documentation. However, finding visual and relevant information in an image is a huge task for Image Retrieval community and a very discussed issue in digital image processing. In fact, image can be extracted from a big collection of images, in the purpose of responding to user's need. Image Retrieval processes based on classical techniques may not be sufficient to user. For several years, great efforts have been devoted to integrate semantic aspect, in order to enhance relevance of the result and ensure high-level content consideration in image. This paper presents a state of the art of Image Retrieval approaches using graph theory due to the growing interest given to graphs in terms of performance, representation and its ability to ingrate semantic aspect. We review a number of recently available graph-based approaches in Image Retrieval aiming to determine factors adding semantic aspect in Image Retrieval system.

Keywords: Image retrieval · Semantic aspect · Graph theory

1 Introduction

Since the advent of information technology, the number of multimedia documents has grown continuously. In fact, more than 80% of company and organization data is in the form of documents. Multimedia documents are characterized by a rich content and complex structures, which complicates access to specific granules in such documents and therefore makes the document retrieval a tedious task. Graph is the most effective representation model, which allow the representation of complex and connected data, such as multimedia document. Comparing two documents structurally, means comparing graphs that represent them. The graph theory could be of great interest in the evaluation of the structural similarity.

In general, comparing two graphs leads to find the better matching between them. The general approaches proposed by graph theory concerning matching

© Springer Nature Switzerland AG 2020
A. El Moataz et al. (Eds.): ICISP 2020, LNCS 12119, pp. 299–307, 2020.
https://doi.org/10.1007/978-3-030-51935-3_32

are: exact matching and approximate matching. In many fields of application, the goal is not to show that two graphs are structurally identical, but it is more interesting to know how similar these graphs are. In such applications, graph similarity based on exact matching is not appropriate. For this purpose, approximate error-tolerant graph matching based on finding the maximum common subgraph or on the calculation of the graph editing distance has been proposed in [8,9,17]. In the context of image, finding visual information granularity in an image, requires the use of special techniques in order to respond to user's need. Indeed, Image Retrieval can be based either on the text or on the visual content. However, a key limitation of traditional image retrieval system is the ignorance of semantic aspect. Several works has used graphs to represent images [2,6,7,32]. In [1], graphs are rich data structures with the ability to represent complex and structured objects.

In this paper, we present the basic concepts of Image Retrieval and image retrieval techniques. Also, a comparative study of image retrieval techniques has been carried out to present their advantages, and drawbacks. The problematic of this work revolves around: the integration of semantic aspect in graph-based Images Retrieval approaches.

The remainder of this paper is organized as follows. Section 2, describes the concept of Image Retrieval and the techniques of Image Retrieval, and presents advantages and limits of using each type of techniques. Section 3 presents a comparative study on graph-based Image Retrieval and outlines the integration of semantic aspect in graph-based approaches. Finally, Sect. 4 discusses our issue focused on graph-based semantic Image Retrieval.

2   Image Retrieval

Information Retrieval (IR) is concerned with the acquisition, structuring, storage, retrieval, and ranking of information [13]. It based on the information needs of user. This task can be applied for all types of data: text, image, video, music track. An Image is the visual representation of an object by different mediums or support.

2.1   Image Retrieval Techniques

Image Retrieval (ImR) is the area studying the way to find images in image collection. Moreover, the issue is to rank the similar images to the user's request. ImR attracts the attention of many researchers in the field of: digital libraries, remote sensing, astronomy, etc. It has been a very active field since the 1970s. ImR systems can be classified into three techniques [20]:

- Text-based Image Retrieval (TBIR), is the system for retrieving images by text queries. The extraction of an image similar to a user query is based on indexation process, which proposes to attach to an image, a set of descriptors. This technique use the textual indexation of an image, its metadata or the textual elements attached to the image. A lot of research has been done on TBIR, but are very ancient due to the great importance given to the other types of ImR. TBIR can be based on annotation [15]. The major drawback of this approach is that for a descriptive annotation, it must be manual, hence, the complexity of the task. One of the first examples of TBIR is [36], it presents a framework that performs annotations to images using text and then uses text-based databases.

- Content-based Retrieval (CBIR): over the last decades, a big interest in images collection has grown with the development of image acquisition devices, storage capacities and the availability of high-quality digitization techniques. Indeed, CBIR is a system based on colors, textures, shapes, and other characteristics (depending on the user's needs). In other words, it consists of extracting visual descriptors and retrieving by visual similarity. This technique responds to many needs in the field of ImR and overcomes the limitations of TBIR. An image can be described by a weighting function that reflects the importance of the features and varies widely according to the system and the objectives. Work [34] presents an effective content-based visual ImR system, by extracting color histogram and spatial information. Work [14] presents CBIR approach using a computational visual attention model, based on saliency regions and energy features of the gray-level co-occurrence and saliency structure histogram. Work [16] presents an approach using local visual attention feature, based on fast and performant salient point detector, and the salient point expansion.

- Semantic-based Image Retrieval (SBIR), is the technique that defines image using semantic terms to determine the significance conveyed in the image. SBIR can be obtained by extracting visual descriptors from the image in order to identify significant and interesting regions of the image, followed by a process for extracting knowledge in order to obtain a semantic description of the image. This technique is performed by several factors. The following section focus on that. Table 1 summarizes the advantages and limits of each techniques.

According to Table 1, we conclude that each technique has its advantages and limitations, therefore the use of TBIR, CBIR and SBIR depends on the objective of the task carried out by the user and the context studied. Several works aim to increase the relevance of the result, to that end, they combined more than one category. Work [25] presents a decisive content based ImR approach for feature fusion in visual and textual images. In [30] a system is proposed to combine textual and visual statistics in a single index vector for content-based retrieval. Work [29] presents a system based on content-based and develops its own ontology module, it contributes to significantly increase the relevance of retrieval results, by enhancing the ranking of images.

Table 1. Advantages and Limits of ImR techniques

Technique	Advantages	Limits
TBIR	• More natural to general user • Superior in terms of quality: well targeted, relevant and precise	• Time consuming, expensive, and subjective • Impractical in large-scale image databases
CBIR	• Easier to integrate into mathematical formulations • Overcome difficulties of TBIR • Less time-consuming	• Not able to capture semantic phenomenon • Unusable for general users
SBIR	• Relevant result • Accurate image interpretation	• Time-consuming • Use of external resources

2.2   Image Representation Models

For several years great effort has been devoted to the study of image representation models, most commonly used are: vectors, strings, trees, and graphs. Vectors are often used in ImR, works [16, 29, 34] model the image as a vector.

String is an ordered set of elements, used in ImR when it becomes important to order elements. The distance between two string are often defined by the editing distance of Levenshtein's chain [33], as in [23]. However, studies on vector-based and string-based approaches are still lacking due to modeling poverty and may not be conventional in all situations to model complex objects. Likewise, tree-based model allows the representation of hierarchical relationships, and not practicable to model complex relations. To solve this issue, many researchers have proposed graph theory in ImR.

Graph-based model permits to represent all possible relations between components, the semantics associated with an arc is not limited to a typing or membership relation. Graphs offer a very rich modeling of the document and their structures. Graph-based Image is represented as a set of components and a set of binary relations between these components. They are widely used in many applications due to their very high expressiveness in terms of structure and semantics. Note that strings and trees are particular graphs.

The application of graph theory to IR is studied in several works due to its advantages in terms of improving the efficiency of the IR engine. Indeed, graph-based measures provide the use of the graph as a semantic representation model for queries and documents and also its exploitation in a semantic document search model [10, 24, 28].

3   Graph-Based Semantic Image Retrieval

3.1   Graph-Based Image Retrieval

In graph data, nodes represent entities and edges represent relationships between nodes. Graph structure has the ability of representing meaning of entities and relationships between entities. This excellent ability makes graph more and more popular in the field of computer. In general, the mathematical theory of graphs could be of great interest in measuring the similarity of objects. In [12], subgraph isomorphism can be used to show the inclusion or the equivalence of two graphs. Below, we give the mathematical definition of a graph:

Definition 1. *A graph G can be defined by a pair (V, E), where V is the set of nodes of G and $E \in V \times V$ represents the set of edges of G (relations between nodes)*

In the following a number of works regarding graph-based ImR.

3.2   Factors Adding Semantic Aspect

Taking into account semantic aspect in ImR, means to retrieve the relevant result with considering the overall signification of the image. SBIR aims to give the adequate interpretation of the image. The purpose of the following table is to determine factors adding semantic aspect in graph-based ImR.

Table 2. Approaches using graphs

Approach	Work context	Graph model	Factor adding semantic aspect
[3]	Out-of-sample image	Anchor graph	– Not ignoring the underlying structure information – One anchor of graph shares one concept
[21]	Generic context	Scene graph	Graph specificity
[11]	Plant identification	Conceptual graph	Graph specificity
[19]	Generic context	Simple graph model	Relevance feedback
[22]	Image search in Wikipedia	Semantic graph	Graph specificity
[26]	Image and all other media	Simple graph model	Context based Image Retrieval using structure of the image
[18]	Generic context	Simple graph model	Query annotation

Table 2 shows image-based approaches, classified according to the graph model and the factors reflecting semantic aspect in graph-based ImR approaches. Those factors contributed to enhance semantic aspect in the ImR system.

In the following, a set of factors is grouped together to clearly identify the factors that implicitly and explicitly influence the semantics of ImR approaches.

To evaluate the retrieval performance, work [5] designs an automatic scheme to simulate the **relevance feedback**. The simulation system automatically classifies a database image as relevant if the image belongs to the same semantic category with the initial query. The work affirms that experimental results show that the relevance feedback technique improves retrieval performance for semantic categories with clear region correspondence.

In [35], relevance feedback was defined as a powerful interactive technique used to improve the performance of ImR systems. With user provided relevant/irrelevant information on the retrieved images, the system can capture the semantic concept of the query more correctly and gradually improve the retrieval precision.

Work [18] presents an intelligent annotation-based ImR system, that introduces concepts and instances, where annotations are stored as RDF triples and can be queried to find images. Annotations at concept level, are enable to create semantic links between concepts and then addresses many challenges.

Relevance feedback and query annotation are techniques that allow the expansion of the query to enhance the query expressing of the user's need. Query annotation is a technique that influences the graph before the retrieval process and the relevance feedback allows the extension of the graph to optimize the retrieval engine.

According to **the underlying structure**, most traditional methods focus on the data features, but, they ignore the underlying structure information, which plays a major role for semantic discovery, especially when the label information is unknown. Many databases have underlying cluster or manifold structure [3].

The context of a node in graph model has an influence on the semantics, but in a lower degree, it implies to take into account the ascendant and descendant nodes, in order to make a general interpretation of the image, such is in [26].

Specific graphs such as: the *semantic graph* and *conceptual graph*, used for the representation of knowledge and reasoning. Work [22] presents CKSGIS, it retrieves automatically an interactive semantic graph of convigned terms that allow users to easily find related images, not limited to a specific search term. In [4,11,27], *conceptual graphs* have been used in semantic representations for ImR. They are very used in graph-based ImR. As well as *scene graph*, it logically structures the spatial representation of a scene graph, such as in [21,31].

4   Discussion and Coclusion

In the last few years, several privileges have been acquired while working with graph-based, due to the tree architecture of graph, its ability to model complex objects (in our case: images), and complex relationships between these objects (e.g.: representation of multiple relationships between the same nodes).

In ImR domain, there are several factors adding semantic aspect, to increase the performance of the retrieval system and to improve the relevance of the result. Based on the approaches presented in Table 2, the main problematic of this work is to know in what extent, graphs integrate semantic aspect in ImR, using not only approaches dealing with semantics in an explicitly way, but also those expressing semantics in an implicitly way. In this work, we were interested in the semantic aspect of graph-based ImR approaches, depending on the context of the study and the objectives pursued.

In this paper, we presented a state of the art of works related to graph-based ImR. In general, the paper reviews the ImR technique (TBIR, CBIR, and SBIR) and image representation models (strings, lists, trees, and graphs). From this we deduce that TBIR and CBIR techniques are not enough to deal with relevant and effective ImR and according to image representation models literature, we deduce that, graph model has a great importance in ImR.

Graph-based ImR permits to represent all the possible relations between the components, the semantics associated with an arc is not limited to a typing or membership relation. An image is represented as a set of components and a set of binary relations between these components.

The main purpose of the paper is to draw attention to graph-based approaches, and its vital role to increase significance of the image and optimally serve the user's interest. We have made an overview of existing approaches on ImR. We have concluded that semantic aspect can be carried from several angles, in accordance with study's context and the objectives pursued. Based on this state of the art, we can conclude that graph-based approaches to ImR can open other leads to improve image-related information retrieval systems.

ImR involves into a promising field, but existing methods of semantic ImR must be adapted. There remain many challenges to overcome in this domain. Our future work will focus on approaches using semantic in IR.

References

1. Idarrou, A., Mammass, D.: Structural clustering multimedia documents: an approach based on semantic sub-graph isomorphism. Int. J. Comput. Appl. **51**(1), 14–21 (2012)
2. Kumar, A., Kim, J., Wen, L., Fulham, M., Feng, D.: A graph-based approach for the retrieval of multi-modality medical images. Med. Image Anal. **18**(2), 330–342 (2014)
3. Bin, X., Jiajun, B., Chen, C., Wang, C., Cai, D., He, X.: EMR: a scalable graph-based ranking model for content-based image retrieval. IEEE Trans. Knowl. Data Eng. **27**(1), 102–114 (2013)
4. Hernández-Gracidas, C., Enrique Sucar, L., Montes-y Gómez, M.: Modeling spatial relations for image retrieval by conceptual graphs. In: Proceedings of the First Chilean Workshop on Pattern Recognition (2009)
5. Li, C.-Y., Hsu, C.-T.: Image retrieval with relevance feedback based on graph-theoretic region correspondence estimation. IEEE Trans. Multimed. **10**(3), 447–456 (2008)
6. Pedronette, D.C.G., Torres, R.D.S.: A correlation graph approach for unsupervised manifold learning in image retrieval tasks. Neurocomputing **208**, 66–79 (2016)

7. Pedronette, D.C.G., Gonçalves, F.M.F., Guilherme, I.R.: Unsupervised manifold learning through reciprocal KNN graph and connected components for image retrieval tasks. Pattern Recognit. **75**, 161–174 (2018)
8. Conte, D., Foggia, P., Sansone, C., Ven-to, M.: Thirty years of graph matching in pattern recognition. Int. J. Pattern Recognit. Artif. Intell. **18**(03), 265–298 (2004)
9. Hidović, D., Pelillo, M.: Metrics for attributed graphs based on the maximal similarity common subgraph. Int. J. Pattern Recognit. Artif. Intell. **18**(03), 299–313 (2004)
10. Boubekeur, F., Boughanem, M., Tamine-Lechani, L.: Semantic information retrieval based on CP-nets. In: 2007 IEEE International Fuzzy Systems Conference, pp. 1–7. IEEE (2007)
11. Gonçalves, F.M.F., Guilherme, I.R., Pedronette, D.C.G.: Semantic guided interactive image retrieval for plant identification. Expert Syst. Appl. **91**, 12–26 (2018)
12. Salton, G., et al.: The smart system-experiments in automatic document processing (1971)
13. Salton, G.: Recent trends in automatic information retrieval. In: Proceedings of the 9th Annual International ACM SIGIR Conference on Research and Development in Information Retrieval, pp. 1–10. ACM (1986)
14. Liu, G.-H., Yang, J.-Y., Li, Z.Y.: Content-based image retrieval using computational visual attention model. Pattern Recognit. **48**(8), 2554–2566 (2015)
15. Abioui, H., Idarrou, A., Bouzit, A., Mammass, D.: Review: automatic image annotation for semantic image retrieval. In: Mansouri, A., El Moataz, A., Nouboud, F., Mammass, D. (eds.) ICISP 2018. LNCS, vol. 10884, pp. 129–137. Springer, Cham (2018). https://doi.org/10.1007/978-3-319-94211-7_15
16. Yang, H.-Y., Li, Y.-W., Li, W.-Y., Wang, X.-Y., Yang, F.-Y.: Content-based image retrieval using local visual attention feature. J. Vis. Commun. Image Represent. **25**(6), 1308–1323 (2014)
17. Bunke, H., Shearer, K.: A graph distance metric based on the maximal common subgraph. Pattern Recognit. Lett. **19**(3–4), 255–259 (1998)
18. Chen, H., Trouve, A., Murakami, K.J., Fukuda, A.: An intelligent annotation-based image retrieval system based on RDF descriptions. Comput. Electr. Eng. **58**, 537–550 (2017)
19. Urban, J., Jose, J.M.: Adaptive image retrieval using a graph model for semantic feature integration. In: Proceedings of the 8th ACM International Workshop on Multimedia Information Retrieval, pp. 117–126. ACM (2006)
20. H'roura, J.: Contributions à l'extraction de descripteureurs sur des données non conventionnelles pour a reconnaissance d'objets 3D. Ph.D. thesis, Université Ibn Zohr (2019)
21. Johnson, J., et al.: Image retrieval using scene graphs. In: Proceedings of the IEEE Conference on Computer Vision and Pattern Recognition, pp. 3668–3678 (2015)
22. Shieh, J.-R., Yeh, Y.-T., Lin, C.-H., Lin, C.-Y., Wu, J.-L.: Collaborative knowledge semantic graph image search. In: Proceedings of the 17th International Conference on World Wide Web, pp. 1055–1056. ACM (2008)
23. Jenni, K., Mandala, S., Sunar, M.S.: Content based image retrieval using colour strings comparison. Procedia Comput. Sci. **50**, 374–379 (2015)
24. Maisonnasse, L., Chevallet, J.P., Berrut, C.: Incomplete and fuzzy conceptual graphs to automatically index medical reports. In: Kedad, Z., Lammari, N., Métais, E., Meziane, F., Rezgui, Y. (eds.) NLDB 2007. LNCS, vol. 4592, pp. 240–251. Springer, Heidelberg (2007). https://doi.org/10.1007/978-3-540-73351-5_21

25. La Cascia, M., Sethi, S., Sclaroff, S.: Combining textual and visual cues for content-based image retrieval on the World Wide Web. In: Proceedings of IEEE Workshop on Content-Based Access of Image and Video Libraries (Cat. No. 98EX173), pp. 24–28. IEEE (1998)
26. Torjmen-Khemakhem, M., Pinel-Sauvagnat, K., Boughanem, M.: Investigating the document structure as a source of evidence for multimedia fragment retrieval. Inf. Process. Manag. **49**(6), 1281–1300 (2013)
27. Mechkour, M., Berrut, C., Chiaramella, Y.: Using conceptual graph frame work for image retrieval. In: International Conference on MultiMedia Modeling (MMM 1995), Singapore, pp. 127–142 (1995)
28. Baziz, M., Boughanem, M., Loiseau, Y., Prade, H.: Fuzzy logic and ontology-based information retrieval. In: Wang, P.P., Ruan, D., Kerre, E.E. (eds.) Fuzzy Logic, vol. 215, pp. 193–218. Springer, Heidelberg (2007). https://doi.org/10.1007/978-3-540-71258-9_10
29. Allani, O., Zghal, H.B., Mellouli, N., Akdag, H.: A knowledge-based image retrieval system integrating semantic and visual features. Procedia Comput. Sci. **96**, 1428–1436 (2016)
30. Unar, S., Wang, X., Wang, C., Wang, Y.: A decisive content based image retrieval approach for feature fusion in visual and textual images. Knowl.-Based Syst. **179**, 8–20 (2019)
31. Schuster, S., Krishna, R., Chang, A., Fei-Fei, L., Manning, C.D.: Generating semantically precise scene graphs from textual descriptions for improved image retrieval. In: Proceedings of the Fourth Workshop on Vision and Language, pp. 70–80 (2015)
32. Sorlin, S., Solnon, C.: Similarité de graphes: une mesure générique et un algorithme tabou réactif (2005)
33. Levenshtein, V.I.: Binary codes capable of correcting deletions, insertions, and reversals. In: Soviet Physics doklady, vol. 10, pp. 707–710 (1966)
34. Li, X., Chen, S.-C., Shyu, M.-L., Furht, B.: An effective content-based visual image retrieval system. In: Proceedings 26th Annual International Computer Software and Applications, pp. 914–919. IEEE (2002)
35. Rui, Y., Huang, T.S., Ortega, M., Mehrotra, S.: Relevance feedback: a power tool for interactive content-based image retrieval. IEEE Trans. Circuits Syst. Video Technol. **8**(5), 644–655 (1998)
36. Rui, Y., Huang, T.S.: A novel relevance feedback technique in image retrieval (1999)

A New Texture Descriptor:
The Homogeneous Local Binary
Pattern (HLBP)

Ibtissam Al Saidi[1]([⊠]), Mohammed Rziza[1]([⊠]), and Johan Debayle[2]([⊠])

[1] LRIT Laboratory, Associated Unit to CNRST (URAC 29), Rabat IT Center,
Faculty of Sciences, Mohammed V University in Rabat, Rabat, Morocco
`ibtissam_alsaidi@um5.ac.ma, mohammed.rziza@gmail.com`
[2] MINES Saint-Etienne, CNRS, UMR 5307 LGF, Centre SPIN, 158 cours Fauriel,
42023 Saint-Etienne Cedex 2, France
`debayle@emse.fr`

Abstract. This paper presents a simple and novel descriptor named Homogeneous Local Binary Pattern (HLBP) for texture analysis. The purpose of this description is to improve the Local Binary Pattern (LBP) approach basing on the impact of criterion homogeneous region using General Adaptive Neighborhood (GAN) principle. HLBP method is generated by using the criterion homogeneity which helps to represent a significant feature based on relationships between neighboring pixels. The main idea of HLBP is to threshold the distance between the current pixel and each of its neighbors with a homogeneity tolerance value which correspond more to the underlying spatial structures consequently allow extracting highly distinctive invariant features of the image. To assess the performance of the our proposed descriptor, we use "Outex" database and compared with the basic (LBPs). The experimental results show that the proposed Homogeneous Local Binary Pattern gives a good performance in term of classification accuracy.

Keywords: Texture classification · Local Binary Pattern (LBP) ·
General Adaptive Neighbor (GAN) · Feature extraction

1 Introduction

Texture wears the basic characteristic that appears in each natural surface, which gives a hardness to determine a specific definition by the scientific community [15]. Texture analysis is omnipresent as one of the most arduous and defying issue in computer founded pattern recognition, due to their similarity in spectral features between the various surfaces. However, until now, it has been considered as the only powerful manner that can present a fundamental discriminate based on various methods of feature extraction used in image processing and recognition. In texture image processing, several applications such as: face recognition [14], object detection [13], image matching [5] segmentation [1] and

© Springer Nature Switzerland AG 2020
A. El Moataz et al. (Eds.): ICISP 2020, LNCS 12119, pp. 308–316, 2020.
https://doi.org/10.1007/978-3-030-51935-3_33

texture classification [9]. In the literature, there exists four main approaches for texture analysis: statistical methods [11], model-based method [2], structural-based approach [8] and spectral-based or filter-based methods [7].

In order to allow an efficient classification of the texture, two main steps are required. The texture representation phase and the classification phase. These make it possible to generate a model using compact and discriminating characteristics extracted from the first phase. To extract features, several approaches have been developed. Specifically, the Local Binary Patterns (LBP) descriptor introduced by [11] has obtained important attention caused by their robustness and performance in front of illumination and scale variation with minimum algorithmic complexity. Based on LBP, many extensions have been defined and studied such as: ILBP [6], DLBP [1], CLBC [17], LDP [16], MRELBP [10] and SSLBP [4].

Recently, GAN-based Minkowski maps proposed by [12] have been employed for texture analysis as a powerful technique that has already proved its effectiveness on gray tone and color images in a local, adaptive and multi-scale way. In [3] this method is applied in combination with LBP descriptors to classify skin lesions. More precisely, the geometrical GAN-based features of each pixel are extracted from the texture images, then the LBP are computed to characterize them. The GAN processes the image locally.

In this work, to obtain better details of the image, we present a new LBP method based on the impact of homogeneous tolerance used in General Adaptive Neighborhood [3, 12] named Homogeneous Local Binary Pattern HLBP. This one allows to overcome the shortcoming of the initial approach by using a thresholding based on a selected criterion such as luminance, contrast, thickness, etc. Pixel criterion values fit within a specified range of homogeneity tolerance which allow to better take into consideration the relationships between the neighboring pixels. Therefore, the neighborhoods of the pixels correspond more to the underlying spatial structures. Consequently, HLBP descriptor obtains more intrinsic characteristics of the image. The proposed descriptor offers robust and powerful discriminative features in low dimension which are used as input of the support vector machine (SVM) classifier. The classical database Outex of texture images is employed to evaluate the method. According to the experimental results, the proposed HLBP gives the best performance in term of classification accuracy.

The following paper is organized as follows: a depth explanation of the specified methods and tools including the new approach of LBP is presented after the introduction. Thereafter, the performance of the descriptor is evaluated using the outex texture image dataset in Sect. 3, followed by discussion and conclusion in the last section.

2   Background

2.1   Local Binary Pattern LBP

LBP is one of the effective approaches to represent the texture characteristics which generate binary code for each pixel by thresholding the neighboring pixels with the central pixel value [11]. For the neighbors that have a value strictly negative from the center value pixel are encoded with 0; otherwise, they are encoded with 1. A concatenation of these binary values presents a binary number that is multiplied with corresponding weights to generate the LBP code as a decimal value employed for labeling the given pixel. This code is computed as follows:

$$LBP_{P,R}(x,y) = \sum_{i=0}^{P-1} s(g_i - g_c)2^i, s(g_i, g_c) = \begin{cases} 1 & g_i \geq g_c \\ 0 & g_i < g_c \end{cases} \tag{1}$$

Where P is the number of neighbor pixels and R is the radius, g_c and g_i are respectively pixel center and pixels neighbors of position i^{th}. The basic LBP is extending to a circular symmetric area [11], due to the limitation that can provide from LBP using square neighbors front of rotating invariant features. The following equations illustrate the rotation invariant uniform LBP ($LBP_{P;R}^{riu2}$)

$$LBP_{P,R}^{riu2} = \begin{cases} \sum_{i=0}^{P-1} s(g_i - g_c)2^i, & ifU(LBP_{P,R}) \geq 2 \\ P+1 & otherwise \end{cases} \tag{2}$$

where the uniform LBP is given by:

$$U(LBP_{P,R}) = |s(g_{P-1} - g_c) - s(g_0 - g_c)| + |s(g_i - g_c) - s(g_{i-1} - g_c)|. \tag{3}$$

3   Proposed Approach

In this section, we will describe in details our new proposed approach HLBP, in order to improve LBP and obtain more intrinsic image characteristics.

3.1   Homogeneous Local Binary Pattern Method

First of all, we observe that the GAN approach is provides homogeneous spatial regions using m tolerance with a specific criterion function (luminance, contrast ...) that make him robust compared to other methods for analyzing gray tone and color images. Therefore, based on this observation, we propose the HLBP descriptor that aims to improve LBP by using the condition of criterion homogeneity of its pixels g_i close to the one of the center g_c. Formally, the $HLBP_{P,R}$ is defined as:

$$HLBP_{P,R,m}(x,y) = \sum_{i=0}^{P-1} s(g_i - g_c)2^i, \tag{4}$$

$$s(g_i, g_c) = \begin{cases} 1 & \{|h(g_i) - h(g_c)| \leq m\} \\ 0 & otherwise \end{cases} \tag{5}$$

With the same notations as LBP, g_c is the center of the pixel, g_i denotes the pixel neighbors at the ith position, P the number of samples and R the radius. $s()$ represents the sign function with a new additional parameter for HLBP: the homogeneity tolerance $m \in \mathbb{R}^+$ with respect to the criterion (luminance, contrast, ...) h which helps to threshold all the neighbor pixels in a specific interval.

To get the HLBP code we need to follow this steps:

1. Firstly, we need to specify the size of the window by choosing the radius R and the sample number of neighbor pixels P.
2. Secondly, we must precise homogeneity tolerance m by considering in our approach the luminance criterion $h \equiv f$ (f is pixel intensity) to get pixels that are adjacent to g_c in relation to the criterion mapping.
3. Finally, calculating the code of homogeneous Local Binary Pattern. In this step, the value of the center is substituted with binary code of the neighbors which are thresholded in relation with the center pixel $|f(g_i) - f(g_c)| \leq m$. Precisely, each pixel is sill by homogeneity tolerance m with its neighbors by subtracting the center pixel value. The resulting that is respect threshold values are encoded with 1, otherwise are encoded with 0. Then, all binary numbers are concatenated to obtain decimal code which label the pixel.

To clarify the proposed approach, in figure "Fig. 1" we illustrate the complete procedure with specific parameters of R and P (8,1). We have an example of local pattern with the center $g_c = 190$ we substitute all neighbors $|g_c - g_i|$ then we compare it with homogeneity tolerance that we choose as $m = 15$ to compose binary code "10001101".

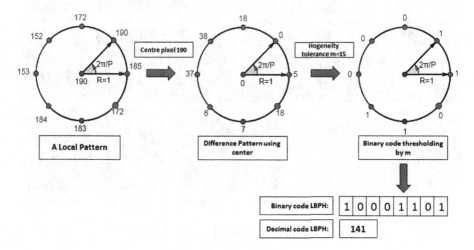

Fig. 1. Example of HLBP of (P = 8, R = 1) and m = 15.

4   Experimental Results

The objective of this section is to figure out the effectiveness of our proposed method in texture classification.

4.1   Texture Dataset

The experiment is realized using one of the most popular databases in texture analysis "Outex TC 0010". This database is composed of 24 classes of rich surface texture (random motifs, irregular patterns, texture regularity, various granularity, ect), each class contains 180 images with size 128*128. In the total, 4320 texture images are provided, with a variety of angle orientation (0.5, 10, 15, 30, 45, 60, 75 and 90). Some samples of each class are illustrated in figure "Fig. 2".

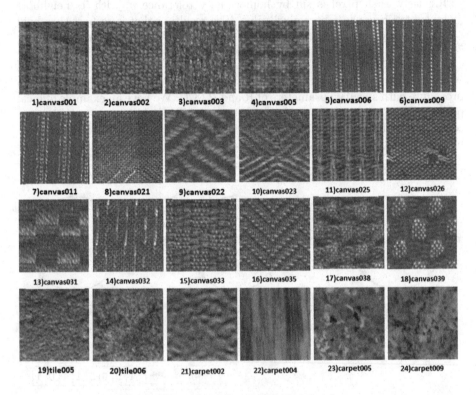

Fig. 2. Example of Outex TC 0010 database.

4.2   Experimental Procedure

In order to test our proposed method, we use the support vector machine SVM as one of the powerful classifiers in literature where 10 cross validation is employed.

In the proposed method, we use variation of two parameters radius R = 1, 2, 3, 4, 5, 6 and homogeneity tolerance m = 5, 10, 15, 20, 30. This variation allows to extract multiscale HLBP. In order to reduce the feature dimension and computational time in this paper, we fix P to 8 neighbors for all R. With the aim to evaluate the performance of our descriptor, we use Outex_TC_00010. Table 1 illustrates the results of the proposed experiments.

Experiment 1

In experiment 1, we compare our method with the texture descriptor based on LBP "General Adaptive Neighborhood-Minkowski Local Binary Pattern Descriptor" presented in [3].

It is clearly remarkable that our proposed descriptor gives the best classification accuracy compared with GANLBP for $m = 15$ we have 92.48%, 98.40% and 98,17% for R = 1, 2 and 3 respectively. Furthermore, they achieve also good results: 98.17%, 98.66% and 97.87% with different parameters of homogeneity tolerance $m = 15, m = 20$ and $m = 30$ respectively. "Fig. 3" illustrates examples of HLBP using different homogeneity tolerances m and radius R. Generally, our approach HLBP gives better performance by capturing more intrinsic characteristics of the image due to the use of sill of homogeneity tolerance m allowing to better take into account the relationships between neighboring pixels.

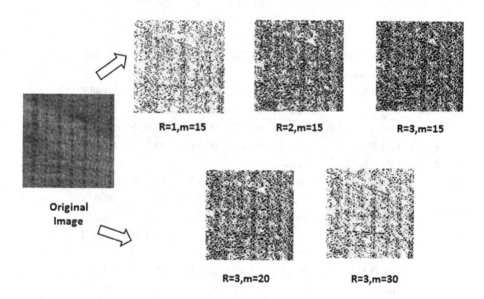

Fig. 3. Example of images HLBP employing different parameter.

Experiment 2

In Table 1 for homogeneity tolerance $m = 15$ our approach outperforms LBP for most of radius. However, we did not stop here since we want to increase robustness of HLBP to surpass LBP precisely, in all multiscale variation (radius and homogeneity tolerance), therefore, we proposed the second experiment. The idea is to combine the histograms of HLBP and LBP.

Table 1. Classification accuracy (%) of the HLBP and HLBP+LBP descriptor for different values of radius R on the Outex-TC-10 dataset.

R	Approaches	Homogeneity tolerance m				
		5	10	15	20	30
1	LBP classic	96.26%				
	GANLBP	79.54%	85.51%	89.10%	90.16%	89.79%
	HLBP	95.69%	95.19%	92.48%	87.96%	69.19%
	GANLBP+LBP	97.80%	98.40%	98.52%	98.73%	98.47%
	HLBP+LBP	99.00%	98.35%	98.59%	**99.05%**	98.59%
2	LBP classic	97.10%				
	GANLBP	82.55%	97.19%	88.73%	90.60%	92.48%
	HLBP	92.24%	98.38%	98.40%	97.73%	95.35%
	GANLBP+LBP	98.89%	97.25%	99.21%	99.12%	99.05%
	HLBP+LBP	99.31%	99.51%	**99.54%**	99.38%	99.35%
3	LBP classic	97.94%				
	GANLBP	79.72%	81.44%	85.37%	88.10%	90.02%
	HLBP	87.01%	96.48%	98.17%	98.66%	97.87%
	GANLBP+LBP	98.68%	99.00%	99.42%	99.10%	99.03%
	HLBP+LBP	99.00%	99.38%	**99.70%**	99.51%	99.49%
4	LBP classic	96.71%				
	GANLBP	78.45%	85.37%	85.51%	88.61%	90.46%
	HLBP	84.93%	94.31%	96.09%	96.00%	96.16%
	GANLBP+LBP	98.80%	99.03%	99.17%	99.00%	99.44%
	HLBP+LBP	99.05%	99.33%	99.54%	99.58%	**99.61%**
5	LBP classic	94.00%				
	GANLBP	80.09	72.63%	80.30%	81.51%	82.78%
	HLBP	80.16%	90.79%	94.47%	94.58%	96.04%
	GANLBP+LBP	97.85%	98.08%	97.73%	97.73%	98.26%
	HLBP+LBP	97.41%	98.31%	98.91%	**98.94%**	98.80%
6	LBP classic	94.12%				
	GANLBP	78.13%	80.42%	79.81%	79.68%	80.28%
	HLBP	79.38%	90.49%	94.42%	95.60%	95.51%
	GANLBP+LBP	97.36%	97.66%	97.94%	97.57%	97.25%
	HLBP+LBP	97.85%	98.66%	**99.14%**	99.07%	98.98%

As expected the performance of other descriptors such as GANLBP+LBP and LBP drops for this dataset owing to loose the structural information compared to HLBP+LBP which keep their performance in front of other methods through a margin that shows its robustness with the homogeneity criterion m.

For $m = 15$ we have 98.59% 99.54% 99.70% 99.54% 98.91% 99.14% in R = (1, 2, 3, 4, 5 and 6) respectively.

5 Conclusion

In this work we have proposed a new texture classification descriptor HLBP that helps to get a robust discriminative information from an image based on the tolerance threshold m which take into consideration the relationship between neighboring pixels. Therefore, the neighborhoods of pixels much more correspond to the underlying spatial structures. Different experiments are carried out using one of the publicly available texture databases Outex_TC_0010 showing the performance and robustness of our approach.

References

1. Arof, H., Deravi, F.: Circular neighbourhood and 1-D DFT features for texture classification and segmentation. IEE Proc.-Vis. Image Sig. Process. **145**(3), 167–172 (1998)
2. Chen, J.L., Kundu, A.: Rotation and gray scale transform invariant texture identification using wavelet decomposition and hidden Markov model. IEEE Trans. Pattern Anal. Mach. Intell. **16**(2), 208–214 (1994)
3. González-Castro, V., et al.: Texture descriptors based on adaptive neighborhoods for classification of pigmented skin lesions. J. Electron. Imaging **24**(6), 061104 (2015)
4. Guo, Z., Wang, X., Zhou, J., You, J.: Robust texture image representation by scale selective local binary patterns. IEEE Trans. Image Process. **25**(2), 687–699 (2015)
5. Heikkilä, M., Pietikäinen, M., Schmid, C.: Description of interest regions with local binary patterns. Pattern Recognit. **42**(3), 425–436 (2009)
6. Jin, H., Liu, Q., Lu, H., Tong, X.: Face detection using improved LBP under Bayesian framework. In: Third International Conference on Image and Graphics (ICIG 2004), pp. 306–309. IEEE (2004)
7. Kokare, M., Biswas, P., Chatterji, B.: Rotation-invariant texture image retrieval using rotated complex wavelet filters. IEEE Trans. Syst. Man Cybern. Part B (Cybern.) **36**(6), 1273–1282 (2006)
8. Lam, W.K., Li, C.K.: Rotated texture classification by improved iterative morphological decomposition. IEE Proc.-Vis. Image Sig. Process. **144**(3), 171–179 (1997)
9. Liao, S., Law, M.W., Chung, A.C.: Dominant local binary patterns for texture classification. IEEE Trans. Image Process. **18**(5), 1107–1118 (2009)
10. Liu, L., Lao, S., Fieguth, P.W., Guo, Y., Wang, X., Pietikäinen, M.: Median robust extended local binary pattern for texture classification. IEEE Trans. Image Process. **25**(3), 1368–1381 (2016)
11. Ojala, T., Pietikainen, M., Maenpaa, T.: Multiresolution gray-scale and rotation invariant texture classification with local binary patterns. IEEE Trans. Pattern Anal. Mach. Intell. **24**(7), 971–987 (2002)
12. Pinoli, J.C., Debayle, J.: General adaptive neighborhood mathematical morphology. In: 2009 16th IEEE International Conference on Image Processing (ICIP), pp. 2249–2252. IEEE (2009)

13. Satpathy, A., Jiang, X., Eng, H.L.: LBP-based edge-texture features for object recognition. IEEE Trans. Image Process. **23**(5), 1953–1964 (2014)
14. Tan, X., Triggs, B.: Enhanced local texture feature sets for face recognition under difficult lighting conditions. IEEE Trans. Image Process. **19**(6), 1635–1650 (2010)
15. Tuceryan, M., Jain, A.K.: Texture analysis. In: Handbook of Pattern Recognition and Computer Vision, pp. 235–276. World Scientific (1993)
16. Zhang, B., Gao, Y., Zhao, S., Liu, J.: Local derivative pattern versus local binary pattern: face recognition with high-order local pattern descriptor. IEEE Trans. Image Process. **19**(2), 533–544 (2009)
17. Zhao, Y., Huang, D.S., Jia, W.: Completed local binary count for rotation invariant texture classification. IEEE Trans. Image Process. **21**(10), 4492–4497 (2012)

Considerably Improving Clustering Algorithms Using UMAP Dimensionality Reduction Technique: A Comparative Study

Mebarka Allaoui[1]([⊠])[iD], Mohammed Lamine Kherfi[2,3][iD],
and Abdelhakim Cheriet[3][iD]

[1] LAGE Laboratory, Kasdi Merbah University Ouargla, Ouargla, Algeria
moubarakaallaoui1994@gmail.com
[2] LAMIA Laboratory, Université du Québec à Trois-Rivières, Trois-Rivières, Canada
[3] Kasdi Merbah University Ouargla, Ouargla, Algeria

Abstract. Dimensionality reduction is widely used in machine learning and big data analytics since it helps to analyze and to visualize large, high-dimensional datasets. In particular, it can considerably help to perform tasks like data clustering and classification. Recently, embedding methods have emerged as a promising direction for improving clustering accuracy. They can preserve the local structure and simultaneously reveal the global structure of data, thereby reasonably improving clustering performance. In this paper, we investigate how to improve the performance of several clustering algorithms using one of the most successful embedding techniques: Uniform Manifold Approximation and Projection or UMAP. This technique has recently been proposed as a manifold learning technique for dimensionality reduction. It is based on Riemannian geometry and algebraic topology. Our main hypothesis is that UMAP would permit to find the best clusterable embedding manifold, and therefore, we applied it as a preprocessing step before performing clustering. We compare the results of many well-known clustering algorithms such ask-means, HDBSCAN, GMM and Agglomerative Hierarchical Clustering when they operate on the low-dimension feature space yielded by UMAP. A series of experiments on several image datasets demonstrate that the proposed method allows each of the clustering algorithms studied to improve its performance on each dataset considered. Based on Accuracy measure, the improvement can reach a remarkable rate of 60%.

Keywords: Dimensionality reduction · UMAP · Clustering · Embedding manifold · Big data analytics · Machine learning · Comparative study

1 Introduction

Clustering is a fundamental pillar of unsupervised machine learning and it is widely used in a range of tasks across disciplines. In past decades, a variety

© Springer Nature Switzerland AG 2020
A. El Moataz et al. (Eds.): ICISP 2020, LNCS 12119, pp. 317–325, 2020.
https://doi.org/10.1007/978-3-030-51935-3_34

of clustering algorithms have been developed [5] such as k-means [6], Gaussian Mixture Models (GMMs) [14], HDBSCAN [1], and hierarchical algorithms [15]. However, these clustering algorithms typically require features to be hand crafted or learned for each dataset and task. Then, those features should be analyzed using feature selection, in order to eliminate redundant or poor quality features. Those requirements are more challenging in the unsupervised setting. Additionally, this process is time-consuming and brittle [17], since the choice of features has a large influence on the performance of the clustering algorithm. In this paper, we formulate the following hypothesis: if we apply an adequate embedding on our raw data, i.e., an embedding which allows to find a good distance preserving manifold, than this could help clustering algorithms in doing their job. One key question was: which embedding technique to apply it to find the best embedding manifold. Many methods exists, including those performing a linear transformation of data like the well-known Principal Component Analysis (PCA) [10]. However, PCA is a linear method and does not perform well in cases where relationships are non-linear. Thankfully, alternative non-linear manifold learning methods exist, and can be categorized by their focus on finding local or global structure. Isomap [7] is well known globally focused method. While T-SNE [8] is considered as locally focused method. More recently manifold learning technique is UMAP [11], UMAP showed better performance to preserve both the local and global structure. In this paper, we will investigate the use of this latter technique: because it outperforms its concurrents [11] and it has proven to be able to exactly meet our needs [16,18]. In this paper, Our main focus was on measuring the improvement achieved by each clustering algorithm thanks to the application of UMAP embedding manifold, and in order to validate our method we conduct a number of experiments on five datasets. We empirically observe that this method allows to the clustering algorithms to be competitive with state-of-the-art techniques. The rest of this paper is organized as follows. We present more details about UMAP technique in Sect. 2. In Sect. 3 we introduce our idea. Section 4 discusses the experimental results in five image datasets. Section 5 concludes our work.

2   UMAP Embedding Technique for Dimensionality Reduction

Uniform Manifold Approximation and Projection (UMAP) is a recently proposed manifold learning method, which seeks to accurately represent local structure and better incorporate global structure [9]. Compared to t-SNE it has a number of advantages. UMAP has been shown to scale well with large datasets, while t-SNE typically struggles with them. UMAP relies on three hypothesis, namely that 1) the data is uniformly distributed on a Riemannian manifold, 2) the Riemannian metric is locally constant 3) the manifold is locally connected. From these assumptions it is possible to represent the manifold with a fuzzy topological structure of high dimensional data points. The embedding manifold is found by searching for a fuzzy topological structure of low dimensional projection of the data. To construct the fuzzy topological structure UMAP represents the data points by a high-dimensional graph. The constructed high-dimensional graph is weighted graph, with edge weights representing the likelihood that two points are connected. UMAP uses exponential probability distribution to compute the similarity between high dimensional data points:

$$p_{i|j} = \exp(-\frac{d(x_i, x_j) - \rho_i}{\sigma_i}) \tag{1}$$

Where $d(x_i, x_j)$ is the distance between the i-th and j-th data points and ρ is the distance between i-th data points and its first nearest neighbor. In cases that the weight of the graph between i and j nodes is not equal to the weight between j and i nodes. UMAP uses a symmetrization of the high-dimensional probability:

$$p_{ij} = p_{i|j} + p_{j|i} - p_{i|j}p_{j|i} \tag{2}$$

As we said above the constructed graph is a likelihood graph, and UMAP needs to specify k the number of nearest neighbor:

$$k = 2^{\sum_i p_{ij}} \tag{3}$$

Once the high-dimensional graph is constructed, UMAP constructs and optimizes the layout of a low-dimensional analogue to be as similar as possible. For modelling distance in low dimensions, UMAP uses probability measure similar to Student t-distribution:

$$q_{ij} = (1 + a(y_i - y_j)^{2b})^{-1} \tag{4}$$

where $a \approx 1.93$ and $b \approx 0.79$ for default UMAP.

UMAP uses binary cross-entropy (CE) as a cost function due to its capability of capturing the global data structure:

$$CE(P, Q) = \sum_i \sum_j [p_{ij} \log(\frac{p_{ij}}{q_{ij}}) + (1 - p_{ij}) \log(\frac{1 - p_{ij}}{1 - q_{ij}})] \tag{5}$$

Where P is the probabilistic similarity of the high dimensional data points, and Q is for the low dimensional data points.

The derivative of the cross-entropy used to update the coordination of the low-dimensional data points to optimize the projection space until the convergence. UMAP applied Stochastic Gradient Descent (SGD) due to its faster convergence and it reduces the memory consumption since we compute the gradients for a subset of the data set.

UMAP has a number of important hyper-parameters that influence its performance. These hyper-parameters are:

- The dimensionality of the target embedding
- The number of neighbor k, choosing small value means the interpretation will be very local and capture fine detail structure. While choosing a large value means the estimation will be based on larger regions, and thus, will missing some of the fine detail structure.

– The minimum allowed distance between points in the embedding space. Lower values of this minimum distance will more accurately capture the true manifold structure, but may lead to dense clouds that make visualization difficult.

3 Our Method

Our method relies primarily on the application of clustering algorithms on embedding manifold extracted by manifold learning methods UMAP [9] due to its success in preserving both the local and the global structure. We chose four of well-known algorithms as clustering algorithms which are represented in k-means [6], HDBSCAN [1], GMM [14] and Agglomerative Clustering [15]. We will show that by augmenting the clustering task with a manifold learning technique which explicitly takes local structure into account, we can increase the quality of clustering performance of the different algorithms. Figure 1 represents the architecture of our method.

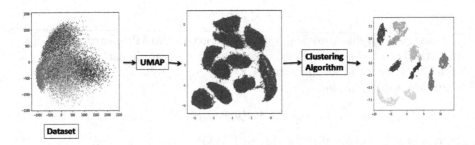

Fig. 1. The structure of our method.

4 Experiments

To assess the improvement of using UMAP with the clustering algorithms studied, we conduct experiments on a range of diverse datasets, including standard datasets widely used to evaluate clustering algorithms.

4.1 Datasets

We conducted our experiments on five diverse image datasets, including standard datasets used to evaluate deep clustering algorithms. Those datasets are MNIST [2], Fashion MNIST [3], USPS [13], Pen Digits [4] and UMIST Face Cropped [12]. Table 1 summarizes the main characteristics of each dataset.

Table 1. Datasets statistics.

Dataset	Number of images	Number classes	Feature vector length
MNIST	20000	10	784
F-MNIST	20000	10	784
USPS	9298	10	256
Pen digits	1797	10	64
UMIST face	575	20	10304

4.2   Evaluation Metrics

In order to validate the performance of unsupervised clustering algorithms, we use the two standard evaluation metrics, accuracy (ACC) and Normalized Mutual Information (NMI).

$$ACC = max_m \frac{\sum_{i=1}^{n} 1\{y_i = m(c_i)\}}{n} \tag{6}$$

$$NMI = \frac{2I(y,c)}{[H(y) + H(c)]} \tag{7}$$

4.3   Results

Figure 2 shows the resulting clusters when using k-means for visualization purposes. We could see that the visualization is better when we apply the algorithm on the UMAP embedded manifold of the five datasets. However, in order to better understand the effectiveness of our method at clustering we will study each clustering algorithm via measuring its own results on the different datasets using the accuracy and NMI, as well as when we apply it on the extracted features by UMAP.

Table 2 and Table 3 show the accuracy and NMI results for the clustering algorithms on five different datasets comparable to the same algorithms applied on embedding manifold of the datasets extracted by UMAP. In both tables, improvement score rows represent the difference between the results of the algorithms and the results after the application of these algorithms on the features extracted by UMAP. By doing so, we can see clearly how UMAP can help the four clustering algorithms and to what extent the results improved. Actually, great results were achieved by the algorithms on embedded data points, where the results are improved by an increase of up to 60% in term of accuracy, and in range of 5% to 48% in term of NMI. What is striking is how UMAP helped HDBSCAN to improve its result by 60 % points on USPS dataset. Also, it had an improvement better than the other algorithms in 2 of the 5 datasets with at least 50% in term of accuracy and over than 38% in term of NMI measure. GMM is improved better than the other, on 3 of the 5 datasets, with percentage over than 34% in term of accuracy, and over than 25% in term NMI measure.

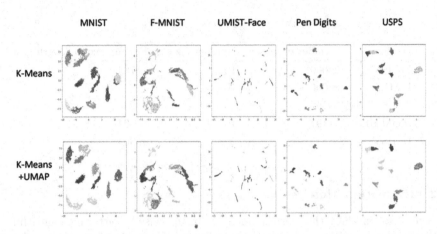

Fig. 2. Visualization of K-Means applied to all five datasets. The first row represents the K-Means visualization of the five datasets themselves, and the second row represents the visualization of K-Means on the UMAP embedded manifold of these datasets.

Table 2. Comparison between the different clustering algorithms on the five datasets according to the accuracy measure.

	MNIST	F-MNIST	UMIST face	Pen digits	USPS
K-means	0.5278	0.4750	0.4348	0.7028	0.6678
UMAP + K-means	**0.9054**	**0.5865**	**0.7409**	**0.8843**	**0.8105**
Improvement score	0.3776	0.1115	0.3061	0.1815	0.1427
Agglomerative	0.5751	0.5766	0.4539	0.7451	0.6834
UMAP + Agglomerative	**0.8918**	**0.5925**	**0.7270**	**0.8737**	**0.9584**
Improvement score	0.3167	0.0159	0.2731	0.1286	0.2740
HDBSCAN	0.2765	0.2140	0.4904	0.5453	0.3529
UMAP + HDBSCAN	**0.7765**	**0.3458**	**0.6730**	**0.9004**	**0.9553**
Improvement score	0.5000	0.1318	0.1826	0.3551	0.6024
GMM	0.4507	0.4579	0.3826	0.4836	0.4802
UMAP+GMM	**0.9159**	**0.5885**	**0.7287**	**0.8748**	**0.6727**
Improvement score	0.4652	0.1306	0.3461	0.3912	0.1925

The accuracy and NMI measures showed us that the studied clustering algorithms in general and HDBSCAN as a particular case had bad results and especially in MNIST and Fashion MNIST datasets. The problem here is all the clustering algorithms tend to suffer from the curse of dimensionality: high dimensional data requires more observed samples to produce much density. If we could reduce the dimensionality of the data more we would make the density more evident and make it far easier for those algorithms to cluster the data. What we need is strong manifold learning, and this is where UMAP can come into play. One of the reasons which help the studied algorithms to perform well on

Table 3. Comparison between the different clustering algorithms on the five datasets according to the NMI measure

	MNIST	F-MNIST	UMIST face	Pen digits	USPS
K-means	0.4774	0.5139	0.6647	0.6998	0.6266
UMAP + K-means	**0.8494**	**0.6377**	**0.8663**	**0.8545**	**0.8602**
Improvement score	0.3720	0.1238	0.2016	0.1547	0.2336
Agglomerative	0.6360	0.6080	0.6673	0.7965	0.7250
UMAP + Agglomerative	**0.8463**	**0.6511**	**0.8764**	**0.8456**	**0.9000**
Improvement score	0.2103	0.0431	0.2091	0.0491	0.1750
HDBSCAN	0.3674	0.2535	0.6933	0.5804	0.4442
UMAP + HDBSCAN	**0.8315**	**0.6323**	**0.8427**	**0.8871**	**0.8923**
Improvement score	0.4641	0.3788	0.1494	0.3067	0.4481
GMM	0.3882	0.5471	0.6160	0.5203	0.4232
UMAP+GMM	**0.8654**	**0.6424**	**0.8648**	**0.8447**	**0.8231**
Improvement score	0.4772	0.0953	0.2488	0.3244	0.3999

the learned manifold is to set the min distance (the hyper-parameter of UMAP) to be 0. And thus make the points packed together densely as well as making cleaner separations between clusters.

Table 4. The execution time before and after applying UMAP on the different clustering algorithms on the five datasets.

Time in second	MNIST	F-MNIST	UMIST face	Pen digits	USPS
K-means	112.13	74.69	17.24	0.94	12.93
UMAP + K-means	**1.22**	**1.20**	**0.33**	**0.26**	**0.57**
Agglomerative	710.08	674.14	6.57	0.48	47.93
UMAP + Agglomerative	**88.31**	**100.14**	**0.03**	**0.28**	**8.51**
HDBSCAN	1603.26	1660.25	17.77	1.14	117.56
UMAP + HDBSCAN	**5.13**	**4.49**	**0.03**	**0.12**	**0.75**
GMM	24.51	26.27	3.49	0.58	25.26
UMAP+GMM	**0.51**	**0.42**	**0.06**	**0.03**	**0.14**

Table 4 gives us the execution time taken for each clustering algorithm on the different datasets compared to the run-time of these algorithms applied to the embedding manifold of the five datasets. We can observe that the run-time is also improved, where it was reduced to a few seconds and sometimes to a few split-seconds, and this is really a good achievement for our method compared to the size of the datasets. Especially for agglomerative and HDBSCAN algorithms, the run-time of HDBSCAN is reduced from over than 26 min until around 5 s in

MNIST and Fashion MNIST datasets. From these results, we demonstrate that these clustering algorithms can now handle large databases well.

5   Conclusion

In this paper, we investigated the use of UMAP technique for dimensionality reduction before applying a number of well-known clustering algorithms on datasets. We showed that it can drastically improve the performance of the studied algorithms, both in terms of clustering accuracy and time. Experimental results indicate that the proposed approach can improve clustering performance obviously, we show how our proposed method can make the mentioned clustering algorithms competitive with the current state-of-the-art clustering approaches. It is also validated by experiments that our method allows to the clustering algorithms considered to deal better on larger data sets.

References

1. Campello, R.J.G.B., Moulavi, D., Sander, J.: Density-based clustering based on hierarchical density estimates. In: Pei, J., Tseng, V.S., Cao, L., Motoda, H., Xu, G. (eds.) PAKDD 2013. LNCS (LNAI), vol. 7819, pp. 160–172. Springer, Heidelberg (2013). https://doi.org/10.1007/978-3-642-37456-2_14
2. Deng, L.: The MNIST database of handwritten digit images for machine learning research [best of the web]. IEEE Sig. Process. Mag. **29**(6), 141–142 (2012)
3. Xiao, H., Rasul, K., Vollgraf, R.: Fashion-MNIST: a novel image dataset for benchmarking machine learning algorithms. arXiv preprint arXiv:1708.07747 (2017)
4. Alpaydin, E., Alimoglu, F.: Pen-based recognition of handwritten digits data set. University of California, Irvine. Machine Learning Repository. Irvine: University of California, **4**(2) (1998)
5. Xu, D., Tian, Y.: A comprehensive survey of clustering algorithms. Annals of Data Science **2**(2), 165–193 (2015). https://doi.org/10.1007/s40745-015-0040-1
6. MacQueen, J.: Some methods for classification and analysis of multivariate observations. In: Proceedings of the Fifth Berkeley Symposium on Mathematical Statistics and Probability, pp. 281–297. University of California Press, Berkeley (1967)
7. Tenenbaum, J.B., De Silva, V., Langford, J.C.: A global geometric framework for nonlinear dimensionality reduction. Science **290**(5500), 2319–2323 (2000)
8. Maaten, L.V.D., Hinton, G.: Visualizing data using t-SNE. J. Mach. Learn. Res. **9**, 2579–2605 (2008)
9. McInnes, L., Healy, J., Melville, J.: Umap: uniform manifold approximation and projection for dimension reduction. arXiv preprint arXiv:1802.03426 (2018)
10. Pearson, K.: LIII. On lines and planes of closest fit to systems of points in space. Lond. Edinb. Dublin Philos. Mag. J. Sci. **2**(11), 559–572 (1901)
11. Pedregosa, F., et al.: Scikit-learn: machine learning in Python. J. Mach. Learn. Res. **12**, 2825–2830 (2011)
12. Graham, D.B., Allinson, N.M.: Characterising virtual eigen signatures for general purpose face recognition. In: Wechsler, H., Phillips, P.J., Bruce, V., Soulié, F.F., Huang, T.S. (eds.) Face Recognition. NATO ASI Series (Series F: Computer and Systems Sciences), vol. 163, pp. 446–456. Springer, Heidelberg (1998). https://doi.org/10.1007/978-3-642-72201-1-25

13. Hull, J.J.: A database for handwritten text recognition research. IEEE Trans. Pattern Anal. Mach. Intell. **16**(5), 550–554 (1994). https://doi.org/10.1109/34.291440
14. Rasmussen, C.E.: The infinite Gaussian mixture model. In: NIPS 1999 Proceedings of the 12th International Conference on Neural Information Processing Systems, pp. 554–560. MIT Press, Cambridge (2000)
15. Madhulatha, T.S.: An overview on clustering methods. J. Eng. **2**(4), 719–725 (2012)
16. McConville, R., Santos-Rodriguez, R., Piechocki, R.J., Craddock, I.: N2D: (Not Too) deep clustering via clustering the local manifold of an auto encoded embedding. arXiv preprint arXiv:1908.05968 (2019)
17. Miao, J., Niu, L.: A survey on feature selection. Procedia Comput. Sci. **91**, 919–926 (2016)
18. Becht, E., et al.: Dimensionality reduction for visualizing single-cell data using UMAP. Nat. Biotechnol. **37**(1), 38 (2019)

Logo Detection Based on FCM Clustering Algorithm and Texture Features

Wala Zaaboub[1,2](✉) Ⓘ, Lotfi Tlig[1] Ⓘ, Mounir Sayadi[1] Ⓘ, and Basel Solaiman[3] Ⓘ

[1] Research Laboratory SIME, ENSIT, Tunis, Tunisia
walazaaboub@gmail.com
[2] National Engineering School of Sfax ENIS, Sfax, Tunisia
[3] ITI, IMT-Atlantique, Technopole Brest, Plouzané, France

Abstract. Logo detection methods usually depend on logo shapes and need for training data or *a-priori* information on the processed images. This limits their effectiveness to real-world applications. In this paper, we tackle these challenges by exploring the textural information. Specifically we propose a novel approach for administrative logo detection based on a fuzzy classification with a multi-fractal texture feature, capable of automatically characterizing texture measures describing logo and non-logo regions. Experimental results, using two real datasets, confirm the feasibility of the proposed method for degraded administrative documents. Extensive comparative evaluations demonstrate the superiority of this approach over the state-of-the-art methods.

Keywords: Logo detection · Texture feature · Fuzzy classification · Logo extraction

1 Introduction

Administrative documents generally contain different salient graphical objects such as logos, signatures, seals, stamps, and bar codes. These objects are considered as a rich source of contextual information, to perfectly treat the problem of document retrieval and classification. Logo is a key visual feature for users to identify the source and the ownership of a document. Recently, many research works have been carried out on the related topics of logo retrieval, detection and recognition for images [1–3]. This paper is about the logo detection for administrative documents retrieval.

The manual identification of logos is a difficult problem, as in organization documents flow is continuously coming and growing rapidly. Consequently, many research papers focus on automatic detection and verification of logos to facilitate more reliable and appropriate documents-based systems [4,5].

However, there are many challenges to have a performing and accurate document logo detection. First, administrative documents commonly contain a mixture of machine printed and handwritten text, stamps, seals, signatures, tables and other elements. Second, logo typically appears as a complex-mixture of text

© Springer Nature Switzerland AG 2020
A. El Moataz et al. (Eds.): ICISP 2020, LNCS 12119, pp. 326–336, 2020.
https://doi.org/10.1007/978-3-030-51935-3_35

and graphic, which may lead to false detection. Third, administrative documents are generally binary images that degrade the document analysis. Fourth, logo has often a considerably disproportionate size compared to other elements of the document, it is often very small in relative to the document image resolution. The low image quality, the presence of degradation and noise in scanned documents and the large intra-class variations of logos make the detection task more difficult.

The logo detection techniques can be categorized into three categories; connected-component-based methods, local descriptor-based methods and block-based methods. This work fall in the last category. Its main purpose is to overcome the problems of the state-of-art methods like; the high temporal complexity of the training process, the requirement of *a-priori* information on processed images, the difficulty of accurately processing documents with a high degree of degradation and of being unable to process correctly on unusual fonts. The proposed logo detection task will be treated as a linear three-class texture segmentation problem (logo, non-logo, and background) based on a fuzzy classifier. This paper is laid out as follows: Sect. 2 presents the proposed texture-based fuzzy approach, Sect. 3 covers the experiments and Sect. 4 concludes the work.

2   Proposed Approach

2.1   Segmentation on Logo and Non-Logo Regions

Pre-processing is a primordial phase in any analysis system that is involved in improving the image quality for further processing especially when the image is corrupted by skew, bad illumination, noise or blur. Our pre-processing phase is composed of skew correction, noise reduction and image down-scaling.

We considered that the combination of luminous intensity descriptors with texture features is a qualified process to have satisfactory image classification result in remote sensing [6]. It is on this basis that this combination will be used for administrative documents segmentation. For the texture feature, our idea is to choose the texture measures describing the different textures of administrative documents (logo/non-logo), and then only the features that quantify these measures are selected, instead of blindly extracting features or performing expensive tests on all possible features.

Texture features qualify measures like: density, coarseness, roughness, linearity and direction. Logo texture is often rough with an irregular structure, unlike the text zone which is in the form of repetitive patterns (alphabetic characters) spatially placed under a rule. In this case, the chosen texture measures describing the different administrative documents textures are the roughness and the regularity. In this work, multi-fractal feature is preferred because it is efficient in expressing these measures. It perfectly characterizes the administrative documents textures (logo, non-logo, and background).

We propose to adapt the multi-fractal analysis method via the wavelet-transform to detect logo regions from administrative documents. Indeed, there are two multi-fractal-based methods: the first one generates the local information that is represented by the singularity exponent (Hölder exponent), and the second method provides the global and the local information that are given by the multi-fractal spectrum. We propose to implement the first method, as our goal is the extraction of local statistical information. What further argues our choice is that this Hölder exponent has the power to characterize the textural roughness with the advantage of multi-scale analysis [7].

The estimated singularity exponent α (1) is defined as follows:

$$\alpha(\overrightarrow{x}) = \lim_{r \to 0} \frac{\log T_\psi^r \mu(\overrightarrow{x})}{\log r} \tag{1}$$

Where μ is the multi-fractal measure, ψ is the wavelet, $\overrightarrow{x}$ is one point of the image, r is an expansion factor and $T_\psi^r \mu$ is the wavelet projection.

Turiel et al. [8] defined the class of wavelet functions adapted to the singularity analysis (for which (1) is verified). For our approach, we choose the Lorentz wavelet (with the order $\gamma = 1$) to estimate the singularity exponents. Its general expression is given by (2):

$$\psi(\overrightarrow{x}) = (1 + |\overrightarrow{x}|^2)^{-\gamma} \tag{2}$$

Using the chosen analysis method and its feature, we generate the image of Hölder exponents. The texture feature expresses a statistical moment from the generated image. We use the feature "Mean" calculated over a sliding window of the Hölder image, and assigned to the central pixel, to create the "image of texture features" that will be used in the classification step.

For luminous intensity feature, we calculate the Fisher Score [9] for every descriptor to evaluate its discrimination ability. The evaluation of this criterion on a set of test images allows the selection of the most appropriate luminous intensity descriptor characterizing the different classes (logo, non-logo, and background). Among the four statistical features ("mean", "variance", "skewness" and "kurtosis"), we choose the "mean" feature having the highest value of the characterization degree.

Using this feature expressing a statistical moment, we generate which is called the "image of intensity features". To create the "image of means" that will be used in the classification, the mean is calculated over a sliding window of the input image and assigned to the central pixel.

The Fuzzy C-Means algorithm FCM [10] is an extension of the K-Means algorithm. In fact, it presents a fuzzy concept in the calculation of the membership degree. Its goal is to build a fuzzy partition of the processed image. The image segmentation with FCM algorithm is realized using an estimation, for each pixel, of a vector of membership degrees to each class. Then, minimization of a particular objective function is required.

In this work, the FCM classification is developed in order to use a vector of features representing each pixel, instead of a pixel-based classification. The chosen features compose the vector used in the classification: the luminous intensity feature "mean" and the texture feature "Hölder exponent". This signature is calculated using a sliding window, with an appropriate size, centered around the corresponding pixel. The image segmentation based on FCM approach using a vector of features follows the steps presented in the following algorithm.

Algorithm of FCM approach using a vector of features
Begin
1. Initialization:
fuzziness parameter $m > 1$ ($m = 1.1$), threshold parameter $\varepsilon > 0$ ($\varepsilon = 0.01$), and matrix $\boldsymbol{V}^{(0)}$ of classes centers. (Size of the matrix: Number of classes × Number of features).
2. Computing the normalized matrix $\boldsymbol{F}$ of the features values. $\boldsymbol{F}$ is calculated using the original image and the image of Hölder exponents (size of F: Number of pixels without edges × Number of features).
3. Computing the partition matrix $\boldsymbol{U}^{(t)}$ of elements u_{ik} (3).

$$
u_{ik} = \left(\sum_{j=1}^{C} \left(\frac{\|F_k - v_i\|}{\|F_k - v_j\|} \right)^{\frac{2}{m-1}} \right)^{-1}
\tag{3}
$$

C: Number of classes, v_i: Vector of the center of class i, and $\boldsymbol{F_k}$: Extracted feature vector of the k-th pixel.
4. Updating the matrix $\boldsymbol{V}^{(t)}$ of classes centers v_i (4).

$$
v_i = \frac{\sum_{k=1}^{N} (u_{ik})^m F_k}{\sum_{k=1}^{N} (u_{ik})^m}
\tag{4}
$$

5. Test if the stop condition (5) is reached, else return to step 3.

$$
\left\| V^{(t)} - V^{(t-1)} \right\| > \varepsilon
\tag{5}
$$

$\boldsymbol{V}^{(t)}$: matrix of class centers calculated in the current iteration, and $\boldsymbol{V}^{(t-1)}$: matrix of class centers calculated in the previous iteration.
End

The main advantage of the algorithm comes from the use of membership degrees. Due to them, the iterative optimization process become more robust, by taking into account overlaps between classes. Thus, it allows to obtain partitions that are more relevant and closer to reality. In addition, these degrees allow to take nuanced decisions for the pixel assignation to a class, which is very interesting for any classification type. Among the other advantages of the algorithm, we can note that its complexity is relatively reduced compared to other non-supervised classification, this makes it exploitable to deal with large problems (big amount of data). This strong point argues our classifier choice for administrative documents, as in organization documents flow is continuously coming and growing rapidly and usually they are high resolution documents. A summarizing schema of the proposed segmentation (logo, non-logo, and background) for administrative documents is shown in Fig. 1.

One of the fundamental issues of implementing classification algorithm is that the system cannot have error-free information about the interest region. In addition, for logos contained near-scattered components, an ordinary extraction algorithm is typically unable to detect all these components. Therefore, as a post-segmentation, the proposed system deployed the morphological operations [11] for a better approach of refining logo region extraction. The idea is to implement both opening and closing operations in order to join separated logo sub-areas and remove the unwanted non-logo regions, which can be sometimes confused with the logo regions. One of the advantages of these two operations is that keep general shape but smooth with respect to object (for opening) or background (for closing).

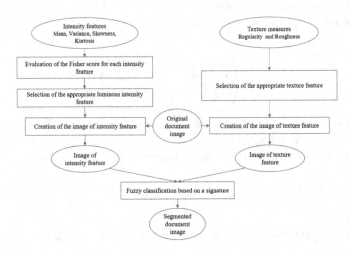

Fig. 1. Proposed schema of the segmentation of documents on: logo, non-logo, and background

2.2 Logo Detection and Extraction

First, the logo is located on the scanned document using bounding box surrounding this zone. Second, it is extracted from the input image taking into consideration the corresponding position of the detected zone in the segmented image. Logo localization is based on the segmented image and specific contextual rules defining shape and location of logos in administrative documents:

- Shape: *The logo is typically a small compact region in relative to the document resolution.*
- Location: *The representational property of the logo requires placing it in very distinct locations on the document, in an isolated region not embedded in the text zone as other graphic components, generally at the document border.*

We should say that these contextual rules differ from the *a-priori* information that depends on the processed documents. These rules are generic and valid for most administrative documents.

3 Experimental Study

In this paper, two datasets are involved to evaluate the performance and effectiveness of the proposed approach. A public document image collection with realistic scope and complexity has a significant effect on the document analysis. We should note that an evaluation using large public datasets is more realistic than that using self-collected datasets which are limited and capture fewer variations.

First, experimental study is realized on the Tobacco-800 dataset [12]; a public subset of the complex document image processing test collection constructed by Illinois Institute of Technology, assembled from 42 million pages of documents (in seven million multi-page TIFF images) released by tobacco companies under the master settlement agreement and originally hosted at University of California. It is a realistic database composed of 1290 documents images collected and scanned in binary format using a wide variety of equipment over time, for document analysis and retrieval. The resolutions of these documents vary significantly from 150 to 300 DPI and the dimensions of images range from 1200 by 1600 to 2500. Tobacco-800 has become the standard public dataset for research works on logos in scanned documents. Ground truth labels of the logos are created using XML documents and only consist on rectangular bounding boxes on logos. In this paper, the experimentation is performed using 432 logos across 35 classes detected from this dataset. Noting that this dataset is commonly used for its significant noise, as result of the binarization and the noise of scanning devices.

Second, experiments are realized on a color document image dataset called Scanned Pseudo-Official Data-Set (SPODS) [13] containing 1080 real world official documents characterized by the presence of logos, signatures, stamps and printed text. It consists of 32 logos, 32 stamps, 32 signatures, and 12 document types. The resolution of these documents is 300 DPI. Groundtruth of this public dataset is available by annotating the different elements in documents.

A close inspection of the experimentation indicates that the proposed system delivers good results on public data and detects almost all the logos from the processed datasets. Examples of correctly detected logos from Tobacco-800 dataset and SPODS are shown in Fig. 2 and Fig. 3 respectively. Here, we can notice the robustness of multi-fractal feature and its ability to find different logos.

In previous works like in [14], the noise was detected as part of the logo. In addition, small logos degraded the recall of the detection. Unlike these works, our system has the power to detect noisy and small logos (at bottom of Fig. 2). Thus, one of the strong points of the proposed approach is that the logo detection is not affected by low image quality with much noise.

The experiments carried out on another type of documents having an adhesion of the logo and the text that do not belong to the logo area in ground truth.

Fig. 2. Examples of correctly detected logos from the Tobacco-800 dataset

Fig. 3. Examples of correctly detected logos from the SPODS dataset

The difficulty of logo detection increases in the case where the gap between logo and text is narrow. We performed our method to show the impact of this adhesion on the logo detection. Unlike the other methods, our system managed to successfully extract logos from Tobacco-800 dataset, as depicted in Fig. 4. Thus, second strong point of our method is that it can deal with logos having close text zones. The experimentation on various scanned documents show the effective ability of the texture-based fuzzy system to locate logos.

The experiment results on SPODS dataset show that the performance of the proposed approach to detect logos is satisfactory. Another strong point of our method is that it can distinguish between logo and stamp having almost the same shape characteristics, as reported in Fig. 5. Thus, the qualitative analysis emphasizes the great results of detection process.

The quantitative analysis of images shows a visible improvement. The results indicate that the proposed system performs superbly in accuracy, precision, recall, and F-measure. By performing our system on the Tobacco-800 dataset, we gain 98.93% on precision, 97.89% on recall and 96.86% on accuracy in document logo detection, with high F-measure value (98.41%). The proposed approach is also successfully tested on SPODS. These results demonstrate the success of the method and highlight the importance of using a multi-fractal texture feature to correctly detect logos.

An effective comparative analysis is that where the environment of experimentation is the same for all the methods. In such case, the comparison of the performance of these methods is meaningful. In this context, we focused on the most recent and relevant methods, and only which are performed on the same Tobacco-800 dataset for automatic logo detection. The proposed logo detection yield the best result as compared to the other works, as illustrated in Table 1.

Fig. 4. Examples of qualitative evaluations on the Tobacco-800 dataset: predicted bounding boxes (in red), and ground truth bounding boxes (in green) (Color figure online)

Fig. 5. Examples of qualitative evaluations on the SPODS dataset: predicted bounding boxes (in red), and ground truth bounding boxes (in green) (Color figure online)

We notice that the evaluation measures are higher than those obtained by the other methods. The considerable improvement in accuracy proofs the potential of the proposed method for detecting logos. Other advantages of our approach over state-of-the-art methods is that it is scalable for the logo scale variation and it does not require *a-priori* knowledge on processed images nor a training step with a huge amount of training data to have a reliable detection. Compared to block-based methods, this technique is simpler and more effective using only one suitable texture feature. According to the obtained results, we can affirm that the proposed approach is more effective than many other state-of-the-art logo detection methods.

Table 1. Comparison of document logo detection performances

Document logo detection techniques	Dataset	Images	Accuracy (%)	Precision (%)
Veershetty and Hangarge [4]	Tobacco-800	200	–	87.80
Naganjaneyulu et al. [5]	Tobacco-800	1151	91.47	98.10
Li et al. [15]	Tobacco-800	1290	86.50	99.40
Pham et al. [16]	Tobacco-800	426	90.05	92.98
Dixit and Shirdhonkar [17]	Tobacco-800	1290	89.52	–
Jain and Doermann [18]	Tobacco-800	435	–	87.00
Le et al. [19]	Tobacco-800	1290	88.78	91.15
Kumar and Ranjith [20]	Tobacco-800	500	91.30	–
Wiggers et al. [21]	Tobacco-800	1290	–	72
Alaei et al. [22]	Tobacco-800	1290	91.50	75.25
Proposed approach	Tobacco-800	1290	96.86	98.93

4   Conclusion

In this work, we present an accurate, effective, and efficient approach for document logo detection. The basic logo detection problem is the difficulty of accurately processing documents with a high degree of degradation and being unable to process correctly on unusual fonts. The traditional approaches to logo detection require *a-priori* information on processed images. Our solution employs the textural information to distinguish between logo and non-logo regions. We designed a complete detection pipeline containing a texture feature extraction and a fuzzy classification. Experiments are carried out on SPODS and Tobacco-800 dataset. Our solution outperforms the methods in the state-of-the-art. However, there exist many unsolved challenges such as the generalization of this method to do logo recognition and other logical document components. We will attempt to address these issues in our future work.

References

1. Soon, F.C., Khaw, H.Y., Chuah, J.H., Kanesan, J.: Hyper-parameters optimisation of deep CNN architecture for vehicle logo recognition. IET Intel. Transp. Syst. **12**, 939–946 (2018)
2. Tuzko, A., Herrmann, C., Manger, D., Beyerer, J.: Open set logo detection and retrieval. In: Proceedings of the 13th International Joint Conference on Computer Vision, Imaging and Computer Graphics Theory and Applications, pp. 284–292. SciTePress, Madeira (2018)
3. Alaei, A., Roy, P.P., Pal, U.: Logo and seal based administrative document image retrieval: a survey. Comput. Sci. Rev. **22**, 47–63 (2016)
4. Veershetty, C., Hangarge, M.: Logo retrieval and document classification based on LBP features. In: Nagabhushan, P., Guru, D.S., Shekar, B.H., Kumar, Y.H.S. (eds.)

Data Analytics and Learning. LNNS, vol. 43, pp. 131–141. Springer, Singapore (2019). https://doi.org/10.1007/978-981-13-2514-4_12

5. Naganjaneyulu, G.V.S.S.K.R., Sai Krishna, C., Narasimhadhan, A.V.: A novel method for logo detection based on curvelet transform using GLCM features. In: Chaudhuri, B.B., Kankanhalli, M.S., Raman, B. (eds.) Proceedings of 2nd International Conference on Computer Vision & Image Processing. AISC, vol. 704, pp. 1–12. Springer, Singapore (2018). https://doi.org/10.1007/978-981-10-7898-9_1

6. Zaaboub, W., Dhiaf, Z.B.: Approach of texture signature determination–application to forest cover classification of high resolution satellite image. In: 6th International Conference of Soft Computing and Pattern Recognition (SoCPaR), pp. 325–330. IEEE (2014)

7. Krim, J., Indekeu, J.: Roughness exponents: a paradox resolved. Phys. Rev. E **48**, 1576 (1993)

8. Turiel, A., Parga, N.: The multifractal structure of contrast changes in natural images: from sharp edges to textures. Neural Comput. **12**, 763–793 (2000)

9. Gu, Q., Li, Z., Han, J.: Generalized fisher score for feature selection. arXiv preprint arXiv:1202.3725 (2012)

10. Miyamoto, S., Ichihashi, H., Honda, K., Ichihashi, H.: Algorithms for Fuzzy Clustering. Springer, Heidelberg (2008). https://doi.org/10.1007/978-3-540-78737-2

11. Soille, P.: Morphological Image Analysis: Principles and Applications. Springer, Heidelberg (2013). https://doi.org/10.1007/978-3-662-05088-0

12. Lewis, D., Agam, G., Argamon, S., Frieder, O., Grossman, D., Heard, J.: Building a test collection for complex document information processing. In: Proceedings of the 29th Annual International ACM SIGIR Conference on Research and Development in Information Retrieval, pp. 665–666. ACM (2006)

13. Nandedkar, A.V., Mukherjee, J., Sural, S.: SPODS: a dataset of color-official documents and detection of logo, stamp, and signature. ICVGIP 2016. LNCS, vol. 10481, pp. 219–230. Springer, Cham (2017). https://doi.org/10.1007/978-3-319-68124-5_19

14. Alaei, A., Delalandre, M.: A complete logo detection/recognition system for document images. In: 2014 11th IAPR International Workshop on Document Analysis Systems (DAS), pp. 324–328. IEEE (2014)

15. Li, Z., Schulte-Austum, M., Neschen, M.: Fast logo detection and recognition in document images. In: 2010 International Conference on Pattern Recognition, pp. 2716–2719. IEEE (2010)

16. Pham, T.A., Delalandre, M., Barrat, S.: A contour-based method for logo detection. In: 2011 International Conference on Document Analysis and Recognition, pp. 718–722. IEEE (2011)

17. Dixit, U.D., Shirdhonkar, M., Automatic logo detection and extraction using singular value decomposition. In: International Conference on Communication and Signal Processing (ICCSP), pp. 0787–0790. IEEE (2016)

18. Jain, R., Doermann, D.: Logo retrieval in document images. In: 2012 10th IAPR International Workshop on Document Analysis Systems (DAS), pp. 135–139. IEEE (2012)

19. Le, V.P., Nayef, N., Visani, M., Ogier, J.M., De Tran, C.: Document retrieval based on logo spotting using key-point matching. In: 2014 22nd International Conference on Pattern Recognition (ICPR), pp. 3056–3061. IEEE (2014)

20. Sharath Kumar, Y.H., Ranjith, K.C.: An approach for logo detection and retrieval in documents. In: Santosh, K.C., Hangarge, M., Bevilacqua, V., Negi, A. (eds.) RTIP2R 2016. CCIS, vol. 709, pp. 49–58. Springer, Singapore (2017). https://doi.org/10.1007/978-981-10-4859-3_5

21. Wiggers, K.L., Britto, A.S., Heutte, L., Koerich, A.L., Oliveira, L.E.S.: Document image retrieval using deep features. In: International Joint Conference on Neural Networks (IJCNN), pp. 1–8. IEEE (2018)
22. Alaei, A., Delalandre, M., Girard, N.: Logo detection using painting based representation and probability features. In: 12th International Conference on Document Analysis and Recognition (ICDAR), pp. 1235–1239. IEEE (2013)

Mathematical Imaging and Signal Processing

Discrete p-bilaplacian Operators on Graphs

Imad El Bouchairi[✉], Abderrahim El Moataz[✉], and Jalal Fadili[✉]

Normandie Univ, ENSICAEN, UNICAEN, CNRS, GREYC, Caen, France
imad.elbouchairi@gmail.com, abderrahim.elmoataz@unicaen.fr,
Jalal.Fadili@greyc.ensicaen.fr

Abstract. In this paper, we first introduce a new family of operators on weighted graphs called p-bilaplacian operators, which are the analogue on graphs of the continuous p-bilaplacian operators. We then turn to study regularized variational and boundary value problems associated to these operators. For instance, we study their well-posedness (existence and uniqueness). We also develop proximal splitting algorithms to solve these problems. We finally report numerical experiments to support our findings.

Keywords: p-bilaplacian · Weighted graphs · Regularization · Boundary value problems · Proximal splitting

1 Introduction

Regularized variational problems and partial differential equations (PDEs) play an important role in mathematical modeling throughout applied and natural sciences. For instance, many variational problems and PDEs have been studied to model and solve important problems in a variety of areas such as, e.g., in physics, economy, data processing, computer vision. In particular they have been very successful in image and signal processing to solve a wide spectrum of applications such as isotropic and anisotropic filtering, inpainting or segmentation.

In many real-world problems, such as in machine learning and mathematical image processing, the data is discrete, and graphs constitute a natural structure suited to their representation. Each vertex of the graph corresponds to a datum, and the edges encode the pairwise relationships or similarities among the data. For the particular case of images, pixels (represented by nodes) have a specific organization expressed by their spatial connectivity. Therefore, a typical graph used to represent images is a grid graph. For the case of unorganized data such as point clouds, a graph can also be built by modeling neighborhood relationships between the data elements. For these reasons, there has been recently a wave of interest in adapting and solving nonlocal variational problems and PDEs on data which is represented by arbitrary graphs and networks. Using this framework, problems are directly expressed in a discrete setting where an appropriate discrete differential calculus can be proposed; see e.g., [4,5] and references therein.

© Springer Nature Switzerland AG 2020
A. El Moataz et al. (Eds.): ICISP 2020, LNCS 12119, pp. 339–347, 2020.
https://doi.org/10.1007/978-3-030-51935-3_36

This mimetic approach consists of replacing continuous differential operators, e.g., gradient or divergence, by reasonable discrete analogues, which makes it possible to transfer many important tools and results from the continuous setting.

Contributions. In this work, we introduce a novel class of p-bilaplacian operators on weighted graphs, which can be seen as proper discretizations on graphs of the classical p-bilaplacian operators [9]. Building upon this definition, we study a corresponding regularized variational problem as well as a boundary value problem. The latter naturally gives rise to p-biharmonic functions on graphs and equivalent definitions of p-biharmonicity [8]. For these two problems, we start by establishing their well-posedness (existence and uniqueness). We then turn to developing proximal splitting algorithms to solve them, appealing to sophisticated tools from non-smooth optimization. Numerical results are reported to support the viability of our approach.

2   Notations and Preliminary Results

Throughout this paper, we assume that $G = (V, E, \omega)$ is a finite connected undirected weighted graph without loops and parallel edges, where V is the set of vertices, E is the set of edges, and the symmetric function $\omega : V \times V \to [0, 1]$ is the weight function. We denote by $(x, y) \in E$ the edge that connects the vertices x and y, and we write $x \sim y$ for two adjacent vertices. For two vertices x, $y \in V$ with $x \not\sim y$ we set $\omega(x, y) = \omega(y, x) = 0$ and thus the set of edges E can be characterized by the support of the weight function ω, i.e., $E = \{(x, y) | \omega(x, y) > 0\}$.

Let $\mathcal{H}(V) \stackrel{\text{def}}{=} \{u : x \in E \mapsto u(x) \in \mathbb{R}\}$ be the vector space of real-valued functions on the vertices of the graph. For a function $u \in \mathcal{H}(V)$ the $\ell^p(V)$-norm of u is given by

$$\|u\|_p = \left(\sum_{x \in V} |u(x)|^p \right)^{\frac{1}{p}}, \quad 1 \leqslant p < \infty, \quad \text{and} \quad \|u\|_\infty = \max_{x \in V} |u(x)|.$$

We define in a similar way $\mathcal{H}(E)$ as the vector space of all real-valued functions on the edges of the graph.

Let $u \in \mathcal{H}(V)$ and $x, y \in V$. The (nonlocal) gradient operator is defined as

$$\nabla_\omega u : \mathcal{H}(V) \to \mathcal{H}(E)$$

$$u \mapsto U, \quad U(x, y) = \sqrt{\omega(x, y)}(u(y) - u(x)), \forall (x, y) \in V.$$

This is a linear antisymmetric operator whose adjoint is the (nonlocal) weighted divergence operator denoted div_ω. It is easy to show that

$$\text{div}_\omega : \mathcal{H}(E) \to \mathcal{H}(V)$$

$$U \mapsto u, \quad u(x) = \sum_{y \sim x} \sqrt{\omega(x, y)}(U(y, x) - U(x, y)), \forall x \in E.$$

For $1 < p < \infty$, the anisotropic graph p-Laplacian operator $\Delta_{\omega,p} : \mathcal{H}(V) \to \mathcal{H}(V)$ is thus defined by

$$\Delta_{\omega,p}u(x) \stackrel{\text{def}}{=} \text{div}_\omega(|\nabla_\omega u|^{p-2}\nabla_\omega u)(x)$$
$$= 2\sum_{y\sim x}(\omega(x,y))^{\frac{p}{2}}|u(y) - u(x)|^{p-2}(u(y) - u(x)), \forall x \in V.$$

Unless stated otherwise, in the rest of the paper, we assume $p \in]1, +\infty[$.

3 p-biharmonic Functions on Graphs

We define p-biharmonic functions on graphs inspired by the way p-harmonic functions were introduced in [8] for networks. Let's consider the following functional

$$\mathcal{F}_d(u; p) \stackrel{\text{def}}{=} \frac{1}{p}\|\Delta_{\omega,2}u\|_p^p. \tag{1}$$

Observe that $\Delta_{\omega,2}$ is the standard Laplacian on a graphs, which is a self-adjoint operator.

Definition 1. *We define the p-bilaplacian operator for a function $u \in \mathcal{H}(V)$ by*

$$\Delta_p^2 u(x) \stackrel{\text{def}}{=} \Delta_{\omega,2}(|\Delta_{\omega,2}u|^{p-2}\Delta_{\omega,2}u)(x), \quad x \in V.$$

Definition 2. *Let $A \subset V$. We say that a function u is p-biharmonic on A if it is a minimiser of the functional $\mathcal{F}_d(\cdot; p)$ among functions in V with the same values in $A^c = V \setminus A$, that is, if*

$$\mathcal{F}_d(u; p) \leqslant \mathcal{F}_d(v; p)$$

for every function $v \in \mathcal{H}(V)$, with $u = v$ in A^c.

Inspired by [8], existence and uniqueness of p-biharmonic functions can be established using standard arguments.

Proposition 1. *Let A subset of V and $u \in \mathcal{H}(V)$. The following assertions are equivalent:*

(i) the function u is p-biharmonic on A.
(ii) the function u satisfies

$$\sum_{x\in V}|\Delta_{\omega,2}(u)(x)|^{p-2}\Delta_{\omega,2}(u)(x)\Delta_{\omega,2}w(x) = 0, \quad x \in A, \tag{2}$$

for every function $w \in \mathcal{H}(V)$, with $w = 0$ in A^c.
(iii) the function u solves

$$\Delta_p^2 u(x) = 0, \quad \text{for all } x \in A.$$

4 p-bilaplacian Dirichlet Problem on Graphs

Consider the following boundary value (Dirichlet) problem

$$\begin{cases} \Delta_p^2 u = 0, & on\ A \\ u = g, & on\ A^c, \end{cases} \tag{3}$$

where $g \in \mathcal{H}(V)$, $p \in]1, +\infty[$, $A \subset V$ and $A^c = V \setminus A$. Observe that since the graph G is connected, there always exists a path connecting any pair vertices in $A \times A^c$. Our goal now is to establish well-posedness of (3). This will be derived using Dirichlet's variational principe (hence the subscript d in $\mathcal{F}_d$), which, in view of Proposition 1, amounts to equivalently studying the minimization problem

$$\min\{\mathcal{F}_d(u; p) : u \in \mathcal{H}_g(V)\}, \tag{4}$$

where $\mathcal{H}_g(V) = \{u \in \mathcal{H}(V) : u = g\ on\ A^c\}$ is the subspace of the functions with a zero "trace".

Theorem 1. *The problem* (3) *has a unique solution.*

Proof. Let $\iota_{\mathcal{H}_g(V)}$ be the indicator function of $\mathcal{H}_g(V)$, i.e. it is 0 on $\mathcal{H}_g(V)$ and $+\infty$ otherwise. By the Poincaré-type inequality established in Lemma 1, we get that $\mathcal{F}_d(\cdot; p) + \iota_{\mathcal{H}_g(V)}$ is coercive. Since this objective is lower semicontinuous (lsc) by closedness of $\mathcal{H}_g(V)$ and continuity of $\mathcal{F}_d(\cdot; p)$, (4) has a minimizer. This together with strict convexity of $\mathcal{F}_d(\cdot; p)$ on $\mathcal{H}_g(V)$ (see Lemma 2) then entails uniqueness.

Lemma 1. *There is* $\lambda = \lambda(\omega, V, A^c) > 0$ *such that*

$$\lambda \sum_{x \in V \setminus A^c} |u(x)|^2 \leqslant \sum_{x \in V} \sum_{y \sim x} \omega(x, y)|u(y) - u(x)|^2 + \sum_{x \in A^c} |g(x)|^p, \tag{5}$$

for all $u \in \mathcal{H}_g(V)$. *Thus* $\mathcal{F}_d(\cdot; p)$ *is coercive on* $\mathcal{H}_g(V)$.

Proof. Set

$$S_0 = A^c;$$
$$S_1 = \{x \in V \setminus S_0 : \exists y \in S_0;\ y \sim x\},$$
$$S_{j+1} = \{x \in V \setminus (\cup_{k=0}^j S_k) : \exists y \in S_j;\ y \sim x\}, \quad j = 1, 2, \dots.$$

Since the graph G is connected, there is $l \in \mathbb{N}$ such that $\{S_j\}_{j=0}^l$ forms a partition of V. By Jensen's inequality, we have

$$\sum_{x \in V} \sum_{y \sim x} \omega(x, y)|u(y) - u(x)|^2 \geqslant \sum_{x \in S_j} \sum_{y \in S_{j-1}} \omega(x, y)|u(y) - u(x)|^2,$$

$$\geqslant \alpha \sum_{x \in S_j} |u(x)|^2 - \beta \sum_{y \in S_{j-1}} |u(y)|^2,$$

$j = 1, \cdots, l$, where $\alpha = \frac{1}{2}\min\{\omega(x,y) : (x,y) \in E\}$ and $\beta = \sum_{x,y \in V} \omega(x,y)$. Since $u = g$ on $S_0 = A^c$ and $\{S_j\}_{j=0}^l$ forms a partition of V, it is easy to see that there exists $\hat{\lambda} = \hat{\lambda}(\omega, V, A^c) > 0$ such that

$$\sum_{x \in V \setminus A^c} |u(x)|^2 \leqslant \hat{\lambda} \sum_{x \in V} \sum_{y \sim x} \omega(x,y)|u(y) - u(x)|^2 + \hat{\lambda} \sum_{x \in A^c} |g(x)|^2.$$

We arrive at the coercivity result by taking $\lambda = \hat{\lambda}^{-1}$.

Let $C_g = 2\sum_{x \in A^c} |g(x)|^2$. We then have from Hölder and Young inequalities

$$\lambda\|u\|_2^2 \leqslant \sum_{x \in V} \sum_{y \sim x} \omega(x,y)|u(y) - u(x)|^2 + C_g$$

$$= -\sum_{x \in V} u(x)\Delta_{\omega,2}u(x) + C_g$$

$$\leqslant \frac{\varepsilon}{2}\|u\|_q^2 + \frac{p}{2\varepsilon}\mathcal{F}_d(u;p)^{2/p} + C_g$$

where $1/p + 1/q = 1$. Since the norms are equivalent in any finite-dimensional vector space, there exists $C(n) > 0$ with $n = |V|$, such that

$$\lambda\|u\|_2^2 \leqslant C(n)^2\frac{\varepsilon}{2}\|u\|_2^2 + \frac{p}{2\varepsilon}\mathcal{F}_d(u;p)^{2/p} + C_g.$$

Choosing $\varepsilon = 2\rho\lambda/C(n)^2$, $\rho \in]0,1[$, we get

$$(1 - \rho)\lambda\|u\|_2^2 \leqslant \frac{pC(n)^2}{4\rho\lambda}\mathcal{F}_d(u;p)^{2/p} + C_g,$$

whence coercivity of $\mathcal{F}_d(\cdot;p)$ follows immediately.

Lemma 2. *The functional $\mathcal{F}_d(\cdot;p)$ is strictly convex on $\mathcal{H}_g(V)$.*

Proof. Assume that $\mathcal{F}_d(\cdot;p)$ is not strictly convex on $\mathcal{H}_g(V)$. Then there exist $u, v \in \mathcal{H}_g(V)$ with $u \neq v$ such that $\tau\mathcal{F}_d(u;p) + (1-\tau)\mathcal{F}_d(v;p) = \mathcal{F}_d(\tau u + (1-\tau)v;p)$ for all $\tau \in]0,1[$. But since the function $t \mapsto t^p$ is strictly convex on $\mathbb{R}^+$ for $p \in]1, +\infty[$, this equality entails that $\Delta_{\omega,2}u = \Delta_{\omega,2}v$ on V, hence on A. Clearly $w = u - v$ satisfies

$$\begin{cases} \Delta_{\omega,2}w = 0, & on\ A \\ w = 0, & on\ A^c. \end{cases}$$

But we know from [8, Theorem 3.11 and Corollary 3.16] that $w = 0$ on V, i.e., $u = v$ on V, leading to a contradiction.

5 p-bilaplacian Variational Problem on Graphs

In this section, we consider the following minimization problem, which is valid for any $p \in [1, +\infty]^1$,

$$\min_{u \in \mathcal{H}(V)} \left\{ E(u;p) \stackrel{def}{=} \frac{1}{2}\|f - Au\|_2^2 + \lambda\mathcal{F}_d(u;p) \right\}, \tag{6}$$

---

[1] Obviously $\lim_{p \to +\infty} \frac{1}{p}\|\cdot\|_p^p = \iota_{\|u\|_\infty \leqslant 1}$.

where $A : \mathcal{H}(V) \to \mathcal{H}(V)$ is a linear operator, $f \in \mathcal{H}(V)$, $\lambda > 0$ is the regularization parameter, and $\mathcal{F}_d(\cdot; p)$ is given by (1). Problems of the form (6) can be of great interest for graph-based regularization in machine learning and inverse problems in imaging; see [7] and references therein. Problem (6) is well-posed under standard assumptions.

Theorem 2. *The set of minimizers of $E(\cdot; p)$ is non-empty and compact if and only if $\ker(A) \cap \ker(\Delta_{w,2}) = \{0\}$. If, moreover, either A is injective or $p \in]1, +\infty[$, then $E(\cdot; p)$ has a unique minimizer.*

Proof. For any proper lsc convex function f, recall its recession function from [11, Chapter 2], denoted f_∞. We have from the calculus rules in [11, Chapter 2] that

$$E_\infty(d; p) = \lambda \left(\tfrac{1}{p} \| \cdot \|_p^p \right)_\infty (\Delta_{w,2}d) + \frac{1}{2} \left(\| f - \cdot \|_2^2 \right)_\infty (Ad).$$

Since $\frac{1}{p} \| \cdot \|_p^p$ and $\| f - \cdot \|_2^2$ are non-negative and coercive, we have from [11, Proposition 3.1.2] that their recession functions are positive for any non-zero argument. Equivalently,

$$E_\infty(d; p) > 0, \quad \forall d \notin \ker(A) \cap \ker(\Delta_{w,2}).$$

Thus $E_\infty(d; p) > 0$ for all $d \neq 0$ if and only if $\ker(A) \cap \ker(\Delta_{w,2}) = \{0\}$. Equivalence with the existence and compactness assertion follows from [11, Proposition 3.1.3].

Let's turn to uniqueness. When A is injective, the claim follows from strict (in fact strong convexity) of the data fidelity term. Suppose now that $p \in]1, +\infty[$. By strict convexity of $\frac{1}{p} \| \cdot \|_p^p$ and $\| f - \cdot \|_2^2$, a standard contradiction argument shows that for any pair of minimizers $u^\star$ and $v^\star$, we have $u^\star - v^\star \in \ker(A) \cap \ker(\Delta_{w,2})$. This yields the uniqueness claim under the stated assumption.

6 Algorithms and Numerical Results

To solve both (3) and (6), we adopt a primal-dual proximal splitting (PDS) framework with an appropriate splitting of the functions and linear operators.

6.1 A PDS for the Boundary Value Problem (3)

Problem (3) is equivalent to (4). The latter takes the form

$$\min_{u \in \mathcal{H}_V} F(\Delta_{w,2}u) + G(u), \quad \text{where} \quad F(u) = \frac{1}{p} \| u \|_p^p, \quad G(u) = \iota_{\mathcal{H}_g(V)}(u). \tag{7}$$

The latter can be solved with the following PDS iterative scheme [3], which reads in this case

$$
\begin{aligned}
u^{k+1} &= \mathrm{proj}_{\mathcal{H}_g(V)}(u^k - \tau \Delta_{w,2}v^k) \\
v^{k+1} &= \mathrm{prox}_{\frac{\sigma}{q} \| \cdot \|_q^q}(v^k + \sigma \Delta_{w,2}(2u^{k+1} - u^k)),
\end{aligned}
\tag{8}
$$

where $\tau, \sigma > 0$, $\text{proj}_{\mathcal{H}_g(V)}$ is the orthogonal projector on the subspace $\mathcal{H}_g(V)$ (which has a trivial closed form), $1/p + 1/q = 1$, and $\text{prox}_{\frac{\sigma}{q}\|\cdot\|_q^q}$ is the proximal mapping of the proper lsc convex function $\frac{\sigma}{q}\|\cdot\|_q^q$. The latter can be computed easily, see [7] for details. Combining [3, Theorem 1], Proposition 1 and Theorem 1, the convergence guarantees of (8) are summarized in the following proposition.

Proposition 2. *If $\tau\sigma\|\Delta_{\omega,2}\|^2 < 1$, then the sequence $(u^k, v^k)_{k\in\mathbb{N}}$ provided by (8) converges to $(u^\star, v^\star)$, where $u^\star$ is a solution to (3), which is unique if $p \in]1, +\infty[$.*

6.2 A PDS for the Variational Problem (6)

For simplicity and space limitation, we restrict ourselves here to the case where A is the identity. In this case, inspired by the work in [6], we use the (accelerated) FISTA iterative scheme [2,10] to solve the Fenchel-Rockafellar dual problem of (6), and recover the primal solution by standard extremality relationships. Our scheme reads in this case

$$
\begin{aligned}
y^k &= v^k + \frac{k-1}{k+b}(v^k - v^{k-1})\\
v^{k+1} &= \text{prox}_{\gamma\frac{\lambda}{q}\|\cdot/\lambda\|_q^q}\left(y^k + \gamma\Delta_{\omega,2}(f - \Delta_{\omega,2}y^k)\right)\\
u^{k+1} &= f - \Delta_{\omega,2}v^{k+1}
\end{aligned}
\tag{9}
$$

where $\gamma \in]0, \|\Delta_{\omega,2}\|^{-2}]$, $b > 2$.

Combining Theorem 2, [6, Theorem 2], [1, Theorem 1.1], the scheme (9) has the following convergence guarantees.

Proposition 3. *The sequence $(u^k)_{k\in\mathbb{N}}$ converges to $u^\star$, the unique minimizer of (6), at the rate $\|u^k - u^\star\|_2 = o(1/k)$.*

6.3 Numerical Results

We apply the scheme (9) to solve (6) in order to denoise a function f defined on a 2-D point cloud. We apply (3) in a semisupervised classification problem which amounts to finding the missing labels of a label function defined on a 2-D point cloud. The nodes of the graph are the points in the 2-D cloud and $u(x)$ is the value at a point/vertex x. We choose the nearest neighbour graph with the standard weighting kernel $\exp(-|\mathbf{x} - \mathbf{y}|)$ when $|\mathbf{x} - \mathbf{y}| \leqslant \delta$ and 0 otherwise, where $\mathbf{x}$ and $\mathbf{y}$ are the 2-D spatial coordinates of the points in the cloud. The original point cloud used in our numerical experiments consists of $N = 2500$ points that are not on a regular grid. For the variational problem, the noisy observation is generated by adding a white Gaussian noise of standard deviation 0.5 to the original data, see Fig. 1(a). For the Dirichlet problem, the initial label function takes the values of the original data on a set of points/vertices where this set corresponds to the boundary data, it is chosen randomly and is equal to $N/4$ of the original points/vertices, see Fig. 1(b).

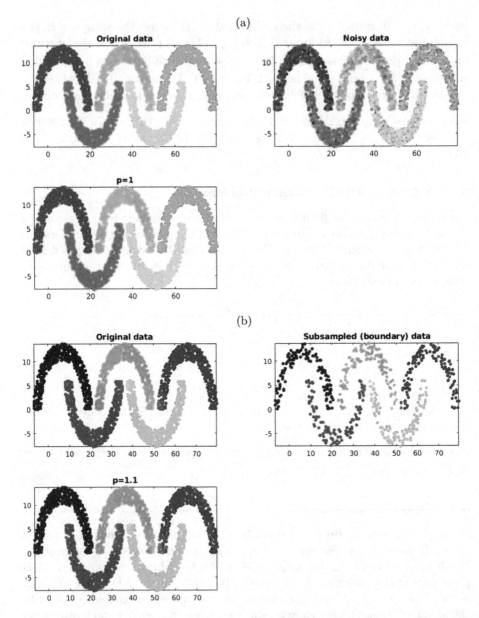

Fig. 1. (a): results for denoising with $p = 1$. (b): results for a semisupervised classification problem with $p = 1.1$. For each setting, we show the original function on the point cloud, its observed version, and the result provided by each of our algorithms.

References

1. Attouch, H., Peypouquet, J.: The rate of convergence of Nesterov's accelerated forward-backward method is actually faster than $1/k^2$. SIAM J. Optim. **26**(3), 1824–1834 (2016)

2. Chambolle, A., Dossal, C.: On the convergence of the iterates of the "fast iterative shrinkage/thresholding algorithm". J. Optim. Theory Appl. **166**(3), 968–982 (2015). https://doi.org/10.1007/s10957-015-0746-4
3. Chambolle, A., Pock, T.: A first-order primal-dual algorithm for convex problems with applications to imaging. J. Math. Imaging Vis. **40**(1), 120–145 (2011). https://doi.org/10.1007/s10851-010-0251-1
4. Elmoataz, A., Lézoray, O., Bougleux, S.: Nonlocal discrete regularization on weighted graphs: a framework for image and manifold processing. IEEE Trans. Image Process. **17**(7), 1047–1060 (2008)
5. Elmoataz, A., Toutain, M., Tenbrinck, D.: On the p-Laplacian and ∞-Laplacian on graphs with applications in image and data processing. SIAM J. Imaging Sci. **8**(4), 2412–2451 (2015)
6. Fadili, M.J., Peyré, G.: Total variation projection with first order schemes. IEEE Trans. Image Process. **20**(3), 657–669 (2010)
7. Hafiene, Y., Fadili, M.J., Elmoataz, A.: Continuum limits of nonlocal p-Laplacian variational problems on graphs. SIAM J. Imaging Sci. **12**(4), 1772–1807 (2019)
8. Holopainen, I., Soardi, P.M.: p-Harmonic functions on graphs and manifolds. Manuscripta Mathematica **94**(1), 95–110 (1997)
9. Katzourakis, N., Pryer, T.: On the numerical approximation of p-biharmonic and ∞-biharmonic functions. Num. Methods Partial Differ. Equ. **35**(1), 155–180 (2019)
10. Nesterov, Y.: A method for solving the convex programming problem with convergence rate $O(1/k^2)$. Dokl. Akad. Nauk SSSR **269**(3), 543–547 (1983)
11. Auslender, A., Teboulle, M.: Asymptotic Cones and Functions in Optimization and Variational Inequalities. Springer, New York (2003). https://doi.org/10.1007/b97594

Image Watermarking Based on Fourier-Mellin Transform

Khadija Gourrame[1,2(✉)], Hassan Douzi[1], Rachid Harba[2],
Riad Rabia[1], Frederic Ros[2], and Mehamed ElHajji[1]

[1] IRF-SIC Laboratory, Ibn Zohr University, Agadir, Morocco
khadija.gourrame@edu.uiz.ac.ma
[2] PRISME Laboratory, Orleans University, Orleans, France

Abstract. Geometric attacks are still challenging issues in image watermarking. In this paper, the robustness of different insertion position and shape of the watermark are evaluated in watermarking scheme based on Fourier-Mellin transform. We propose diagonal, rectangular, and circular insertion of the mark. The robustness of these techniques against geometric deformations such as rotation, scaling and translation (RST) is evaluated. Results show that the circular insertion performs better for translation and scaling attacks, while the diagonal insertion is better for rotations and RST attacks. The last point makes the diagonal insertion to be preferred in industrial applications since the combination of RST attacks often occurs in many applications such as printing the image on a physical support, and scanning it (print-scan attack).

Keywords: Image watermarking · Synchronization · Fourier-Mellin transform · RST

1 Introduction

Robustness of image watermarking techniques is an active research area for many years. Yet, there is no existing watermarking method that achieves a complete robustness to all common watermarking attacks including geometric distortions. In general, image watermarking techniques are implemented in spatial domain or transformed domains [1]. The most used transforms are discrete cosine transform (DCT), discrete wavelet transform (DWT), and discrete Fourier transform (DFT). Although DCT and DWT are very robust to common signal processing attacks such as compression and signal filtering, they are weak to resist geometric distortions [2]. However, DFT has the advantage of being more resistant to geometric attacks [3, 4].

Geometric distortions are one of the major issues in industrial watermarking applications where the so-called print-scan occurs (print-scan means that the image is printed on a physical support, and then scanned). RST (rotation, scaling and translation) deformations are typically produced by print-scan process. Geometric attacks cause a loss of synchronization between the extracted watermark and the embedded one. Thus, many schemes are proposed in the literature to solve the synchronization issue.

© Springer Nature Switzerland AG 2020
A. El Moataz et al. (Eds.): ICISP 2020, LNCS 12119, pp. 348–356, 2020.
https://doi.org/10.1007/978-3-030-51935-3_37

In our team, Riad et al. have developed a robust watermarking method in the Fourier domain for print-scan process using a Fourier method for identity (ID) images [4]. The object was to increase the security of official documents as for example ID cards or passports. In that case, the print-scan attack produces rotations and translations of small amplitudes. To be rotation invariant, the watermark vector is embedded in the magnitude of the Fourier transform of the original image, as a circle of radius r around the center of the transformed image in the medium frequency domain, translation problems are being naturally handled by Fourier capabilities. Regarding scaling deformations, in [3, 4], they proposed to resize the image to its original size before the watermark detection. However, this Fourier method is not robust to scale deformations in case if the size of original image is unknown. To overcome the scaling problem, Fourier log-polar approach or Fourier-Mellin transform [5, 6] is proposed as an RST invariant method. Converting the cartesian coordinates of Fourier magnitude into log-polar coordinates, transforms the rotation and scaling operations into translation ones, so that the whole process becomes invariant to RST. Many watermarking techniques are proposed in literature using FMT. Many research works try to reduce the steps in Fourier-Mellin transform as in [6]. Log-polar transform (LPT) is used as a partial step of FMT technique for proposing robust image watermarking [7], however this proposition is designed for non-blind watermarking method. Fourier-Mellin moments are used as image features for watermark registration and extraction for copyright authentication [8]. Quaternion Discrete Fourier transform (QDFT) combined with LPM represents new version of FMT in recent years [9], this method is robust against signal processing attacks for small image blocks with QDFT. However, for geometric attacks, the robustness was secured with the help of a template, which limits and reduces the range of RST attacks. After the O. Ruanaidh and T. Pun [5] work, the followed proposed method in literature try to avoid to go through all the steps to get an RST invariant domain, which it compromise on the robustness of the method in the first place. Early studies show that considering shape, position and other parameters related to the watermark are improving on the robustness of the method [4, 10].

In this paper, different insertion watermarking schemes based on Fourier-Mellin transform (FMT) are studied. These techniques are tested against geometric deformation such as translation, rotation and scaling (RST). The methods associate watermark insertion in different geometrical shapes and positions in invariant RST domains based on Fourier-Mellin transform. The presented techniques are circular insertion [4], diagonal insertion [10] and rectangular insertion. The robustness of the presented methods are compared and discussed.

The paper is organized as follows. Section 2 presents the theoretical background. Section 4 describes the watermarking method based on FMT and it shows results of the robustness test of the three tested methods under RTS attacks, Sect. 4 gives a conclusion.

2 Fourier-Mellin Transform - Mathematical Background

In this section, geometric properties of the Fourier transform are presented. The properties of the Fourier-Mellin transform are also explained which show that it is naturally resistant to RST.

2.1 Fourier Transform

The Fourier transform $F(u, v)$ of an image $f(x, y)$ is:

$$f(x, y) \overset{FT}{\leftrightarrow} \left(F(u, v) = \int_{-\infty}^{\infty} \int_{-\infty}^{\infty} f(x, y) e^{-j2\pi(ux + vy)} dx dy \right). \tag{1}$$

Ft is invariant to translations in term of magnitude part:

$$f(x - a, y - b) \overset{FT}{\leftrightarrow} e^{-j2\pi(au + bv)} F(u, v), \tag{2}$$

where a and b are translation parameters. Rotation of an image with angle θ in spatial domain is converted into a rotation with the same angle in Fourier domain:

$$\begin{aligned} f(x \cos(\theta) - y \sin(\theta), x \sin(\theta) + y \cos(\theta)) & \overset{FT}{\leftrightarrow} \\ F(u \cos(\theta) - v \sin(\theta), u \sin(\theta) + v \cos(\theta)). & \end{aligned} \tag{3}$$

Scaling in spatial domain with a factor a is converted into a scaling with the inverse factor in Fourier domain:

$$f(ax, by) \overset{FT}{\leftrightarrow} \frac{1}{|ab|} F\left(\frac{u}{a}, \frac{v}{b}\right). \tag{4}$$

Properties in (2) and (3) demonstrate that the magnitude of the Fourier transform naturally resists to rotations and translations, but not to scaling changes. It is therefore naturally adapted to RT attacks.

2.2 Fourier-Mellin Transform

Fourier-Mellin transform of image $f(r, \theta)$(in polar coordinates) is:

$$FM(u, v) = \frac{1}{2\pi} \int_0^\infty \int_0^{2\pi} f(r, \theta) r^{-jv-1} e^{-ju\theta} dr d\theta. \tag{5}$$

With a set of integration by substitution, FMT can be expressed as FT of an image in log polar coordinates [5]:

$$FM(u, v) = \frac{1}{2\pi} \int_0^{2\pi} \int_{-\infty}^{\infty} f'(\rho, \theta) e^{-j(v\rho + u\theta)} d\rho d\theta = F\left\{ f'(\rho, \theta) \right\}. \tag{6}$$

The geometric properties of the Fourier-Mellin transform are related to the geometric properties of both the log-polar transform and the Fourier transform:

Log-polar transform converts rotation and scaling into translation as expressed in the following equations:

$$\left(x',y'\right) \leftrightarrow (r, \theta + \varphi) \leftrightarrow (\rho, \theta + \varphi), \tag{7}$$

$$(ax, by) \leftrightarrow (ar, \theta) \leftrightarrow (\log(ar), \theta) = (\rho + \log(a), \theta), \tag{8}$$

where (x', y') are cartesian coordinates after rotation with φ. (r, θ) and (ρ, θ) are the polar and log-polar coordinates respectively.

Fourier transform is invariant to translation therefore Fourier transform of an image in log-polar coordinates is invariant to rotation and scaling. As a result, Fourier-Mellin transform is invariant to rotation and scaling.

3 Fourier-Mellin Method's Comparison

Fourier-Mellin domain for watermarking is described first by O. Ruanaidh and T. Pun [5]. Figure 1. illustrates the process of obtaining the rotation, scaling and translation (RST) transformation invariant from a digital image. This is explained in the following. Imagine that the input image at the bottom of Fig. 1 is subject to a RST attack. As seen in Fig. 1, a series of 3 transforms is going to be applied to this image. For the first FT, the magnitude of the FT of the image is invariant to translation, while a rotation and a scale changes are still present. When the LPT is applied (the first part of the FM transform), the remaining rotation and scale are transformed into translations. All these properties were recalled in Sect. 2 of this document. For the last FT (the second part of the FMT), the magnitude of the FT is now invariant to translations (known from the first FT), but it is also invariant to rotation and scale. It results in a RST invariant domain.

The watermark takes the form of two dimensional spread spectrum signal in the RST transformation invariant domain. First, Fourier transform (FFT) is applied and then followed by a Fourier-Mellin transform (FMT- A log-polar mapping (LPM) followed by a Fourier transform). The invariant coefficients selected for their robustness to image processing are marked using a spread spectrum signal. The inverse mapping is computed as an inverse Fourier transform (IFFT) followed by an inverse Fourier-Mellin transform (IFMT- An inverse log-polar mapping (ILPM) followed by an inverse FFT).

The embedding process of the three methods consists on: First, luminance values of the cover work are transformed to the RST domain Fig. 1. Then the watermark W as a pseudo-random sequence of N binary elements is inserted in the mid frequencies magnitude coefficients along a circle of radius r, which can be expressed with this equation:

$$M_W = M_f + \alpha \times W, \tag{9}$$

where M_W is the mid frequencies in RST domain Fig. 1, M_f is the original frequencies before the watermark insertion, and α is the strength parameter.

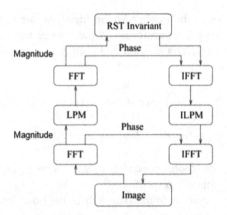

Fig. 1. RST invariant watermarking scheme based on Fourier-Mellin transform [5].

The choice of α is related to the invisibility of the watermark. The watermark is spread over all the image pixels taking into account the peak signal-to-noise ratio (PSNR) metric. Hence, an adaptive strength α is determined to obtain the desired value of PSNR, in general equal to 40 dB [3, 11]. Finally, the watermarked image is obtained by applying the inverse process of FMT Fig. 1. The blind detection needs only the tested image and the watermark W. First, the FMT is applied to the luminance of the image. Then, the coefficients are extracted from the RST domain. the normalized correlation is computed between the extracted coefficients and the sequence W of the watermark [2].

3.1 Circular Insertion of the Watermark

See Fig. 2 and Table 1.

 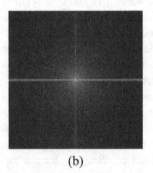

(a)   (b)

Fig. 2. (a) Watermarked image and (b) circular watermark insertion in the magnitude of FMT.

Table 1. PSNR and correlation values of circular insertion

PSNR	40.4007 dB
Correlation value	0.9750
Correlation without watermark	0.0822

3.2 Diagonal Insertion

See Fig. 3 and Table 2.

(a)

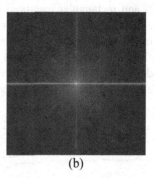
(b)

Fig. 3. (a) Watermarked image and (b) diagonal watermark insertion in the magnitude of FMT.

Table 2. PSNR and correlation values of diagonal insertion

PSNR	40.3173 dB
Correlation value	0.9182
Correlation without watermark	0.0346

3.3 Rectangular Insertion

See Fig. 4 and Table 3.

Fig. 4. Rectangular watermark insertion in the magnitude of FMT.

Table 3. PSNR and correlation values of rectangular insertion

PSNR	40.267 dB
Correlation value	0.8923
Correlation without watermark	0.2989

3.4   Comparison Results

Comparison results are presented in Fig. 5. Three methods of invisible watermarking based on Fourier-Mellin are tested. As explained earlier, those methods are circular, diagonal and rectangular insertions. Watermarked and unwatermarked image of each method are tested under four types of geometric distortions: Diagonal translations with pixels in the range [0,100], rotations between 1 and 19°, scaling with factors in the range [0.88, 1.24] and combination of the above geometric distortions RST.

Under translation attack, the three tested methods give better result with the higher performance of the circular insertion method. Diagonal insertion shows better

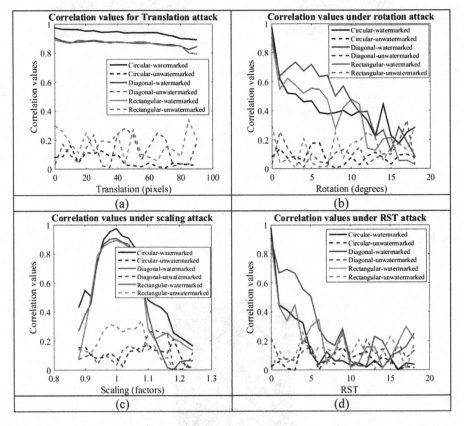

Fig. 5. Correlation values of the three tested methods for watermarked and unwatermarked images with (a) translation attack, (b) rotation attack and (c) scaling attack and (d) RST attack.

performance against rotation attack compared with circular and diagonal insertion method. However, circular insertion shows higher performance against scaling attack. Under a combination of the attacks RST the performance of the three methods is similar with a slight outperform of diagonal insertion.

4 Conclusion

In this paper, the robustness of different insertion position and shape of the watermark were evaluated in a watermarking scheme based on the Fourier-Mellin transform (FMT). The diagonal, rectangular and circular insertions were tested against geometric deformation such as rotation, scaling and translation (RST). Results show that the circular insertion performs better for translations and scaling attacks, while the diagonal insertion is better for rotations and RST attacks. This last point makes that the diagonal insertion should be preferred in industrial applications when printing the image on a physical support, and scanning it (print-scan process or print-scan attack). Future works concerned the protection of official documents containing an Identity image (passport or ID cards) using watermarking methods. In that case, it is expected in the future that the detection of the mark in the ID document will be assessed using a smartphone (print-cam process). The geometric attack is stronger than the RST attack. It is a perspective attack, a RST attack combined to a tilling of the optical axis of the camera that takes the image. In a first step, we will test the robustness of the FMT watermark insertion with various insertion position and shape to the perspective attack. In a second step, we will develop an android version of the method.

References

1. Barni, M., Cox, I., Kalker, T., Kim, H.-J. (eds.): Digital Watermarking, IWDW 2005. LNCS, vol. 3710. Springer, Heidelberg (2005). https://doi.org/10.1007/11551492
2. Cox, I., Miller, M., Bloom, J., Fridrich, J., Kalker, T.: Digital Watermarking and Steganography. Morgan Kaufmann, Boston (2007)
3. Poljicak, A., Mandic, L., Agic, D.: Discrete Fourier transform-based watermarking method with an optimal implementation radius. J. Electron. Imaging **20**(3), 033008 (2011)
4. Riad, R., Harba, R., Douzi, H., Ros, F., Elhajji, M.: Robust Fourier watermarking for id images on smart card plastic supports. Adv. Electr. Comput. Eng. **16**(4), 23–30 (2016)
5. Ruanaidh, J.J.O., Pun, T.: Rotation, scale and translation invariant spread spectrum digital image watermarking. Sig. Process. **66**(3), 303–317 (1998)
6. Lin, C.-Y., Wu, M., Bloom, J.A., Cox, I.J., Miller, M.L., Lui, Y.M.: Rotation, scale, and translation resilient watermarking for images. IEEE Trans. Image Process. **10**(5), 767–782 (2001)
7. Zheng, D., Zhao, J., El Saddik, A.: RST-invariant digital image watermarking based on log-polar mapping and phase correlation. IEEE Trans. Circuits Syst. Video Technol. **13**(8), 753–765 (2003)
8. Shao, Z., Shang, Y., Zhang, Y., Liu, X., Guo, G.: Robust watermarking using orthogonal Fourier-Mellin moments and chaotic map for double images. Sig. Process. **120**, 522–531 (2016)

9. Ouyang, J., Coatrieux, G., Chen, B., Shu, H.: Color image watermarking based on quaternion Fourier transform and improved uniform log-polar mapping. Comput. Electr. Eng. **46**, 419–432 (2015)

10. Kim, B.-S., et al.: Robust digital image watermarking method against geometrical attacks. Real-Time Imaging **9**(2), 139–149 (2003)

11. Cheddad, A., Condell, J., Curran, K., Mc Kevitt, P.: Digital image steganography: survey and analysis of current methods. Sig. Process. **90**(3), 727–752 (2010)

A New Sparse Blind Source Separation Method for Determined Linear Convolutive Mixtures in Time-Frequency Domain

Mostafa Bella$^{(\boxtimes)}$ and Hicham Saylani$^{(\boxtimes)}$

Laboratoire d'Électronique, Traitement du Signal et Modélisation Physique,
Faculté des Sciences, Université Ibn Zohr, BP 8106, Cité Dakhla, Agadir, Morocco
mostafa.bella@edu.uiz.ac.ma, h.saylani@uiz.ac.ma

Abstract. This paper presents a new Blind Source Separation method for linear convolutive mixtures, which exploits the sparsity of source signals in the time-frequency domain. This method especially brings a solution to the artifacts problem that affects the quality of signals separated by existing time-frequency methods. These artifacts are in fact introduced by a time-frequency masking operation, used by all these methods. Indeed, by focusing on the case of determined mixtures, we show that this problem can be solved with much less restrictive sparsity assumptions than those of existing methods. Test results show the superiority of our new proposed method over existing ones based on time-frequency masking.

Keywords: Blind source separation · Linear convolutive mixtures · Sparsity · Time-frequency masking · Bin-wise clustering · Determined mixtures

1 Introduction

Blind Source Separation (BSS) aims to find a set of N unknown signals, called sources and denoted by $s_j(n)$, knowing only a set of M mixtures of these sources, called observations and denoted by $x_i(n)$. This discipline is receiving increasing attention thanks to the diversity of its fields of application. Among these fields, we can cite those of audio, biomedical, seismic and telecommunications. In this paper, we are interested in so-called linear convolutive (LC) mixtures for which each mixture $x_i(n)$ is expressed in terms of the sources $s_j(n)$ and their delayed versions as follows:

$$x_i(n) = \sum_{j=1}^{N} \sum_{q=0}^{Q} h_{ij}(q) \cdot s_j(n-q) = \sum_{j=1}^{N} h_{ij}(n) * s_j(n), \qquad i \in [1, M], \qquad (1)$$

© Springer Nature Switzerland AG 2020
A. El Moataz et al. (Eds.): ICISP 2020, LNCS 12119, pp. 357–366, 2020.
https://doi.org/10.1007/978-3-030-51935-3_38

where:

- $h_{ij}(q)$ represents the impulse response coefficients of the mixing filter linking the source of index j to the sensor of index i,
- Q is the order of the longest filter,
- the symbol "$*$" denotes the linear convolution operator.

Indeed, in the field of BSS, the case of LC mixtures is still of interest since the performance of existing methods is still modest compared to the particular case of linear instantaneous mixtures for which $Q = 0$. BSS methods for LC mixtures can be classified into two main families. The so-called temporal methods that deal with mixtures in the time domain and the so-called frequency methods that deal with mixtures in the time-frequency (TF) domain. The performance of the former is generally very modest and remains very restrictive in terms of assumptions compared to the latter. Indeed, based mostly on the independence of source signals, most efficient methods are compared to frequency ones only for very short filters (i.e. Q low), and generally require over-determined mixtures (i.e. for $M > N$) [12,16]. Based mostly on the sparsity of source signals in the TF domain, the frequency methods have shown good performance in the determined case (i.e. for $M = N$) or even under-determined case (i.e. for $M < N$), and this despite increasing the filters length [4,8,9,13–15]. These frequency methods start by transposing the Eq. (1) into the TF domain using the short time Fourier transform (STFT) as follows:

$$X_i(m, k) = \sum_{j=1}^{N} H_{ij}(k) \cdot S_j(m, k), \quad m \in [0, T - 1], \quad k \in [0, K - 1], \quad (2)$$

where:

- $X_i(m, k)$ and $S_j(m, k)$ are the STFT representations of $x_i(n)$ and $s_j(n)$ respectively,
- K and T are the length of the analysis window[1] and the number of time windows used by the STFT respectively[2],
- $H_{ij}(k)$ is the Discrete Fourier Transform of $h_{ij}(n)$ calculated on K points.

Among most efficient and relatively more recent frequency methods, we can mention those based on TF masking [2,4,6–9,13–15]. The sparsity is often exploited by these methods by assuming that the source signals are W-disjoint orthogonal, i.e. not overlapping[3] in the TF domain. The principle of these methods is to estimate a separation mask, denoted by $M_j(m, k)$ and specific to each source $S_j(m, k)$, which groups the TF points where only this source is present.

---

[1] Assuming that the length K of the analysis window used is sufficiently larger than the filters order Q (i.e. $K > Q$).

[2] It should be noted however that the equality in Eq. (2) is only an approximation. This equality would only be true if the discrete convolution used was circular, which is not the case here. We also note that this STFT is generally used with an analysis window different than the rectangular window [2,4,6–9,13–15].

[3] Which means, in each TF point at most one source is present.

The application of the estimated mask $M_j(m, k)$ to one of the frequency observations $X_i(m, k)$ allows us to keep from the latter only the TF points belonging to the source $S_j(m, k)$, and then separate it from the rest of the mixture. Depending on the procedure used to estimate the masks, we distinguish between two types of BSS methods based on TF masking. The so-called *full-band* methods [2,4,6,9] for which the masks are estimated integrally using a clustering algorithm that processes all frequency bins simultaneously, and the so-called *bin-wise* methods [7,8,13–15] for which the masks are estimated using a clustering algorithm that processes only one frequency bin at a time.

Among the most popular *full-band* methods we can cite those proposed in [4,9] which are based on the clustering of the level ratios and phase differences between the frequency observations $X_i(m, k)$ to estimate the separation masks. However, this clustering is not always reliable, especially when the order Q of the mixing filters increases [4]. Moreover, when the maximum distance between the sensors is greater than half the wavelength of the maximum frequency of source signals involved, a problem called *spatial aliasing* is inevitable [4]. The *bin-wise* methods [7,8,13–15] are robust to these two problems. However, these methods require the introduction of an additional step to solve a permutation problem in the estimated masks, when we pass from one frequency bin to another, which is a classical problem that is common to all *bin-wise* BSS methods.

However, all of these BSS methods based on TF masking (*full-band* and *bin-wise*) suffers from artifacts problem which affect the quality of the separated signals and due to the fact that the W-disjoint orthogonality assumption is not perfectly verified in practice. Indeed, being introduced by the TF masking operation, these artifacts are more and more troublesome when the spectral overlap of source signals in the TF domain becomes important. In [11] the authors proposed a first solution to this problem which consists of a cepstral smoothing of spectral masks before applying them to the frequency observations $X_i(m, k)$. An interesting extension of this technique, which was proposed in [3], consists in applying cepstral smoothing not to spectral masks but rather to the separated signals, i.e. after applying the separation masks. Knowing that these two techniques [3,11] were have only been validated on a few *full-band* methods, in [5] we have recently proposed to evaluate their effectiveness using a few popular *bin-wise* methods. However, these two solutions could only improve *one particular type of artifact* called *musical noise* [3,5,11]. In the same sense, in order to avoid the artifacts caused by the TF masking operation, we propose in this paper a new BSS method which also exploits the sparsity of source signals in the TF domain for determined LC mixtures. Indeed, by focusing on the case of determined mixtures, we show that we can avoid TF masking and also relax the W-disjoint orthogonality assumption. Note that the case of determined mixtures was also addressed in [1], but with an assumption which is again very restrictive and which consists in having at least a whole time frame of silence[4] for each of the source signals. Thus, our new method makes it possible to carry out the sep-

---

[4] Of length greater or equal to the length K of the analysis window used in the calculation of the STFT.

aration while avoiding the artifacts introduced by the operation of TF masking, with sparsity assumptions much less restrictive than those of existing methods.

We begin in Sect. 2 by describing our method. Then we present in Sect. 3 various experimental results that measure the performance of our method compared to existing methods, then we conclude with a conclusion and perspectives of our work in Sect. 4.

2   Proposed Method

The sole sparsity assumption of our method is the following.

Assumption: For each source $s_j(n)$ and for each frequency bin k, *there is at least one TF point (m, k) where it is present alone*, i.e:

$$\forall j, \ \forall k, \ \exists \ m \ / \ S_j(m, k) \neq 0 \ \& \ S_i(m, k) = 0, \ \forall i \neq j \tag{3}$$

Thus, if we denote by E_j the set of TF points (m, k) that verify the assumption (3), called **single-source points**, then the relation (2) gives us:

$$X_i(m, k) = H_{ij}(k) \cdot S_j(m, k), \quad \forall (m, k) \in E_j. \tag{4}$$

Our method proceeds in two steps. The first step, which exploits the probabilistic masks used by Sawada et al. in [14,15], consists in identifying for each source of index "j" and each frequency bin "k" the index "m_{jk}" such that the TF point (m_{jk}, k) best verifies the Eq. (4), then in estimating the separating filters, denoted $F_{ij}(k)$ and defined by:

$$F_{ij}(k) = \frac{X_i(m_{jk}, k)}{X_1(m_{jk}, k)} = \frac{H_{ij}(k) \cdot S_j(m_{jk}, k)}{H_{1j}(k) \cdot S_j(m_{jk}, k)} = \frac{H_{ij}(k)}{H_{1j}(k)}, \quad i \in [2, M] \tag{5}$$

The second step consists in recombining the mixtures $X_i(m, k)$ using the separating filters $F_{ij}(k)$ in order to finally obtain an estimate of the separated sources. The two steps of our method are the subject of Sects. 2.1 and 2.2 respectively.

2.1   Estimation of the Separating Filters

Since the proposed treatment in this first step of our method is performed independently of the frequency, we propose in this section to simplify the notations by omitting the frequency bin index "k". So using a matrix formulation, the Eq. (2) gives us:

$$\mathbf{X}(m) = \sum_{j=1}^{N} \mathbf{H}_j \cdot S_j(m), \tag{6}$$

where $\mathbf{X}(m) = [X_1(m, k), ..., X_M(m, k)]^T$, $\mathbf{H}_j = [H_{1j}(k), ..., H_{Mj}(k)]^T$ and $S_j(m) = S_j(m, k)$. During this first step, we proceed as follows:

1. Each vector $\mathbf{X}(m)$ is normalized and then whitened as follows:

$$\widetilde{\mathbf{X}}(m) = \frac{\mathbf{X}(m)}{||\mathbf{X}(m)||} \quad \text{and} \quad \mathbf{Z}(m) = \frac{\mathbf{W}\widetilde{\mathbf{X}}(m)}{||\mathbf{W}\widetilde{\mathbf{X}}(m)||}, \tag{7}$$

where $\mathbf{W}$ is given by $\mathbf{W} = \mathbf{D}^{-\frac{1}{2}}\mathbf{E}^H$, with $\mathbb{E}\{\widetilde{\mathbf{X}}(m)\widetilde{\mathbf{X}}^H(m)\} = \mathbf{E}\mathbf{D}\mathbf{E}^H$.

2. Each vector $\mathbf{Z}(m)$ is modeled by a complex Gaussian density function of the form [14]:

$$p(\mathbf{Z}(m)|\mathbf{a}_j, \sigma_j) = \frac{1}{(\pi\sigma_j^2)^M} \cdot \exp\left(-\frac{||\mathbf{Z}(m) - (\mathbf{a}_j^H\mathbf{Z}(m)).\mathbf{a}_j||^2}{\sigma_j^2}\right) \tag{8}$$

where $\mathbf{a}_j$ and σ_j^2 are respectively the centroid (with unit norm $||\mathbf{a}_j|| = 1$) and the variance of each cluster C_j. This density function $p(\mathbf{Z})$ can be described by the following mixing model:

$$p(\mathbf{Z}(m)|\theta) = \sum_{j=1}^{N} \alpha_j \cdot p(\mathbf{Z}(m)|\mathbf{a}_j, \sigma_j), \tag{9}$$

where α_j are the mixing ratios and $\theta = \{\mathbf{a}_1, \sigma_1, \alpha_1, ..., \mathbf{a}_N, \sigma_N, \alpha_N\}$ is the parameter set of the mixing model.

Then, an iterative algorithm of the type *expectation-maximization* (*EM*) is used to estimate the parameter set θ, as well as the posterior probabilities $P(C_j|\mathbf{Z}(m), \theta)$ at each TF point, which are none other than the probabilistic masks used in [14].

In the *expectation step*, these posterior probabilities are given by:

$$P(C_j|\mathbf{Z}(m), \theta) = \frac{\alpha_j p(\mathbf{Z}(m)|\mathbf{a}_j, \sigma_j)}{p(\mathbf{Z}(m)|\theta)}. \tag{10}$$

In the maximization step, the update of centroid $\mathbf{a}_j$ is given by the eigenvector associated with the largest eigenvalue of the matrix $\mathbf{R}_j$ defined by:

$$\mathbf{R}_j = \sum_{m=0}^{T-1} P(C_j|\mathbf{Z}(m), \theta) \cdot \left\{\mathbf{Z}(m)\mathbf{Z}^H(m)\right\}. \tag{11}$$

The parameters σ_j^2 and α_j are updated respectively via the following relations:

$$\sigma_j^2 = \frac{\sum_{m=0}^{T-1} P(C_j|\mathbf{Z}(m), \theta) \cdot ||\mathbf{Z}(m) - (\mathbf{a}_j^H\mathbf{Z}(m)).\mathbf{a}_j||^2}{M \cdot \sum_{m=0}^{T-1} P(C_j|\mathbf{Z}(m), \theta)} \tag{12}$$

$$\alpha_j = \frac{1}{T} \sum_{m=0}^{T-1} P(C_j|\mathbf{Z}(m), \theta). \tag{13}$$

However, since the *EM* algorithm used in [14,15] is sensitive to the initialization[5], we propose in our method to initialize the masks with those obtained

---

[5] Which is done randomly in [14,15] and can lead to terrible performance.

by a modified version of the MENUET method [4]. Indeed, we replaced, in the clustering step for the estimation of the masks, the *k-means* algorithm used in [4] by the fuzzy c-means *(FCM)* algorithm used in [13], in order to have probabilistic masks.

3. After the convergence of the *EM* algorithm, the classical permutation problem between the different frequency bins is solved by the algorithm proposed in [15], which is based on the inter-frequency correlation between the time sequences of posterior probabilities $P(C_j|\mathbf{Z}(m), \theta)$ in each frequency bin k. In the following we denote these posterior probabilities by $P(C_j|\mathbf{Z}(m, k))$.

4. Unlike the approach adopted in [14,15] which consists in using all the TF points of the estimated probabilistic masks $P(C_j|\mathbf{Z}(m, k))$, we are interested in this step only in identifying one **single-source** TF point for each source of index "j" and for each frequency bin "k", therefore a single time frame index that we denote by "m_{jk}", which best verifies our working assumption (4). We then define this index m_{jk} as being the index "m" for which the presence probability of the corresponding source is maximum[6]:

$$m_{jk} = \underset{m}{\arg\max}\, P(C_j|\mathbf{Z}(m, k)), \quad m \in [0, T-1] \qquad (14)$$

5. After having identified these "best" **single-source** TF points (m_{jk}, k), we finish this first step of our method by estimating the separating filters $F_{ij}(k)$ defined in (5) by:

$$F_{ij}(k) = \frac{X_i(m_{jk}, k)}{X_1(m_{jk}, k)} = \frac{H_{ij}(k)}{H_{1j}(k)}, \quad i \in [2, M] \qquad (15)$$

2.2 Estimation of the Separated Sources

In this section, for more clarity, we provide the mathematical bases for the second step of our method for two LC mixtures of two sources, i.e. for $M = N = 2$. The generalization to the case $M = N > 2$ can be derived directly from this in an obvious way. In this case, the mixing Eq. (1) gives us:

$$\begin{cases} x_1(n) = h_{11}(n) * s_1(n) + h_{12}(n) * s_2(n) \\ x_2(n) = h_{21}(n) * s_1(n) + h_{22}(n) * s_2(n) \end{cases} \qquad (16)$$

As we pass to the TF domain, we get:

$$\begin{cases} X_1(m, k) = H_{11}(k) \cdot S_1(m, k) + H_{12}(k) \cdot S_2(m, k) \\ X_2(m, k) = H_{21}(k) \cdot S_1(m, k) + H_{22}(k) \cdot S_2(m, k) \end{cases} \qquad (17)$$

We use the separating filters $F_{ij}(k)$, with $i = 2$ and $j = 1, 2$, estimated in the first step to recombine these two mixtures as follows:

$$\begin{cases} X_2(m, k) - F_{22}(k) \cdot X_1(m, k) = \widetilde{S}_1(m, k) \\ X_2(m, k) - F_{21}(k) \cdot X_1(m, k) = \widetilde{S}_2(m, k) \end{cases} \qquad (18)$$

---

[6] Note however that in practice, only the indices "m" with an energy $\|\mathbf{X}(m)\|^2$ which is not negligible are concerned by the Eq. (14).

Since we have $F_{21}(k) = \frac{H_{21}(k)}{H_{11}(k)}$ and $F_{22}(k) = \frac{H_{22}(k)}{H_{12}(k)}$, based on the Eq. (15), we get after all simplifications have been made:

$$\begin{cases} \widetilde{S}_1(m,k) = \frac{H_{21}(k).H_{12}(k) - H_{22}(k).H_{11}(k)}{H_{12}(k)} \cdot S_1(m,k) \\ \widetilde{S}_2(m,k) = \frac{H_{22}(k).H_{11}(k) - H_{21}(k).H_{12}(k)}{H_{11}(k)} \cdot S_2(m,k) \end{cases} \tag{19}$$

In order to ultimately obtain the contributions of sources in one of the sensors, we propose to add a post-processing step (as in [1]) which consists in multiplying the signals $\widetilde{S}_j(m,k)$ by filters, denoted by $G_j(k)$, as follows:

$$G_j(k) \cdot \widetilde{S}_j(m,k) = Y_j(m,k), \qquad j \in \{1,2\}, \tag{20}$$

where $\quad G_1(k) = \dfrac{1}{F_{21}(k) - F_{22}(k)} \quad$ and $\quad G_2(k) = \dfrac{1}{F_{22}(k) - F_{21}(k)}. \tag{21}$

After all the simplifications are done, we get:

$$\begin{cases} Y_1(m,k) = H_{11}(k) \cdot S_1(m,k) \\ Y_2(m,k) = H_{12}(k) \cdot S_2(m,k) \end{cases} \tag{22}$$

By denoting $y_j(n)$ the inverse STFT of $Y_j(m,k)$ we get:

$$\begin{cases} y_1(n) = h_{11}(n) * s_1(n) \\ y_2(n) = h_{12}(n) * s_2(n) \end{cases} \tag{23}$$

These signals are none other than the contributions of source signals $s_1(n)$ and $s_2(n)$ on the first sensor (see the expression of the mixture $x_1(n)$ in (16)).

3 Results

In order to evaluate the performance of our method and compare it to the most popular *bin-wise* methods known for their good performance, that is the method proposed by *Sawada* et al. [15] and the *UCBSS* method [13], we performed several tests on different sets of mixtures. Each set consists of two mixtures of two real audio sources, which are sampled at 16 KHz and with a duration of 10 s each, using different filter sets. Generated by the toolbox [10], which simulates a real acoustic room characterized by a reverberation time denoted by RT_{60}[7], the coefficients $h_{ij}(n)$ of these mixing filters depend on the distance between the two sensors (microphones), denoted as D and on the absolute value of the difference between directions of arrival of the two source signals, denoted as $\delta\varphi$. For the calculation of the STFT, we used a 2048 sample Hanning window (as analysis window) with a 75% overlap. To measure the performance we used two of the most commonly used criteria by the BSS community, called *Signal to*

---

[7] RT_{60} represent the time required for reflections of a direct sound to decay by 60 dB below the level of the direct sound.

Distortion Ratio (SDR) and *Signal to Artifacts Ratio* (SAR) provided by the *BSSeval* toolbox [17] and both expressed in decibels (*dB*). The SDR measures the global performance of any BSS method, while the SAR provides us with a specific information on its performance in terms of artifacts presented in the separated signals.

For each test we evaluated the performance of the three methods, in terms of SDR and SAR, over **4 different realizations** of the mixtures related to the use of different sets of source signals cited above. Thus, the values provided below for SDR and SAR represent the average obtained over these 4 realizations[8].

In the first experiment, we evaluated the performance as a function of the parameters D and $\delta\varphi$ for an acoustic room characterized by $RT_{60} = 50\,\mathrm{ms}$. Table 1 groups the performance for $D \in \{0.3\,\mathrm{m}, 1\,\mathrm{m}\}$ and $\delta\varphi \in \{85°, 55°, 30°\}$, where the last column for each value of the parameter D represents the average value of SDR and SAR over the three values $\delta\varphi_i$ of $\delta\varphi$.

Table 1. SDR (dB) and SAR (dB) as a function of D and $\delta\varphi$ for $RT_{60} = 50$ ms.

Method	Performance	$D = 0.3\,\mathrm{m}$				$D = 1\,\mathrm{m}$			
		$\delta\varphi_1$	$\delta\varphi_2$	$\delta\varphi_3$	Mean	$\delta\varphi_1$	$\delta\varphi_2$	$\delta\varphi_3$	Mean
Sawada	SDR	11.75	11.76	12.25	11.92	12.46	12.20	11.88	12.18
	SAR	12.16	12.17	12.72	12.35	12.74	12.59	12.34	12.56
UCBSS	SDR	5.02	8.68	5.82	6.51	8.65	8.71	10.73	9.36
	SAR	7.71	9.99	8.30	8.67	9.67	9.88	11.73	10.43
Proposed method	SDR	16.55	17.40	17.17	**17.04**	16.17	15.08	16.03	**15.76**
	SAR	17.74	18.56	18.57	**18.29**	17.83	16.13	17.81	**17.26**

According to Table 1, we can see that our method is performing better than the other two methods, and this over the 4 realizations of mixtures tested. Indeed, the proposed method shows superior performance over these two methods by about 5 dB for $D = 0.3\,\mathrm{m}$ and 3.5 dB for $D = 1\,\mathrm{m}$ in terms of SDR. This performance difference is even more visible in terms of SAR, which confirms that the artifacts introduced by our method are less significant than those introduced by the other two methods.

In our second experiment we were interested in the behavior of our method with regard to the increase of the reverberation time while fixing the parameters D and $\delta\varphi$ respectively to $D = 0.3\,\mathrm{m}$ and $\delta\varphi = 55°$. Table 2 groups the performance of the three methods in terms of SDR, for RT_{60} belonging to the interval $\{50\,\mathrm{ms}, 100\,\mathrm{ms}, 150\,\mathrm{ms}, 200\,\mathrm{ms}\}$[9].

According to Table 2, we can see again that the best performance is obtained by using our method whichever the reverberation time. However, we note that

---

[8] We have indeed opted for these 4 realizations instead of only one in order to approach as close as possible to a statistical validation of our results.

[9] I.e. the mixing filters length ($Q + 1 = f_s \cdot RT_{60}$) varies from 800 coefficients (for $RT_{60} = 50\,\mathrm{ms}$) to 3200 coefficients (for $RT_{60} = 200\,\mathrm{ms}$).

Table 2. SDR (dB) as a function of RT_{60} for $D = 0.3$ m and $\delta\varphi = 55°$.

Method	RT_{60}			
	50 ms	100 ms	150 ms	200 ms
Sawada	11.76	11.42	9.26	7.65
UCBSS	8.68	5.12	3.83	3.04
Proposed method	**17.40**	**13.60**	**11.02**	**8.12**

this performance is degraded when RT_{60} increases. This result, which is common to all BSS methods, is expected and is mainly explained by the fact that the higher the reverberation time, the less the assumption (here of sparseness in the TF domain) assumed by these methods on source signals is verified.

4   Conclusion and Perspectives

In this paper, we have proposed a new Blind Source Separation method for linear convolutive mixtures with a sparsity assumption in the time-frequency domain that is much less restrictive compared to the existing methods [1,2,4,6–9,13–15]. Indeed, by focusing on the case of determined mixtures, we have shown that our method avoids the problem of artifacts at the separated signals from which suffers most of these methods [2,4,6–9,13–15]. According to the results of the several tests performed, the performance of our new method, in terms of SDR and SAR, is better than that obtained by using the method proposed by *Sawada* et al. [15] and the *UCBSS* method [13], which are known for their good performance within existing methods. Nevertheless, considering that these results were obtained over 4 different realizations of the mixtures and only for some values of the parameters involved, a larger statistical performance study including all these parameters is desirable to confirm this results. Furthermore, it would be interesting to propose a solution to this problem of artifacts also in the case of under-determined linear convolutive mixtures.

References

1. Albouy, B., Deville, Y.: Alternative structures and power spectrum criteria for blind segmentation and separation of convolutive speech mixtures. In: Fourth International Conference on Independent Component Analysis and Blind Source Separation (ICA2003), pp. 361–366, April 2003
2. Alinaghi, A., Jackson, P.J., Liu, Q., Wang, W.: Joint mixing vector and binaural model based stereo source separation. IEEE/ACM Trans. Audio Speech Lang. Process. **22**(9), 1434–1448 (2014)
3. Ansa, Y., Araki, S., Makino, S., Nakatani, T., Yamada, T., Nakamura, A., Kitawaki, N.: Cepstral smoothing of separated signals for underdetermined speech separation. In: 2010 IEEE International Symposium on Circuits and Systems (ISCAS), pp. 2506–2509 (2010)

4. Araki, S., Sawada, H., Mukai, R., Makino, S.: Underdetermined blind sparse source separation for arbitrarily arranged multiple sensors. Sig. Process. **87**(8), 1833–1847 (2007)

5. Bella, M., Saylani, H.: Réduction des artéfacts au niveau des sources audio séparées par masquage temps fréquence en utilisant le lissage cepstral. In: Colloque International TELECOM 2019 and 11emes JFMMA, pp. 58–61, June 2019

6. Ito, N., Araki, S., Nakatani, T.: Permutation-free convolutive blind source separation via full-band clustering based on frequency-independent source presence priors. In: 2013 IEEE International Conference on Acoustics, Speech and Signal Processing, pp. 3238–3242, May 2013

7. Ito, N., Araki, S., Nakatani, T.: Modeling audio directional statistics using a complex bingham mixture model for blind source extraction from diffuse noise. In: 2016 IEEE International Conference on Acoustics, Speech and Signal Processing (ICASSP), pp. 465–468, March 2016

8. Ito, N., Araki, S., Yoshioka, T., Nakatani, T.: Relaxed disjointness based clustering for joint blind source separation and dereverberation. In: 2014 14th International Workshop on Acoustic Signal Enhancement (IWAENC), pp. 268–272, September 2014

9. Jourjine, A., Rickard, S., Yilmaz, O.: Blind separation of disjoint orthogonal signals: demixing N sources from 2 mixtures. In: 2000 IEEE International Conference on Acoustics, Speech, and Signal Processing, Proceedings, vol. 5, pp. 2985–2988, June 2000

10. Lehmann, E.A., Johansson, A.M.: Prediction of energy decay in room impulse responses simulated with an image-source model. J. Acoust. Soc. Am. **124**(1), 269–77 (2008)

11. Madhu, N., Breithaupt, C., Martin, R.: Temporal smoothing of spectral masks in the cepstral domain for speech separation. In: 2008 IEEE International Conference on Acoustics, Speech and Signal Processing, pp. 45–48, March 2008

12. Pedersen, M.S., Larsen, J., Kjems, U., Parra, L.C.: A survey of convolutive blind source separation methods. In: Springer Handbook of Speech Processing. Springer, November 2007

13. Reju, V.G., Koh, S.N., Soon, I.Y.: Underdetermined convolutive blind source separation via time-frequency masking. IEEE Trans. Audio Speech Lang. Process. **18**(1), 101–116 (2010)

14. Sawada, H., Araki, S., Makino, S.: A two-stage frequency-domain blind source separation method for underdetermined convolutive mixtures. In: 2007 IEEE Workshop on Applications of Signal Processing to Audio and Acoustics, pp. 139–142, October 2007

15. Sawada, H., Araki, S., Makino, S.: Underdetermined convolutive blind source separation via frequency bin-wise clustering and permutation alignment. IEEE Trans. Audio Speech Lang. Process. **19**(3), 516–527 (2011)

16. Saylani, H., Hosseini, S., Deville, Y.: Blind separation of convolutive mixtures of non-stationary and temporally uncorrelated sources based on joint diagonalization. In: Elmoataz, A., Mammass, D., Lezoray, O., Nouboud, F., Aboutajdine, D. (eds.) ICISP 2012. LNCS, vol. 7340, pp. 191–199. Springer, Heidelberg (2012). https://doi.org/10.1007/978-3-642-31254-0_22

17. Vincent, E., Gribonval, R., Fevotte, C.: Performance measurement in blind audio source separation. IEEE Trans. Audio Speech Lang. Process. **14**(4), 1462–1469 (2006)

Proposed Integration Algorithm to Optimize the Separation of Audio Signals Using the ICA and Wavelet Transform

Enrique San Juan[1], Ali Dehghan Firoozabadi[2(✉)], Ismael Soto[1], Pablo Adasme[1], and Lucio Cañete[1]

[1] Electrical Engineering Department, Universidad de Santiago de Chile, Santiago, Chile
{enrique.sanjuan, ismael.soto, pablo.adasme, lucio.canete}@usach.cl
[2] Department of Electricity, Universidad Tecnológica Metropolitana, Av. Jose Pedro Alessandri 1242, 7800002 Santiago, Chile
adehghanfirouzabadi@utem.cl

Abstract. In the present work, an integration of two combined methodologies is developed for the blind separation of mixed audio signals. The mathematical methodologies are the independent component analysis (ICA) and the discrete Wavelet transform (DWT). The DWT optimizes processing time by decreasing the amount of data, before that signals are processed by ICA. A traditional methodology for signal processing such as Wavelet is combined with a statistical process as ICA, which assumes that the source signals are mixed and they are statistically independent of each other. The problem refers to very common situations where the human being listens to several sound sources at the same time. The human brain being able to pay attention to the message of a particular signal. The results are very satisfactory, effectively achieving signal separation, where only a small background noise and a attenuation in the amplitude of the recovered signal are noticed, but that nevertheless the signal message is identified in such a way.

Keywords: Voice recognition · Wavelet transform · Voice processing · Independent component analysis

1 Introduction

In the last two decades the wavelet transform has positioned itself as a powerful tool for digital signal processing complementing the Fourier transform and in some cases more powerful, since the wavelet transform performs an analysis both in time and in the frequency at the same time. On the other hand, the independent component analysis (ICA) provides an analysis perspective of the signals from the statistical point of view. In relation to the combined use of the Wavelet transform and the ICA, some related publications are shown below.

In [1], an adaptive hybrid algorithm based on DWT and ICA is proposed to remove noise from images obtained by means of magnetic resonance imaging, where this

© Springer Nature Switzerland AG 2020
A. El Moataz et al. (Eds.): ICISP 2020, LNCS 12119, pp. 367–376, 2020.
https://doi.org/10.1007/978-3-030-51935-3_39

combined technique is compared to conventional techniques, such as DWT, undecimated discrete wavelet transforms (UDWT) and ICA. In [2], the idea of energy efficiency state identification is proposed and the monitoring strategy of energy efficiency state is established for a metal cutting process. A combined application method of continuous wavelet transform (CWT) and fast independent component analysis (FICA) is proposed for feature extraction of low or high energy efficiency state. On the other hand in [3] an electrocardiogram (ECG) signal noise elimination method is proposed based on wavelet transformation and ICA. First, two-channel ECG signals are acquired. We decompose these two ECG signals by wavelet and adding the useful wavelet coefficients separately, obtaining two-channel ECG signals with fewer interference components. Second, these two-channel ECG signals are processed and a channel signal constructed to perform an additional process with ICA, obtaining the separate ECG signal. In [4], a possible solution to the problem is proposed. Forensic speaker verification systems show severe performance degradation in the presence of noise when the signal to noise ratio (SNR) is low. Also use a combination of feature warped Mel frequency Cepstral coefficients (MFCCs) and feature warped MFCC extracted from the DWT of the enhanced speech signals as the feature extraction. In [5], a method for separation of mixed signals using ICA and Wavelet transform is proposed. This problem is solved using DWT based parallel architecture, which is a combined system consisting of two sub-over complete ICA. One process takes the high-frequency wavelet part of observations as its inputs and the other process takes the low-frequency part. Then, the final results are generated by merged these results. In [6], authors proposed a new wavelet based ICA method using Kurtosis for blind audio source separation. In this method, the observations are transformed into an adequate representation using wavelet packets decomposition and Kurtosis criterion.

2 Independent Component Analysis (ICA)

There is an essentially statistical method for signals which simulates the way that human brain is able to differentiate a particular signal when it is listening to several signals at the same time (Fig. 1). We use the premise of the statistical independence from the origin signals. It means the signals are mixed with a noise which has a probability function and subsequently through ICA to obtain the original independent signals.

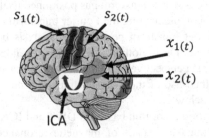

Fig. 1. The human brain according to ICA

2.1   Operation of the ICA Method

Usually when there is a social meeting and two persons are talking, the brain of the person who is listening only concentrates on the voice of the person with whom he is talking and keeping his main characteristics of the signal [7]. The human brain is able to isolate and separate the different sounds that it is listening in order to concentrate only on the voice that is interested (Fig. 2). This ability to separate and recognize a sound source from a noisy environment is known as the Cocktail Party effect technique. This acoustic phenomenon of human psychology was initially proposed by Colin Cherry [8].

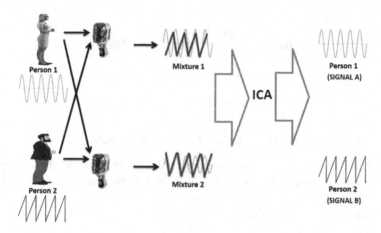

Fig. 2. Cocktail Party effect applied with ICA

In the following example there are two speakers **A** and **B** that each of them is speaking independently. The signals are separated by a mathematical process where two acoustic signals s_a and s_b are generated. Also, The signals are mixed and next to them there is a microphone that record these signals independently [9]. The ICA method takes these signals (generated independently) which are mixed with random noise and then using a statistical process (ICA) to recover the two original signals separately (Fig. 2).

This process can be observed in greater detail when the signals **A** and **B** (Fig. 3) are mixed linearly through a linear function $s = 0.7 * A - 0.11 * B$ to generate two new mixed signals M_A and M_B [10].

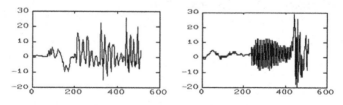

Fig. 3. Independent signals A and B

Once they are separated, the ICA method is applied to the processed signals M_A and M_B. When the process is finished, it is recovered with the two independent signals except for a very low noise factor called scalar which can be positive or negative (Fig. 4).

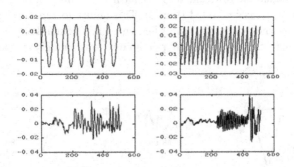

Fig. 4. Recovered signal

2.2 Formulation of the ICA Method for Discrete Signals

Let's consider $x = [x_1, x_2, \ldots x_n]$ as a vector of n components that are the samples of a signal. Also, we assume this vector is produced as a linear combination of n independent denoted signals $s = [s_1, s_2, \ldots s_n]$.

This can be expressed in Eq. 1:

$$
\begin{aligned}
x_1 &= a_{1,1}s_1 + a_{1,2}s_2 + \ldots + a_{1,m}s_m \\
x_2 &= a_{2,1}s_1 + a_{2,2}s_2 + \ldots + a_{2,m}s_m \\
x_m &= a_{m,1}s_1 + a_{m,2}s_2 + \ldots + a_{m,m}s_m
\end{aligned}
\tag{1}
$$

where the coefficients a_n determine a matrix A which is known as the mixing matrix, vector x is called the vector of mixtures and vector s is the vector of the independent components to be processed. Then, it can be described using matrix notation as:

$$
x = As
\tag{2}
$$

In practice, only the vector of mixtures x that is generated when sampling a dialogue event for example is known. Therefore, the ICA method consists in applying the algorithm that allows to find a mixed desiccation matrix W in such a way that $y = Wx$ is a good approximation of the vector s ($\mathbf{y} \cong \mathbf{s}$).

3 Wavelets Process

The wavelet transform has as main characteristics that is low computational and algorithmic complexity and the high probability of recovering small details in the inverse transformation [11]. These are very interesting characteristics for human voice signals. In addition, the process for the compression of the sampled data is the most efficient to optimize the computation time of the ICA method [12]. Another reason why wavelet was selected for use in research is its conduction in the areas of medicine, especially for the separation of brain signals where have good results [13] in combination with ICA. The mother wavelet that is selected for this work is the HAAR because it is appropriate to look for the points of inflection and discontinuities in the signals [14].

3.1 HAAR Wavelet Transform

The HAAR wavelet functions provides an optimal balance between the resolutions in time and frequency spaces. It is more efficient tool to compute the wavelet transform. This wavelet is the most used since it is reduced to calculate averages (sums) and changes (differences) between the data. Every mother wavelet has an associated scale function φ. In the case of the HAAR wavelet, the scale function is represented as:

$$\varphi_{u,v}(t) = \begin{cases} 1 & \text{if } u \leq x < v \\ 0 & \text{otherwise } o \end{cases} \tag{3}$$

The mother scale function is defined as:

$$\varphi_{0,1} = \varphi_{0,\frac{1}{2}} + \varphi_{\frac{1}{2},1} \tag{4}$$

Using the same scale function, the basic HAAR wavelet function is considered as:

$$\varphi_{0,1} = \varphi_{0,\frac{1}{2}} + \varphi_{\frac{1}{2},1} \tag{5}$$

These two previous definitions multiplied by the coefficients 1 and 2 allow to generate the wavelet coefficients with the data of our sample in a way using only averages and differences between the data. If we have a vector s of m data with $m = 2n$, we calculate the first level of wavelet coefficients as:

$$\begin{aligned} c^1 &= \left[\frac{s_1 + s_2}{2}, \frac{s_3 + s_4}{2}, \ldots, \frac{s_{m-1} + s_m}{2}, \frac{s_1 - s_2}{2}, \frac{s_3 - s_4}{2}, \ldots, \frac{s_{m-1} - s_m}{2} \right] \\ &= \frac{1}{2} [s_1 + s_2, s_3 + s_4 + \cdots s_{m-1} + s_m, s_1 - s_2, s_3 - s_4, \ldots, s_{m-1} - s_m] \end{aligned} \tag{6}$$

It is shown that the vector c^1 is divided into two parts. The first part corresponds only to the averages of two elements of s and the second section corresponds to the differences two of the elements of s.

For the next wavelet level, the first section of the vector c^1 is taken and the previous procedure is repeated. The second part of the vector is only multiplied by scale factor. We notice the elements of c^1 as the next level of wavelet coefficients will be the vector c^2 contains in the first quarter, the averages of the elements of s in the second quarter, and the differences of the averages of s and the second half only the differences of the elements of s.

$$c^1 = \left[c_1^+, c_2^+, \ldots, c_m^+, c_1^-, c_2^- \ldots, c_m^-\right] \tag{7}$$

The procedure can be repeated up to n times, which is the number of subdivisions in half that can be made from the vector s. Therefore, in each subsequent level sums and differences of each sub-vector of averages are calculated. In the following, the algorithm for calculating the coefficients of the wavelet HAAR is presented.

3.2   Inverse Wavelet HAAR

It is possible to return by performing simple addition and subtraction operations again in terms of the mother wavelet function and the mother scale function, when calculating the wavelet coefficients up to some level t. To obtain the original data again, it can be expressed through:

$$\frac{1}{2}\left(\varphi_{0,1} + \psi_{0,1}\right) = \varphi_{0,\frac{1}{2}}, \frac{1}{2}\left(\varphi_{0,1} - \psi_{0,1}\right) = \varphi_{\frac{1}{2},0} \tag{8}$$

In terms of the data generated by performing a wavelet level (c^1), we obtained the data as:

$$c + 1 = s_1 + s_{12}, \ c - 1 = s_1 - s_2 \tag{9}$$

The original data is also recovered as:

$$s_1 = C_1^+ + C_1^-, s_1 = C_1^+ - C_1^- \tag{10}$$

This process can be repeated with all the data of a c_t vector until all the data of the previous wavelet level is recovered.

3.3   Data Compression Process Through Wavelet

The simple way is known as the wavelet HAAR allows to calculate wavelet coefficients. It is possible to perform a compression of the data without losing relevant information of them [15]. As was explained, in the HAAR wavelet, additions and subtractions are made. As several c_t levels of wavelet coefficients are calculated, the energy of the signal s is concentrated in the first section of the vector c_t and to the left of the vector. The result of the vector c_8 shows (8 wavelet levels) the coefficients decay approach to zero (Fig. 5).

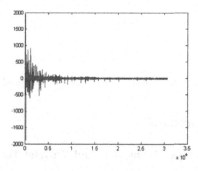

Fig. 5. Density functions of the mixed signals.

4 Implementation Between the ICA and Wavelet Methods

Once these two techniques are described (wavelet transform and the ICA technique), they must combined in an optimal way for a better compression of the sampled data and trying to avoid the loss of information in the communication between these processes. To obtain the two virtual data streams, an arithmetic mean between data is applied by means of such a way to generate two equal data streams in their creation and an independent requirement to be able to use the ICA process. At the time of processing the information through ICA and in order to reduce the amount of sampled data that have to be calculated by the ICA algorithm, the HAAR wavelet algorithm was applied to further reduce the data that ICA needs.

4.1 Process of Generation of Data Trains

This process must be done using the sample taken from a data source and must create two equal samples, but linearly independent between them (Fig. 6).

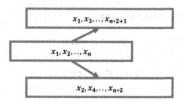

Fig. 6. Mixing process of the implemented algorithm

For this process we should look for the sampled elements that are missing in each of the data streams. These coefficients are searched using a simple arithmetic mean between the previous and above to generate the data stream. When generating the two sampled data streams, they are independent and then processed by the wavelet transform and this generates coefficients that will be taken by the ICA method as shown in Fig. 7.

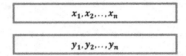

Fig. 7. Arrangements that are created independently

5   Results and Discussions

5.1   Computational Analysis Using ICA and Wavelet Techniques

The algorithm complexity evaluation technique is used to calculate the number of required operations. It focuses on classifying the complexity of the processes and performing the algorithm of the ICA method. It determines that the order is 120**m**6**n**6 where **n** is the number of signals received and **m** the number of iterations of the algorithm. The number of operations necessary for the data reduction process using wavelet HAAR is of the order **mn** (6t−12*2t), where **t** is the number of wavelet levels processed [16]. The calculation of the total number of operations for the wavelet analysis was made, but it is negligible compared to the operations necessary for ICA. These results give a linear polynomial as explained above. They show the potential of wavelet analysis to generate good performance, reducing the computational cost.

5.2   Graphic Interface

For this work two algorithms were implemented, one for signal separation using ICA and another wavelet supported by the platform of a graphical interface that allows to visualize and compare part of the compression process of audio signals. The related process is illustrated through a graph to determine the similarity of the signals. However, to compare the transformations and their application in each reading it is more practical to develop each of the steps and display them visually through a graphic.

5.3   Process Time Between Algorithms

The result of the algorithms (the running time in micro second) are presented in Table 1. As seen in this Table, the used time by wavelet is better in larger files and in smaller files the ICA is better. It can be deduced that for a large data the ICA has more calculations to perform, but if the compression of the data with the wavelet technique is used, it has fewer calculations.

Table 1.   The result of the algorithms (ICA and ICA+wavelet)

ICA	ICA+WAVELET	Files sizes
1995 µs	2334 µs	10 seg.
4123 µs	4256 µs	20 seg.
6234 µs	5232 µs	30 seg

6 Conclusions

The ICA method is a very effective tool for the blind separation of signals, especially if it is combined with other techniques such as the wavelet transform. In the present study, this statistical mathematical model was developed to separate audio signals using the ICA method, which are generated independently. So that these signals do not have a statistical dependency. This process is supported by the wavelet transform to decrease the processing data. The integration of these methods is performed for the separation of audio signals in which the processing time is optimized using previously the wavelet technique before the use of the ICA algorithm. Because the input signals to the ICA algorithm must be statistically independent, these signals are artificially generated. In this way, the problem is faced where the human being listens to several sound sources at the same time, having the ability to pay attention to the signal coming from one in particular. These signals, which are transported through the ear canal, can be separated and identified by the human being. However, the developed application gives a clear example of the versatility of the wavelet technique properly combined with ICA.

Acknowledgment. The authors acknowledge the financial support of Projects: FONDECYT No. 11180107, FONDECYT Postdoctorado No. 3190147 and Dicyt Project 061713AS.

References

1. Rai, H.M., Chatterjee, K.: Hybrid adaptive algorithm based on wavelet transform and independent component analysis for denoising of MRI images. Measurement **144**, 72–82 (2019)
2. Cai, Y., Shi, X., Shao, H., Yuan, J.: Energy efficiency state identification based on continuous wavelet transform-fast independent component analysis. J. Manuf. Sci. Eng. **141** (2), 021012 (2019)
3. Liu, M., Liu, M., Dong, L., Zhao, Z., Zhao, Y.: ECG signals denoising using wavelet transform and independent component analysis. In: International Conference on Optical Instruments and Technology: Optoelectronic Imaging and Processing Technology, p. 962213 (2015)
4. Al-Ali, A.K.H., Dean, D., Senadji, B., Baktashmotlagh, M., Chandran, V.: Speaker verification with multi-run ICA based speech enhancement. In: 11th International Conference on Signal Processing and Communication Systems (ICSPCS), Gold Coast, pp. 1–7 (2017)
5. Chao, M., Xiaohong, Z., Hongming, X., Dawei, Z.: Separation of mixed signals using DWT based overcomplete ICA estimation. In: Fourth International Conference on Computational Intelligence and Communication Networks, Mathura, pp. 416–419 (2012)
6. Mirarab, M.R., Dehghani, H., Pourmohammad, A.: A novel wavelet based ICA technique using Kurtosis. In: 2nd International Conference on Signal Processing Systems, Dalian, pp. V1-36–V1-39 (2010)
7. Zhao, Z., Xie, T., Zhao, Z., Kong, X.: A blind source separation method based on the time delayed correlations and the wavelet transform. In: 8th International Conference on Signal Processing, Beijing, China, vol. 1 (2006)

8. Mirarab, M.R., Sobhani, M.A., Nasiri, A.A.: A new wavelet based blind audio source separation using Kurtosis. In: 3rd International Conference on Advanced Computer Theory and Engineering, Chengdu, China, pp. V1-550–V1-553 (2010)
9. Hamada, T., Nakano, K., Ichijo, A.: Wavelet-based underdetermined blind source separation of speech mixtures. In: International Conference on Control, Automation and Systems, Seoul, South Korea, pp. 2790–2794 (2007)
10. Yilmaz, O., Rickard, S.: Blind separation of speech mixtures via time-frequency masking. IEEE Trans. Sig. Process. **52**(7), 1830–1847 (2004)
11. Debnath, L.: Wavelet Transforms and Their Applications. Birkhäuser, Boston (2002)
12. Osowski, S., Linh, T.H.: ECG beat recognition using fuzzy hybrid neural network. IEEE Trans. Biomed. Eng. **48**, 1265–1271 (2001)
13. Prasad, G.K., Sahambi, J.S.: Classification of ECG arrhythmias using multi-resolution analysis and neural networks. In: IEEE Conference on Convergent Technologies (Tecon 2003), Bangalore, India, pp. 227–231 (2003)
14. Abou-Elseoud, A., Starck, T., Remes, J., Nikkinen, J., Tervonen, O., Kiviniemi, V.J.: The effect of model order selection in group PICA. Hum. Brain Mapp. **31**(8), 1207–1216 (2010)
15. Addison, P.S.: The Illustrated Wavelet Transform Handbook. IOP Publishing, Bristol (2002)
16. Carmona, R., Huang, W.L.: Practical Time-Frequency Analysis Gabor and Wavelet Transforms with an Implementation in S. Academic Press, USA (1998)

ECG Signal Analysis on an Embedded Device for Sleep Apnea Detection

Rishab Khincha(✉) ⓘ, Soundarya Krishnanⓘ, Rizwan Parveenⓘ,
and Neena Goveasⓘ

Department of Computer Science, BITS Pilani Goa Campus, Sancoale 403726, India
khincharishab@gmail.com, soundaryak4898@gmail.com,
{rizwanp,neena}@goa.bits-pilani.ac.in

Abstract. Low cost embedded devices with computational power have the potential to revolutionise detection and management of many diseases. This is especially true in the case of conditions like sleep apnea, which require continuous long term monitoring. In this paper, we give details of a portable, cost-effective and customisable Electrocardiograph(ECG) Signal analyser for real time sleep apnea detection. We have developed a data analysis pipeline using which we can identify sleep apnea using a single lead ECG signal. Our method combines steps including dataset extraction, segmentation, signal cleaning, filtration and finally apnea detection using Support Vector Machines (SVM). We analysed our proposed implementation through a complete run on the MIT-Physionet dataset. Due to the low computational complexity of our proposed method, we find that it is well suited for deployment on embedded devices such as the Raspberry Pi.

Keywords: ECG · Embedded device · Signal processing · Machine learning · Sleep apnea

1 Introduction

Sleep apnea is a condition that occurs when the upper airway becomes blocked during sleep, reducing or completely stopping airflow, or if the brain fails to send the signals needed to breathe. This is a condition that could prove life-threatening if not monitored continuously and automatically.

Sleep apnea diagnosis and treatment is difficult, as the diagnosis involves overnight polysomnography (PSG), during which a medical expert is required to work overnight [1]. Thus, for medical conditions like sleep apnea, setting up automated monitoring and detection of anomalies is a crucial requirement. Medical practitioners use expensive hospital based equipment and in-patient monitoring for detection of anomalies. In addition to being expensive, this kind

NG acknowledges grant of a GPU card for research by NVIDIA. NG acknowledges assistance from ICTP as part of their Senior Research Associate program.

© Springer Nature Switzerland AG 2020
A. El Moataz et al. (Eds.): ICISP 2020, LNCS 12119, pp. 377–384, 2020.
https://doi.org/10.1007/978-3-030-51935-3_40

of monitoring in an artificial setting for conditions like sleep apnea may not give accurate results. The cost and the discomfort may result in many patients avoiding treatment until the condition becomes advanced and life threatening. Furthermore, these devices are not customisable by users in order to add more features or modify existing ones.

Electrocardiography (ECG) signals are one of the most feature rich and non-intrusive ways to detect various cardiac disorders [2]. ECG sensors and cost effective and powerful embedded devices are now freely available. The only thing lacking are mechanisms which can lead to real time analysis of data captured using these devices. Wang et. al. presented an AI based mechanism to detect sleep apnea from a single line ECG [3]. In this work, we propose a simpler algorithm to detect sleep apnea, that uses classifiers to enable execution with required speed on a device such as a Raspberry Pi. With its low cost and portability, any user can buy a device and monitor sleep apnea in the natural setting of their homes. In addition, our device will also be customisable for the users to add additional software to make the most of the device. In this work, we present a data analysis pipeline for raw ECG signals to be used for real time on-device detection of this condition. We find that our data analysis pipeline works with good accuracy with the required real-time efficiency. Our device can have utility for many other use cases requiring continuous monitoring and detection ranging from driver-fatigue detection and health anomaly detection in soldiers deployed in inhospitable locations.

In this work, we study the problem of development of a data analysis pipeline which can be executed on a resource constrained embedded device. In Sect. 3, we describe the steps that are part of a complete end to end pipeline starting from raw signals to an alert based actuation mechanism.

2   ECG Signals

2.1   ECG Signal Features

ECG signals consist of five types of waves, i.e., the P, Q, R, S and T waves. The parts of a signal are shown in Fig. 1, and the medical relevance of these parts is summarised in Table 1.

Table 1. Features of ECG signals

	Time from	Time to	What it indicates	Conditions associated
RR interval	R peak	Next R peak	ECG signal period	Sleep apnea
P-R interval	Start of P wave	Start of QRS complex	Impulse time from sinus node to ventricles	Heart blockage
Q-T interval	Start of Q wave	End of T wave	Ventricular (De/Re)polarization	Cardiac death
QRS complex	Q, R, S waves		Ventricular repolarization	Electrolyte imbalance, Drug toxicity

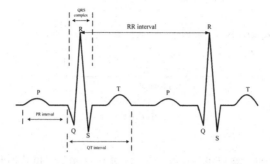

Fig. 1. Parts of an ECG signal

Irregular heartbeats lead to abnormal waves which can be traced from the patient's ECG signals. These irregular rhythms of the heart can be used to detect sleep apnea and other important medical conditions [4]. Rhythmic cycles during a heartbeat, especially the RR intervals in ECG signals have been reported to be associated with sleep apnea.

ECG signals can now be easily measured using commercially available embedded devices [5]. Long term monitoring of these signals is now possible using newly designed innovative power efficient devices, sensors and electrodes which enable ease of use [6]. These efforts have led to the possibility of designing cost effective and customised ECG signal capturing and analysis devices. The need now is to automate the analysis of the collected signals to detect various anomalies.

2.2 ECG Signals Dataset

MIT-Physionet database is the standard database used by researchers around the world for studies involving ECG signals [7]. We have used the Apnea-ECG Database for our work. [8]. The dataset contains 70 records, out of which only 35 have apnea annotations. Recordings vary in length from slightly less than 7 h to nearly 10 h. Each recording includes a continuous digitised ECG signal, a set of apnea annotations (derived by human experts), and a set of machine-generated QRS annotations. The digitised ECG signals (16 bits per sample) are recorded at the rate of 100 samples per second. A subset of the dataset was used for training. 17125 min of ECG recordings, at a rate of 100 Hz was used for training. Out of this, about 6514 min are apneatic, which gives us a well balanced training set. The training set has a 6:1 male to female ratio.

3 Data Pipeline

Raw signals as obtained from any device have several imperfections which need to be dealt with before being used to detect any anomalies. The digitised ECG signals have been recorded at the rate of 100 samples per second. These signals are then processed through the pipeline shown in Fig. 2.

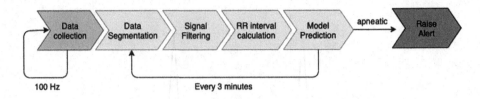

Fig. 2. Pipeline for our device

The data is first segmented into blocks of customisable duration in order to generate features. The signal is then filtered and cleaned as shown in Sect. 3.2 to remove respiratory artefacts and allow for suppression of non-R peaks. The RR intervals are calculated from the filtered signal and various data features are generated. Finally, the SVM model is used to predict whether the patient was apneatic in the last 3 min or not, and an alert is raised if true.

3.1 Data Segmentation

The dataset is segmented into blocks comprising of a few minutes each to perform analysis. Further, we discard the first few and the last few samples in every block to remove edge irregularities, if any. Each block can be customised to accommodate samples for a few minutes starting from one minute long blocks, since the MIT-Physionet dataset has annotations for every minute. A smaller block size gives more frequent predictions while a larger block size gives better statistical insights. In this work, we have used a block size of 3 min which gives us an average of approximately 200 RR intervals for feature generation. In our pipeline, the block duration is customisable to suit the needs of the user.

3.2 Signal Cleaning and Filtering

ECG Signal capturing devices like Holter Monitors continuously collect data over large periods of time. Part of this data acquired over long periods of time may be corrupted due to patient movement, sensor placement, and interference from other sources. As discussed by Mardell et al. [9], algorithms that are computationally less intensive are heavily dependant on the quality of signals provided. Drift artefacts caused due to breathing usually lie around 0.5 Hz while motion artefacts caused due to human motion lie around the 5 Hz range. Filtering using a high pass filter of any frequency above 5 Hz removes these artefacts. Since our analysis uses RR intervals, we found that a high pass filter of 20 Hz gave the best suppression of non-R peak segments of the ECG signal.

Figure 3 is a sample three minute segment of the ECG data without filtering which shows some of the artefacts mentioned above. This signal is then filtered through a high pass filter of 20 Hz causing the artefacts to be removed as shown in Fig. 4.

RR intervals are defined as $RR(i) = R(i+1) - R(i), i = 1, 2, ..., n-1$ where $R(i)$ is the time at which the ith R peak occurs. Thireau et al. [10] have shown

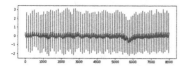

Fig. 3. ECG segment

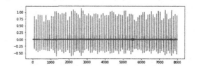

Fig. 4. ECG segment after filtering

with a 95% confidence interval that the R peaks are at least two times the standard deviation above the mean value of the signal. Our pipeline uses this result for R peak detection, thus using minimal computational power and running time.

When calculating the RR intervals, it is possible that the algorithm discussed fails to detect some of the R peaks, resulting in some abnormally long RR intervals. [11]. To avoid this, we remove the abnormal RR intervals by replacing them with the mean RR interval. The final result of filtering before the intervals are extracted helps us easily single out the R peaks by suppressing all the other peaks, as can be seen in Fig. 6. The features generated from the RR intervals, as proposed by Chazal et al. [12] and Isa et al. [13] are seen to be most effective in detecting conditions of sleep apnea. Our model uses the following features: mean, standard deviation and median of RR intervals, NN50 measures (number of pairs of successive RR intervals that differ by more than 50 ms), and the inter quartile range of the RR interval distribution (Fig. 5).

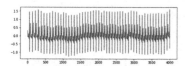

Fig. 5. ECG segment

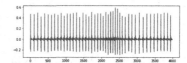

Fig. 6. ECG segment after suppression

3.3 Apnea Detection

Figure 7 shows a sample segment of a subject showing normal breathing pattern, while Fig. 8 shows a similarly sized segment showing sleep apnea. When a person is unable to breathe, it increases the body's 'fight-or-flight' stress response, making the heart beat faster, and thus, makes the RR intervals shorter.

We first attempt a comparative study of some algorithms that require computational power lesser than the one provided by an embedded device like the Raspberry Pi to determine the optimal one for the problem at hand. The comparative results are shown in Table 2.

As seen from Table 2, Support Vector Machines gave an accuracy of 87.23% with a good precision and recall. This algorithm, with an F1 score of 0.897 works well in identifying sleep apnea. Figure 9 depicts the confusion matrix, i.e.

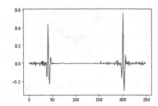
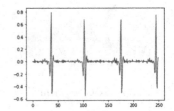

Fig. 7. Non-apneatic segment   **Fig. 8.** Apneatic segment

Table 2. Comparison of different algorithms

Algorithm	Accuracy	Precision	Recall
SVM	87.23%	89.34%	90.14%
Random forest	81.80%	80.39%	86.32%
KNN	84.11%	86.14%	85.62%
Decision tree	85.62%	83.56%	82.18%

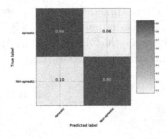

Fig. 9. Confusion Matrix for SVM with a Radial Basis Function (RBF) kernel

the True label vs. the Predicted label for our trained SVM model. We find that the probability of a false negative, where an apneatic episode is failed to be detected is 0.06.

3.4 Timings

Considering the scenario in which blocks of ECG signal data are being appended into a file constantly by the capturing device, our code reads the data written in every block duration and feeds it to our pipeline. The pipeline processes these signals and makes a prediction about the presence or absence of sleep apnea, which is then written onto a file.

We ran 584 h of ECG data from the MIT-Physionet dataset through this pipeline on a Raspberry Pi 3 Model B and a 2.5 GHz Intel Core i7-6500U with 8 GB RAM laptop. Table 3 shows the time statistics for running all the blocks through the pipeline, including the reading and the writing times.

Table 3. Comparison of running time on a laptop vs. a Raspberry Pi

	Signal length (hours)	Mean runtime (s)	Standard deviation (s)
Laptop	584	0.229	0.026
Raspberry Pi	584	2.43	0.0953

3.5 Results

We find that our proposed pipeline can perform end to end analysis of ECG data within 2.5 s on a Raspberry Pi for a 3 min segment when simulated for a 584 h dataset. This gives us an idle time of about 177.5 s every 180 s. This ensures that all the data does not need to be transmitted for analysis. Our proposed pipeline can identify apneatic episodes and only necessary information about the sleep apnea episodes needs to be sent. In addition, the apneatic episode related data can be transmitted within a few seconds using cellular transmission or a Wi-Fi based sensing network as shown by Yang et al. [14].

A comprehensive solution to the identification and treatment of sleep apnea includes actuation of devices attached to the individual. This will ideally result in halting of the episode. Some of the previously proposed mechanisms include vibration of a device [15] and physical stimulation of muscles [16]. We find that using our pipeline executing on a Raspberry pi, there is idle processor time available which can be used to raise alerts and actuate devices as required.

4 Conclusion

In this paper we have introduced a software pipeline suitable for real-time sleep apnea detection using ECG signals which can run on a resource constrained embedded device. Our method combines data segmentation, signal filtering and feature generation from RR intervals to predict occurrence of sleep apnea. We find that our pipeline successfully predicts conditions of sleep apnea on a Raspberry Pi with an execution time of only two percent of the time duration of the data. Our proposed system is portable, customisable and would be an ideal solution for a continuous monitoring device for sleep apnea. In the future, we plan to include a more detailed analysis of the ECG signals, actuation mechanisms, comparison with the user's historical data and incorporation of additional sensors for measuring other body parameters. This opens up the possibility of a portable, cost-effective, customisable health monitoring device.

References

1. Almazay L., Elleithy, K., Faezipour, M.: Obstructive sleep apnea detection using SVM-based classification of ECG signal features. In: 34th Annual International Conference of the IEEE EMBS San Diego, California USA, 28 August–1 September (2012)

2. Benazza-Benyahia, A., Ben Jeb, S.: Multiresolution based reference estimation for adaptive ECG signals denoising. In: International Conference on Image and Signal Processing ICISP, Morocco, vol. 2, pp. 875–882 (2001)

3. Wang, X., Cheng, M., Wang, Y., et al.: Obstructive sleep apnea detection using ECG-sensor with convolutional neural networks. Multimed. Tools Appl. (2018). https://doi.org/10.1007/s11042-018-6161-8

4. Huang, R., Zhou, Y.: Disease classification and biomarker discovery using ECG data. https://doi.org/10.1155/2015/680381

5. Sreekesh, S., Abhimanyu, Z., Keerthi, G., Goveas, N.: Customizable Holter monitor using off-the-shelf components. In: ICACCI, pp. 2302–2306. IEEE Explore (2016). https://ieeexplore.ieee.org/document/7732396

6. Braojos, R., et al.: Ultra-low power design of wearable cardiac monitoring systems. In: Conference Proceedings of Annual Design Automation Conference, vol. 1, pp. 1–6 (2014)

7. Goldberger, A.L., et al.: PhysioBank, physiotoolkit, and physionet: components of a new research resource for complex physiologic signals. Circulation 101(23), e215–e220 (2003)

8. Penzel, T., Moody, G.B., Mark, R.G., Goldberger, A.L., Peter, J.H.: The Apnea-ECG database. In: Computers in Cardiology, vol. 27, pp. 255–258 (2000)

9. Imtiaz, S.A., Mardell, J., Saremi-Yarahmadi, S., Rodriguez-Villegas, E.: ECG artefact identification and removal in mHealth systems for continuous patient monitoring. Healthc. Technol. Lett. 3(3), 171–176 (2016). https://doi.org/10.1049/htl.2016.0020

10. Thireau, J., Zhang, B.L., Poisson, D., Babuty, D.: Heart rate variability in mice: a theoretical and practical guide. Exp. Physiol. 93(1), 83–94 (2008)

11. Karlsson, et al.: Automatic filtering of outliers in RR intervals before analysis of heart rate variability in Holter recordings: a comparison with carefully edited data. BioMed. Eng. (2012). https://doi.org/10.1186/1475-925X-11-2

12. Chazal, P., Penzel, T., Heneghan, C.: Automated detection of obstructive sleep apnea at different time scales using the electrocardiogram. Inst. Phys. Publ. 25(4), 967–983 (2004)

13. Isa, M., Fanany, S., Jatmiko, M.I., Aniati, W.A.: Feature and model selection on automatic sleep apnea detection using ECG, pp. 357–362 (2010). https://doi.org/10.13140/RG.2.1.1624.7124

14. Yang, Z., Zhou, Q., Lei, L., Zheng, K., Xiang, W.: An IoT-cloud based wearable ECG monitoring system for smart healthcare. J. Med. Syst. 40(12), 1–11 (2016). https://doi.org/10.1007/s10916-016-0644-9

15. Scarlata, S., Rossi Bartoli, I., Santangelo, S., Giannunzio, G., Pedone, C., Antonelli Incalzi, R.: Short-term effects of a vibrotactile neck-based treatment device for positional obstructive sleep apnea: preliminary data on tolerability and efficacy. J. Thorac. Dis. 8(7), 1820–1824 (2016). https://doi.org/10.21037/jtd.2016.04.69

16. Spicuzza, L., Caruso, D., Di Maria, G.: Obstructive sleep apnoea syndrome and its management. Therap. Adv. Chronic Dis. 6(5), 273–285 (2015). https://doi.org/10.1177/2040622315590318

Author Index

Printed in the United States
By Bookmasters